SORTALS AND THE SUBJECT-PREDICATE DISTINCTION

The problem of the subject-predicate distinction has featured centrally in much of modern philosophy of language and philosophical logic, and the distinction is taken as basic or fundamental in modern philosophical logic. Michael Durrant, whilst explicitly not denying that the subject-predicate distinction as a distinction is ultimate, seeks to demonstrate that the distinction should not be taken as basic or fundamental and argues that the reason for it being held to be fundamental is a failure to acknowledge the category and role of the sortal.

A sortal is a symbol which furnishes us with a principle for distinguishing and counting particulars (objects), and which does so in its own right relying on no antecedent principle or method of so distinguishing or counting. This book explores sortals and their relationship to the subject-predicate distinction; arguing that the nature of sortal symbols has been misconstrued in much modern writing in the philosophy of logic by failing to distinguish sortals from names and predicates; contending that this misconstruction has led to a failure to appreciate what makes the subject-predicate distinction possible; demonstrating logical difficulties which then follow; and expounding an account of sortal symbols which seeks to be immune from the difficulties. Exploring and challenging aspects of the work of Frege, Russell, Geach, Quine, Evans and Strawson, amongst others, Durrant also provides a new challenge to certain popular presuppositions employed in many areas of contemporary philosophical debate, and offers important insights for those studying across philosophical logic, philosophy of language, and metaphysics and epistemology, in particular.

Michael Durrant is Reader in Philosophy and Honorary Senior Research Fellow in Philosophy, University of Wales, Cardiff, UK; Stephen Horton is Research Assistant, University of Wales, Cardiff, UK.

Sortals and the Subject-Predicate Distinction

MICHAEL DURRANT
Reader in Philosophy, University of Wales and Honorary Senior Research Fellow, Cardiff University

Edited, with a postscript by
STEPHEN HORTON
Research Associate, Cardiff University

Ashgate
Aldershot • Burlington USA • Singapore • Sydney

Published by
Ashgate Publishing Limited
Gower House
Croft Road
Aldershot
Hants GU11 3HR
England

Ashgate Publishing Company
131 Main Street
Burlington, VT 05401-5600 USA

Ashgate website: http://www.ashgate.com

British Library Cataloguing in Publication Data
Durrant, Michael
Sortals and the subject-predicate distinction
1.Predicate (Logic) 2.Language and logic 3.Semantics (Philosophy)
I. Title II. Horton, Stephen
121.6'8

Library of Congress Control Number: 00-134436

ISBN 0 7546 1378 X

Printed and bound by Athenaeum Press, Ltd.,
Gateshead, Tyne & Wear.

Contents

	Editor's Preface	*xi*
	Acknowledgements	*xiii*
	Introduction to the Author by Roger Trigg	*xv*
	Introduction: Aims and Plan of Campaign	1
	A. Principal Aims	1
	B. Sortals, Names, Descriptions, Predicables, Substantival General Terms, Predicates	5
	C. Plan of Campaign	9
I	Sortals, Names and Predicables	13
	A. Sortals: a Detailed Investigation	13
	B. That Sortals are a Distinct Category of Symbol in that they are *not* Reducible to Names or Predicables/Predicates	24
	C. Considerations on a General Thesis which may arise from the claim that Sortals are *not* Reducible to Predicables/Predicates	31
	D. Summary of Chapter	36
II	Sortals and Identification	40
	Part I: The Priority of Sortal Identification over both Naming/Referring and Predication	40
	A. Of the Priority of Sortal Identifying over Naming/Referring	40
	B. A Case in which Referring *may* be taken as prior to Identifying	43
	C. Consideration of a general attack on my thesis that Sortal Identification is prior to Naming/Referring	44
	D. Some general arguments in favour of the thesis that Sortal Identification is prior to Describing or Predicating	49
	Part II: The Contribution of Gareth Evans	52
	E. An introduction to Evans: His concept of 'Demonstrative Identification'	52
	F. That Sortal Identification is logically prior to Demonstrative Identification	53
	G. The Case against Demonstrative Identification (1): the *Misconstruction* of Demonstrative Sortal Phrases	53

H. The Case against Demonstrative Identification (2): Consideration of a stronger and weaker thesis as regards its base 56

I. That Demonstrative Identification (in Evans' sense) rests on, presupposes, Descriptive Identification in his sense and indeed on Sortal Identification 62

J. In that Demonstrative Identification is to be regarded as a mode of *Identification*, as opposed to a mode of Reference, it is reducible to Sortal Identification, in the sense of 'Sortal "Picking-Out"', 'Sortal "Discrimination"' 69

K. Of the relationship of 'Descriptive Identification' to 'Sortal Identification' and 'Predicate Identification' 72

L. Summary of Conclusions from Part I and Part II 75

Part III: Thought and Speech 76

M. Of Evans' general thesis that Identification in Thought is *prior* to Identification in Speech 76

III Sortals and the Subject-Predicate Distinction 88

A. Strawson's Objections to Geach's Explanation of the Subject-Predicate Distinction 88

B. Strawson's 'Mediating' Criterion for the Subject-Predicate Distinction 94

C. An Attempted Account of what distinguishes an expression occurring in the role of Logical Subject 102

D. An Attempted Account of what distinguishes an expression occurring in the role of Logical Predicate 103

E. A more Precise Account of the Relation between Sortals and the Subject-Predicate Distinction 108

F. A Contrast with Quine 111

G. The Issue of a *Criterion* for the Subject-Predicate Distinction 123

H. An Account of the Relations between Referring, Identifying and Existence 125

I. Explanations of the Subject-Predicate Distinction in the sense of Answers to Two Questions 136

IV General Consequences of the Failure to Acknowledge the Category of the Sortal 148

A. The Attempted Reduction of Sortals to Predicates and its consequences 148

B. The consequences of failure to recognise the Sortal as a Principle of Counting and Distinction in its own right *and equally* those of the failure to recognise principles of Counting and Distinction in their own right formed from 'Matter' terms 151

C. Some Elucidations and Expansions 153
D. Summary of the General Consequences of the failure to acknowledge the Category of the Sortal 156

V Frege: Sortals as 'Concepts' 159
A. That Frege fails to recognise both the definite and indefinite form of the Sortal as a *Distinct* and *Primary* Category of general symbol (expression) 159
B. Some consequences and difficulties which arise from Frege's failure to recognise the Sortal as a *Distinct* and *Primary* Category of general symbol (expression) 165
C. Is Frege committed to a doctrine of absolutely simple names and correspondingly, absolutely simple particulars, as a consequence of his failure to acknowledge the Category of the Sortal? 169

VI Russell: Sortals as 'Descriptions' 181
A. Russell's basic Categories of Symbol 181
B. That no case can be made for Russell's *implicit* recognition of Sortals in his Theory of Descriptions 183
C. Particular Difficulties in Russell's position 187
D. Consequences of Russell's failure to acknowledge the Category of the Sortal 190
E. That Russell is committed to a doctrine of *absolutely* simple, basic Names and concomitantly to a doctrine of *absolutely* simple or 'pure' particulars 191

VII Geach: Sortals, Substantival General Terms and General Names 194
A. Geach's recognition of the category of Substantival General Term does not entail recognition of the category of the Sortal as either basic or distinctive 194
B. Geach's reduction of Substantival General Terms to Names or Predicables (Predicates) and its difficulties 195
C. Further criticisms of Geach's thesis that the introduction of Proper Names merely presupposes the introduction of Names 200
D. The disastrous Consequences of reducing Substantival General Terms and Sortals to Names or Predicables (Predicates) 204

VIII Strawson: Sortals - Failure to Recognise Their True Nature; His Dual Position 209
A. The Claim that Strawson recognises the true nature of the Sortal 209

B. That Strawson regards predicate expressions as a kind of *Name*, Sortal expressions as predicates and hence as a kind of *Name* 210
C. That Strawson also treats Sortal expressions as *Predicative* 215
D. A Summary Statement of Strawson's 'Dual Position' 216
E. Some *General* Difficulties for Strawson's 'Category' criterion for the Subject-Predicate Distinction 217

IX Strawson: Sortals and Sortal Instantiation 220
A. That for Strawson one *can* raise the Question: 'What Instances a Sortal Universal?'; that Strawson is *not* committed to either a Doctrine of 'Pure' or 'Relatively Pure' Basic Particulars 220
B. The Case for saying that Strawson is committed to either an *unintelligible* and/or *impossible* Account or *no* Account of what constitutes an Instance of a Sortal 225
C. Conclusion 228

X Strawson: Further Consequences of Failure to Recognise the Nature of the Sortal 229
A. That to treat Sortals as Predicates (indirectly) leads to the Search for a Basis for the Introduction of 'Particulars in General' 229
B. A Second Consequence of the Failure to recognise the Nature of the Sortal: the Search for a *Basic Notion of an 'Instance'* or *Explanation of the Notion of an 'Individual Instance'* 250
C. A Third Consequence of the Failure to recognise the Nature of the Sortal: the Search for an *Underlying Basis* for the Subject-Predicate Distinction in the sense of the Search for a certain type of *Empirical* proposition 252

XI Quine: Sortals and Canonical Notation 256
A. That Quine fails to recognise the Category of the Sortal and analyses Sortals out in favour of the Variables of Quantification and Predicates 256
B. Consequences of the Failure to recognise the Category of the Sortal 263
C. A Suggested Implication of my Thesis Rejected 268
D. Quine's Aims and Claims for the Theory of Canonical Notation 269

Conclusion 275

Postscript by Stephen Horton 280
A. Introduction 280
B. Theories of Reference 280
C. Formal Logic 285
D. Comparative Philosophy: The Analytic/Continental Divide 288
E. Conclusion 301

Bibliography *307*
Bibliography of Other Writings *311*
Index of Names *315*

Editor's Preface

It is not my intention here to offer a detailed description of the content of this work as an introduction is provided by the author. Rather, the purpose of this preface is to briefly explain something of the history of the writing of this book.

Work on this book originally commenced in the late 1970s and continued up until the mid-to-late 1980s, whereupon it was interrupted by a series of events at University College, Cardiff. The net result of these occurrences led to the author becoming Head of the then Philosophy Department. This was a very traumatic period in the history of University College, Cardiff, which meant that the vast majority of the author's time and effort was taken up working for the continued existence of the Philosophy Department. Thus, owing to the overwhelming responsibility of his position, the author was unable to continue with his work on this book. Having set this work aside, it became increasingly difficult for him to find the time, until now, to return to it.

However, whilst having been forced to abandon his original efforts, the work itself was in its advanced stages with some eleven chapters and an introduction completed. Given the already immense amount of effort expended, the quality of the work already achieved and its continued philosophical relevance, funding has recently been made available by Prof. Christopher Norris for myself, acting as editor, to assist the author in the completion of this book.

With regard to the text itself, the Introduction and chapters I to XI inclusive largely mirror, with some amendments and alterations, their original content and format. Chapter XII, the concluding chapter to the book, is a recent addition. Also included is a postscript, written by the editor, which is intended to provide a bridging link between the original work herein and contemporary philosophical discussion. There is also the addition of a bibliography, an index and a bibliography of other writings by the author.

Stephen Horton

Editor's Preface

Acknowledgements

I am grateful to the Vice-Chancellor of Cardiff University; to Professor David Skilton, Head of the School of English, Communications and Philosophy and especially Professor Christopher Norris for making available some research funding to Dr. Stephen Horton for editing this book. My debt to Dr. Horton is enormous; in some instances he has had to work from hand-written manuscripts and in other instances from amateurishly typed text; he has done all the word-processing for and preparation of the final text.

I am indebted to the editor of the *Canadian Journal of Philosophy* and other journals for permission to reproduce articles or sections of articles which originally appeared in their journals.

I should like to extend my great thanks to Professors Robin Attfield and Christopher Norris for their initial and continued encouragement to complete, revise and submit for publication a text originally written in the early-mid 1980s but never completed owing to increasingly demanding Administrative Duties. My thanks are equally extended to Professors Roger Trigg and Basil Mitchell for their unswerving support.

I retired from my teaching post at Cardiff in September 1999 after 37 years and I dedicate this work to the staff and students of Philosophy at Cardiff with whom it has been a privilege and joy to work for so many years.

Michael Durrant

Introduction to the Author

This book forms a fitting climax to Michael Durrant's career in Philosophy over some 40 years. It concerns one of his major interests in the discipline, but a feature of his work has been that he has not allowed himself to be confined to the narrow embrace of one part of philosophical thought. There is at the present day a great danger of over-specialisation in the subject, so that practitioners in one part of it have little idea of, or concern for, work in other parts. This can be particularly stultifying when developments in, say, philosophical logic are not allowed to illuminate philosophical thinking in very different areas.

Michael Durrant's teaching and research have allowed him to integrate thinking from disparate elements in philosophy. A good example of this is the way in which his work in logic has been able to fertilise his thinking in the philosophy of religion. Two books, *The Logical Status of God* (London: Macmillan, 1973) and *Theology and Intelligibility* (London: Routledge and Kegan Paul, 1973) bear witness to this. He has also made substantial contributions to the study of ancient philosophy, a recent example of which is his edition of *Aristotle's* De Anima *in Focus* (London and New York: Routledge, 1993).

Philosophical logic, ancient philosophy and the philosophy of religion, separately and in combination, make a powerful repertoire for any philosopher. In his publications and in his teaching, Michael Durrant has been able to offer a balanced, but rigorous approach to his chosen subject matter. He has, as a result, earned the gratitude of colleagues and pupils alike. For 37 years his work has been centred in that part of the University of Wales which is now known as Cardiff University. He has at times played a vital role in its administration both as Dean of Theology, and as Head of the Philosophy Department.

Nevertheless this has not prevented him being known on a wider stage. He has been a visiting professor at the University of Nebraska in the United States, but perhaps more significantly has played a leading role in the formation of two important learned societies, devoted to the philosophy of religion. He was a member of the executive committee of the British Society for the Philosophy of Religion at its inception, and was also one of the founding members of the European Society for the Philosophy of Religion. This comprises academics from both philosophy and theology in a wide range of European countries. He became the first British President of the Society, and it was partly because of his hard work that it has been able to encourage the development of the philosophy of religion as a rigorous and important discipline in European universities.

As this book shows, Michael Durrant is concerned with the importance of logical thinking. He has never seen this is an end in itself, but has striven to make connections with other issues of major importance. Moreover, he has not allowed himself to be trapped in a narrow academic world. He has played an important part in the life of his local community, and his love of music, shown particularly through his organ playing, has served to remind others that of the life of the scholar need not be dry and arid.

Roger Trigg
University of Warwick

Introduction: Aims and Plan of Campaign

A. Principal Aims

My principal aims are: (I) To show that the nature of sortal symbols has been misconstrued and misunderstood in much modern classical writing in the philosophy of logic; (II) To contend that this misconstruction has led certain writers to fail to appreciate what makes possible the subject-predicate distinction; (III) To present and demonstrate the further consequences of this misconstruction, which generate logical difficulties of a fundamental type; (IV) (a) To illustrate the specific form which these general consequences assume in the writing of selected modern writers; and (b) to exhibit the distinctive difficulties arising for each of these writers; (V) To vindicate my own account of sortal symbols as immune from such difficulties and illuminative of the fundamental nature of systems of discourse.

I give the following as a preliminary account of what I shall mean by a 'sortal'. A sortal is a symbol which furnishes us with a principle for distinguishing and counting particulars and which does so in it own right relying on no antecedent principle or method of so distinguishing and counting.[1] Grammatically a sortal takes the form of a common noun which: (i) takes the indefinite article in its own right; (ii) takes the plural form in its own right. We have as examples: 'man'; 'apple'; 'house'; 'dog'; 'digit'.

A.1. The misconstruction I have alluded to above comprises the construction of the definite form of a sortal ('The so-and-so') as either: (a) a proper name; an 'object word';[2] (b) a definite description;[3] and the indefinite form ('a so-and-so') as either (c) an indefinite description;[4] (d) a general name;[5] (e) a predicative symbol, a 'concept' word;[6] (f) more crudely, a predicate.[7] The misunderstanding, I claim, arises out of such misconstruction. Whilst at one level, the level or aspect of language as a symbolic system pertinent to the expression of thought,[8] I concur with Geach[9] (following Frege), that 'names and predicables are absolutely distinct' and hence mutually exclusive, it is my special claim that at such a level such categories of symbol are not mutually *exhaustive.* We need to recognise the category of the sortal as a *distinct* category of symbol *not* reducible to names or predicables but as being logically prior to, as underlying the introduction of such categories themselves and indeed as underlying the introduction of 'logically proper names' and that class of symbol in favour of which all singular symbols are held by Quine to be eliminable, namely pronouns.

As at one level or aspect of language there has been failure to recognise the sortal as a distinct category of symbol, so also at the level or aspect of language as (i) the role of symbols, function of symbols in discourse; (ii) the use of symbols in speech, i.e. at the level or aspect of language as discourse or speech;[10] there has, I claim, been a corresponding failure to recognise (a) a distinctive mode of language role or function; (b) a distinctive mode of speech activity which I shall call 'Sortal Identification', i.e. the identification of particulars in the sense of (1) the picking out, discrimination of an object by means of the definite form of a sortal; (2) the specification of an object by such means; (3) the specification of an object by means of the indefinite form of the sortal. Rather, such sortal identification, at the singular level, has been reduced to either naming or definite describing ('Descriptive Identification');[11] 'Identifying Reference':[12] at the general level such sortal identifying has been reduced to 'predicating',[13] and thus, I claim, there has been failure to recognise (i) a mode of language role or function; (ii) a mode of speech activity;[14] which is prior to, which underlies *both* naming and predicating or referring and predicating (whether naming/referring is carried out by 'ordinary proper names', 'logically proper names' or 'pronouns' in favour of which singular terms are allegedly eliminable).

Whilst holding the subject-predicate distinction explicated either in terms of categories of *term*[15] or categories of *role*[16] to be an ultimate distinction - a thesis I do not dispute - I claim that such writers have mistakenly regarded such terms, such roles, as *themselves basic* resting upon the introduction of no *other* terms or roles. They have offered an account of 'logical subject' *either* in terms of 'subject term' and then in terms of either *names* (including ordinary proper names)[17] or 'logically proper names' (demonstratives)[18] or pronouns; *or* an account of a symbol occurring in the role of logical subject in terms of naming or referring (or an account of our use of a symbol to perform such a role in terms of naming or referring) and (a) treated ordinary proper names or demonstratives (logically proper names) or pronouns[19] as *either* basic categories of singular symbol resting upon the introduction of no prior category of singular symbol *or* resting upon the introduction of predicates only; (b) treated naming/referring as either (i) ultimately basic forms of language role or function, basic forms of speech activity, resting on no prior category of language function or speech activity or (ii) resting on the introduction of 'predicating' or 'describing' alone. Such writers have offered an account of 'logical predicate' either in terms of 'predicate term' or an account of a symbol occurring in the role of logical predicate in terms of 'predicating' and treated 'predicates' or 'predicables' as either (i) a basic category of symbol resting on no prior category of symbol other than that of the 'proper name' or (ii) a basic symbol resting on no other category of symbol at all; treated 'predicating' as a basic category of language function or speech activity resting on no prior function or activity other than that of naming or again on no such function or activity at all.

In this way I claim such writers have failed to appreciate what underlies the subject-predicate distinction in the sense of what makes this distinction *possible* in its typical manifestation or employment, i.e. in a language in which we speak of particulars in the sense of *objects*. It is to be noted that it is such a context which has, at least by implication, been held to be typical of the manifestation of the subject-predicate distinction. As far as I have been able to determine there is no treatment of the subject-predicate distinction in connection with propositions in which particulars introduced by reference to symbols (terms) formed from what Aristotle would call 'matter' terms, Quine calls 'mass' terms, and Geach calls 'substantival general' terms, feature as typical. I shall be concentrating considerable attention on 'matter' terms in Chapter I and the issue of the relation between such terms and sortals will be an endemic feature of this book. In relation to what I have termed as the typical manifestation of the subject-predicate distinction, it is my case that at *one* level or aspect language - what I have called the level or aspect of language as a symbolic system for the expression of thought - for the expression of thought concerning particulars in the sense of objects - what makes the subject-predicate distinction *possible* in such a system is indeed the presence of sortal symbols: at another level or aspect, at the level or aspect of mode of language role; at the level of speech concerning objects, what makes the subject-predicate distinction possible, what underlies the distinction in this sense, is sortal identification.

It follows from the thesis I am advocating that the definite form of the sortal 'The so-and-so'; 'The *F* which ....', may not primarily occur in, be used in the role of logical subject. What may so occur or be used in such a role is a *name* or its equivalent. This consequence is in accordance with tradition as exemplified in the writings of Frege;[20] initially Russell;[21] Geach, and again, initially, Quine.[22] I shall be considering objections to the thesis that the definite form of a sortal may never occur in or be used in such a role in Chapter III. It further follows from my thesis that I reject the account of 'logical subject' in terms of 'identifying'[23] or 'identifying reference'[24] as set forth by Strawson. On my thesis 'identifying' and in particular 'sortal identifying' is prior to 'referring' in a way to be explained in Chapter II and thus that 'identifying reference' is a confused notion.

It is further to be observed, that in that 'traditional logicians' such as those alluded to, have indeed failed to recognise the category of the sortal and sortal identification, they could not have analysed the notion of 'logical subject' in terms of 'identifying' granted my case to be advocated in Chapter II that other modes of 'identifying' either rest on or are reducible to sortal identification. Indeed Strawson himself could only offer such an account in that he is able to recognise the category of the sortal and sortal identification as a quite distinct category and mode, which I shall claim is not the case.[25] In crude terms, he reduces sortal identification to descriptive identification. Again, even if it could be held that 'traditional' logicians *did* recognise the category of the sortal and sortal identification, this is not *sufficient* for claiming that they could offer (as opposed to would offer) an analysis of 'logical subject' in terms of 'identifying'. To offer such an analysis they would further (a) have to subscribe to the thesis that proper names of their essence have a sense - and notably Russell and Quine do *not*;[26] (b) have to subscribe to the thesis that a proper name is identical with its sense, which is open to strong objection.[27] Strawson, I shall hold, makes a *recommendation* as to the analysis of the notion of 'logical subject'; he cannot be held to be offering a *descriptive* analysis.

A.2. Failure to recognise the true nature of sortals in that sortals have been construed as predicates leads, I hold, to further unacceptable consequences:

(1) Either (A) having to introduce a doctrine of 'absolutely simple' basic names or 'singular terms' (or a more sophisticated version of such a doctrine) and, correspondingly, having to vouchsafe a doctrine of 'absolutely simple individuals', 'pure particulars' or having to introduce a doctrine of 'relatively simple' basic names or singular terms and correspondingly a doctrine of 'relatively simple', 'relatively basic' individuals in a way to be explained in Chapter IV; Or (B) having to advocate either an unintelligible doctrine or no doctrine of what constitutes an instance of a sortal, what instantiates a 'sortal universal'.

(2) Failing to appreciate that *necessarily* to be a particular, an individual in the sense of a particular *object*, is to be specifiable by reference to a sortal (to fall under a sortal) - as opposed to holding that to be a particular (object) is *necessarily* to be specifiable by reference to a sortal (to fall under a sortal).

(3) (indirectly)[28] to the search for a basis for the introduction of 'particulars in general', a basis for 'particular introduction' which I shall claim is: (a) unsuccessful in fact; (b) misconceived in principle, and subsequently to the search for a basic notion of 'an instance' independently of the notion of a 'sortal', which I shall also claim is misconceived;

(4) (indirectly) to the search for an underlying basis for the subject-predicate distinction in the sense: (i) of a search for some one type of *empirical* proposition which underlies the distinction; (ii) of a search for a certain type of empirical proposition which contains no reference to sortals and is formed independently of any reference to sortals. I shall argue that both such searches are misconceived.

As regards (i), I hold that such a search is misguided in principle (misconceived) since: (a) it would make a *logical* distinction rest upon a matter of fact; (b) it fails to appreciate that the subject-predicate distinction *is* a logical distinction, a transcendental distinction, which manifests itself in modes of discourse or speech; it is not merely a *general* distinction. We thus cannot hope to search for: (i) an *empirical* basis for the distinction; (ii) some *one* type of empirical proposition which will form such a basis. All we may hope to search for is a basis for the distinction as it is manifested in a mode of discourse e.g. as manifested in a mode of discourse or a way of speaking in which we (in general) speak of 'objects'; or again e.g. in which we speak of individuals in the sense of 'parcels of matter'.

As regards (ii) I hold such a search to be misconceived: (a) on the grounds just advocated; (b) on the grounds that a sortal has to be invoked in order to *articulate* the proposed basis (*see* Chapter X).

One may further comment here that to hold that we may search for some *one* type of proposition of a certain *status* (e.g. an empirical proposition) as ultimately underlying the subject-predicate distinction, as an ultimate basis for the distinction, is to treat a mode of discourse or way of speaking as *itself* something *within* a mode of discourse or way of speaking, which is absurd. It commits an error parallel to treating a class as a member of itself. The subject-predicate distinction manifests itself *in* a mode of discourse, *in* a way of speaking.

In the light of these comments it may now be asked how I may claim that such writers as I have alluded to have failed to appreciate what underlies the subject-predicate distinction in the sense of what makes the distinction in *general possible.* In my earlier writing however I have made no such general claim. I have limited my claim to the effect that such writers have failed to appreciate what makes the subject-predicate distinction possible in its *typical* manifestation or employment,[29] viz., in a language, area of discourse, way of speaking in which we speak of particulars in the sense of *objects.* There is a sense in which we may raise the question: 'What makes the subject-predicate distinction possible?' but *this* question is strictly to be construed as: 'What makes it possible for the subject-predicate distinction to be manifested in language L, in mode of discourse D, in way of speaking S?', or alternatively: 'What makes it possible for this distinction to be employed in language L etc.?' And *this* question I shall seek to answer by reference to sortals and in this way offer an 'explanation' of the subject-predicate distinction (*see* Chapter III).

Having set out and offered some elucidation of some of my aims, it is now necessary for me to explain why I speak of the sortal as a category of *symbol* and not simply as a category of *expression.*

B. Sortals, Names, Descriptions, Predicables, Substantival General Terms, Predicates

B.1. To so speak is not to deny that a sortal is a category of expression, but it is unsatisfactory to speak of a sortal merely as a category of expression. To so speak does not bring out the point that the understanding of a sortal as set forth in my preliminary account in A above has a necessary reference to role or function. A sortal is a logical classification of an expression determined (a) partly by reference to grammatical criteria; it has the grammatical form of a common noun exhibiting the features of admitting the indefinite article and plural form in its own right; (b) partly by reference to semantic criteria - it is not analysable into parts which themselves have meaning but it itself has meaning; (c) partly by reference to what may be termed 'logical criteria', namely, it is a complete symbol; (d) partly (indeed, relatively speaking, principally) by reference to function. A sortal is the logical classification of an expression *qua* furnishing in its own right a principle of distinction and counting for particulars in the sense of objects. It is in this regard a symbol in that a symbol is not merely an expression but an expression *qua* performing a certain general role.[30]

We may, in this context, observe the following:

A *name*, a proper name, singular name, is a logical classification of an expression determined:

(a) partly by reference to grammatical criteria - it does *not* admit of the indefinite article or the plural form;[31]
(b) partly by reference to semantic criteria - it is not analysable into parts which themselves have meaning, though whether a proper name itself has meaning has been a matter of considerable dispute;[32]
(c) partly by reference to 'logical criteria'; a proper name has a complete sense in its own right;[33]
(d) partly (indeed principally) by reference to function - a proper name stands for, or, to refer to Quine, 'purports to stand for' a single object.

A *general name* again is a logical classification of an expression determined:

(a) partly by reference to grammatical criteria - it has a grammatical form of a common noun which admits of the indefinite article and the plural form;
(b) partly by reference to semantic criteria viz. it is not analysable into parts which themselves have meaning, but itself has meaning;
(c) partly by reference to logical criteria - a general name has a complete sense in its own right;
(d) partly (indeed principally) by reference to its role or function; a general name stands for or 'purports to stand for' several objects.

A *definite description* in Russell's sense is again a logical classification of an expression determined:

(a) partly by reference to grammatical criteria - it is a phrase beginning with a definite article, of the form 'The so-and-so';
(b) partly by reference to semantic criteria - the parts of a definite description, as opposed to those of a proper name, do have meaning;
(c) partly by reference to logical criteria - a definite description does not have a complete sense in its own right; it is an 'incomplete symbol';[34]
(d) partly (indeed principally) by reference to its function - a definite description is reducible to a set of predicates which hold true or false of just one object.[35] Crudely, a definite description, as the locution suggests, describes just one object.

An *indefinite description* in Russell's sense is equally a logical classification of an expression determined:

(a) partly by reference to grammatical criteria - it is a phrase of the form '*a* so-and-so';
(b) partly by reference to semantic criteria - an indefinite description has meaning and in that it is comprised of a complex phrase its parts also have meaning (e.g. 'a man who sits at the receipt of custom');
(c) partly by reference to logical criteria - it too is an 'incomplete symbol';
(d) partly (and indeed principally) by reference to function; an indefinite description is reducible to a predicate or set of predicates which hold true or false of more than one object. Crudely, an indefinite description describes more than one object.

A *predicable*, as introduced by Geach[36] is a logical classification of an expression determined:

(a) partly by reference to grammatical criteria - it has as its grammatical form a verb in the indicative mood or a verb in such a mood together with a complement;
(b) partly by reference to semantic criteria; predicables have a sense, but not a complete sense;
(c) partly by reference to logical criteria: a predicable is a fundamentally incomplete expression, a functional expression writable as '---- is *F*' or '---- *F*'s';
(d) partly, and indeed crucially, by reference to function: a predicable is a functional expression which is a *potential predicate*. Geach writes: '... it has sense only as contributing toward the sense of a proposition; not at all by itself'.[37]

A *substantival general term*, again as introduced by Geach,[38] a 'mass term' (Quine), a 'matter' term (Aristotle) is a logical classification of expression determined:

(a) partly by reference to grammatical criteria - such an expression is a common noun which does not admit of the indefinite article or plural form in its own right and which is typically a common noun for a material substance (cf. Strawson: 'a material name');
(b) partly by reference to semantic criteria - it is not analysable into parts which have a meaning but it itself has meaning;
(c) partly (and indeed principally) by reference to function - it supplies a 'criterion of continued identity'.

Geach writes:

> I had here best interject a note on how I mean this term "criterion of identity". I maintain that it makes no sense to judge whether *x* and *y* are 'the same' or

> whether *x* remains 'the same' unless we add or understand some general term - 'the same *F*'. That in accordance with which we thus judge as to identity, I call a *criterion* of identity: this agrees with the etymology of "criterion".[39]

He later adds: 'Countability is a sufficient condition for our considering as substantival a term in respect of which we count things' and this is so:

> because we (logically) cannot count *A*'s unless we know whether the *A* we are now counting is the same *A* as we counted before. But it is not necessary, in order that "the same *A*" shall make sense, for the question 'How many *A*s?' to make sense; we can speak of the same gold as being first a statue and then a great number of coins but "How many golds?" does not make sense; thus 'gold' is a substantival term, though we cannot use it for counting.[40]

In so far then as the accounts of all types of expression set out above[41] have a necessary reference to function or role, they are symbols. A symbol is an expression in a role or *qua* performing a certain role and I justify my speaking of sortal *symbols* by claiming parity with these other cases. It may be asked why I speak in terms of symbols as opposed to *terms* - singular terms, count terms, substantival general terms etc. My reasons for this are: (i) 'term' seems to be rather too restrictive in the sense that a 'term' is an item of language where there is no clear distinction introduced between language and the use of language (speech) or where there is resistance to making language and speech alternatives;[42] (ii) the use of 'term' in some philosophical writing carries with it a conflation of (a) a category of symbol or expression and (b) the role which such a symbol or expression plays on an occasion - as, for example, in Strawson's notion of 'subject term', 'predicate term'.[43]

It will be noted that in the above list of symbols I have included *predicable* and not *predicate*. This is for the reason that *predicate* introduced as a class of symbol is strictly a confusion. A predicate is not strictly a class of symbol but a role a class of symbol plays on an occasion. As Geach has it:

> a *predicable* is an expression that gives us a proposition about something if we attach it to another expression that stands for what we are forming the proposition about; the predicable then becomes a *predicate* and the other expression becomes its subject; I shall call such a proposition a *predication*.[44]

Even though I have restrained from introducing 'predicate' as a class of symbol for reasons just offered, it will be necessary at stages throughout this book to so speak since some of the theses to be discussed: (i) cannot strictly or even intelligibly be discussed in terms of predicables; (ii) common philosophical parlance demands that they be so discussed. In spite of Geach's insights it has been, and still continues to be, quite common to speak of predicates as a class of symbol, expression or 'term'.

C. Plan of Campaign

I now turn to my plan of campaign. I shall first develop my preliminary account of what I understand as a 'sortal symbol', arguing for the thesis that such symbols are *not* to be construed as either (i) a type of name; (ii) a type of predicable or predicate. Rather, as I have already indicated, I shall argue that they are required for the introduction of such categories of symbol. I shall distinguish between principles of counting and distinction furnished by sortals and such principles formed from 'matter' or 'mass' terms arguing that neither principle may be reduced to the other. I shall then consider a general thesis which might be thought to arise from the arguments advocated in support of my case that sortals are not reducible to predicates or predicables - namely, that at a general theoretical level, sortals are required in order for us to indicate or specify the range of values which may coherently be substituted for an individual variable in a predicative function. I shall reject this strong thesis but nevertheless hold that at such a general theoretical level sortals *are* implicitly required such that whilst in one sense we may embrace Russell's doctrine of 'an essentially undetermined variable "*x*", there is an important sense in which we may *not* (Chapter I).

In parallel fashion I shall argue that at the level or aspect of language as discourse (speech), identification by reference to sortals is logically prior to naming/referring on the one hand and predicating on the other and further that: (a) 'descriptive identification' as introduced by, e.g. Gareth Evans in *The Varieties of Reference* is either reducible to sortal identification or may be regarded as a *quasi* form of sortal identification; (b) 'demonstrative identification' as introduced by Evans rests on sortal identification and may not be regarded as *sui generis* in the way he holds (Chapter II).

I shall then turn to the subject-predicate distinction itself. I shall distinguish between: (a) *explication* of the distinction; (b) an *explanation* of the distinction in the sense of an answer to the question: 'What differentiates an expression occurring in subjective role from an expression occurring in predicative role?'; (c) a *criterion* for the distinction; (d) an *explanation* of the distinction in the sense of answers to the questions: (i) 'What makes the subject-predicate distinction *possible*?'; (ii) 'What *underlies* the distinction?'. I shall offer Geach's 'explanation' as an explication; an importantly revised version of Strawson's 'new' or 'mediating' criterion as given in *Individuals* as an explanation in sense (b); a criterion derived from my own explanation, having first considered Quine's 'explanation' as a possible criterion; an explanation in sense (d) specifically in terms of the presence of sortals in language, more generally in terms of the presence of symbols which supply principles of counting and distinction in their own right (Chapter III).

Next, I shall attempt to work out the *general* consequences of failure to recognise the category of the sortal in that sortals are reduced to 'predicates'; these general consequences I have initially set out at the beginning of A.2. above. This exercise will occupy Chapter IV.

In the rest of the book I shall:

(1) Seek to establish that the writers I have selected do in fact fail to recognise the category of the sortal;
(2) Attempt to work out the peculiar difficulties arising out of such failure;
(3) Illustrate how the general consequences set out in Chapter IV have particular exemplification in the writings of selected authors.

I shall do this first by reference to Frege (Chapter V) and then by reference to Russell (Chapter VI); Geach (Chapter VII); Strawson (Chapters VIII, IX and X) and finally by reference to Quine (Chapter XI). It will be noted that three chapters are devoted to Strawson. My justification for this is the considerable contribution he in particular has made to the issues I am discussing and an undeniable fascination I personally have with his writing which has presented a constant challenge and source of philosophical reflection. After all, Strawson presents us with no less that five accounts of criteria for the subject-predicate distinction in his various works and has avidly sought to establish an ultimate basis for this distinction. Indeed, I hold that a discussion of Strawson's work in this area to be valuable and also illuminating in its own right but particularly so in the context I provide. My penultimate chapter on Quine will draw special attention to: (a) what is required as the basis for his doctrine of 'canonical notation'; (b) his list of 'basic constructions'. I shall claim that 'identification' must be admitted to his list. My final chapter will take the form of a survey of what I have argued in the course of the book and the drawing of conclusions from this survey vindicating my own account of sortals and their importance in our understanding of the nature of systems of discourse.[45]

Notes

1. P.F. Strawson, *Individuals* (London: Methuen, 1959), p. 168; my position even so is to be clearly distinguished from that of Strawson; cf. ch. I, pp. 21-24.
2. Frege.
3. Russell.
4. Russell.
5. Geach.
6. Frege.
7. Strawson, *op. cit.*
8. cf. ch. II, p. 84.
9. In *Reference and Generality* (Ithaca, New York: Cornell University Press, 1962), ch. 2.
10. The question of whether identification in speech is prior to identification in thought will be discussed explicitly in relation to Gareth Evans's position in *The Varieties of Reference*, ed. John McDowell (Oxford: Clarendon Press, 1982), *see* ch. II, sec. M.
11. cf. my discussion of Evans, ch. II.
12. Strawson.
13. Strawson, *op. cit.*

14. I have introduced both (i) and (ii) since some writers, e.g. Geach wish to speak of the role or function of an expression rather than in terms of the use of it or our use of it.
15. Strawson, *op. cit.*
16. Geach. Geach is to be excluded from this general attack since he does hold that the introduction of proper names rests on the introduction of substantival general terms which he does not construe as predicates but he fails to recognise sortals as a distinct category of symbol or 'term' (*see* ch. VII).
17. Frege. Fregean names include Russellian Definite Descriptive phrases and other phrases beginning with the definite article.
18. Russell.
19. Quine.
20. Frege treats such definite forms as *names*, but this does not affect the present point.
21. As is well known, Russell in his final analysis holds that what may occur in such a role is a 'logically proper name': Quine in his final analysis reduces names to pronouns - in logic the variables of quantification.
22. cf. n. 21 above.
23. Strawson in 'Singular Terms and Predication', repr. in P. F. Strawson, *Logico-Linguistic Papers* (London: Methuen, 1971), pp. 53-74.
24. Strawson in *Individuals*.
25. *See* ch. VIII.
26. Frege's position is different (*see* ch. V); Geach clearly does hold such a doctrine.
27. For example, Saul Kripke in 'Identity and Necessity', in Stephen P. Schwartz (ed.), *Naming, Necessity and Natural Kinds* (Ithaca, New York and London: Cornell University Press, 1977), pp. 66ff., esp. pp. 89-94. Russell (e.g.) comes to treat ordinary proper names as 'veiled descriptions' but in that he does this he would reject an account of 'logical subject' in terms of 'identifying'. An expression in the role of logical subject is a 'logically proper name' (a demonstrative); an account of 'logical subject' is to be given in terms of *naming*.
28. The use of 'indirectly' here will be illustrated in Chapter IV and fully explained in Chapter X, section A.1. in reference to Strawson's search in *Individuals* and 'Particular and General'. Again the use of 'indirectly' in relation to (4) above will be explained in Chapter X, section C.
29. For those who wish to speak of *our* employing the subject-predicate distinction.
30. Russell, it is to be noted, speaks in terms of *symbols*; *see* his *Introduction to Mathematical Philosophy* (London: Allen and Unwin, 1930), ch. XVI.
31. Frege, *see* ch. V.
32. Russell holds that logically proper names have no meaning other than what they stand for, i.e. they have no 'nominal essence' but logically proper names are a specially introduced class of symbol. Locke held that there is no 'nominal essence' of proper names - a thesis challenged by Anscombe in G. E. M. Anscombe and Peter T. Geach, *Three Philosophers* (Oxford: Basil Blackwell, 1961).
33. cf. Russell and Geach.
34. *See* Russell's analysis of definite descriptions in (e.g.) *Introduction to Mathematical Philosophy*, ch. XVI; Lecture VI of his 'Lectures on the Philosophy of Logical Atomism', repr. in R.C. Marsh (ed.), *Logic and Knowledge* (London: Allen and Unwin, 1956), pp. 177-281.
35. *See* note 34 above.
36. *Reference and Generality*, ch. 2.

37. *Ibid.*, p. 32.
38. *Ibid.*, p. 39.
39. *Ibid.*
40. *Ibid.*, pp. 39-40.
41. And here one could add, e.g. 'Rigid Designator'.
42. In spite of these worries I shall sometimes find it necessary or convenient, on account of the context of discussion, to use 'term'. For further comments on (i) *see* my discussion in connection with Gareth Evans' point in sec. M of ch. II.
43. *See* Geach's criticism of Strawson in his review of *Subject and Predicate in Logic and Grammar*. I have had difficulty in tracing the precise publication details of this review and am grateful to Professor Geach for his help. The original review, from which I take my point, was published in the TLS soon after the book was published (1974). A more carefully reasoned critique was published in the *Festschrift* for Strawson, Zak van Streaton (ed.), *Philosophical Subjects* (Oxford: OUP, 1980).
44. *Reference and Generality*, p. 25.
45. Given relatively recent developments in 'Discourse Theory', I have to make it clear that my use of 'discourse' pre-dates such developments. Roughly, my use of 'discourse' corresponds to that use of 'discourse' as put forward and understood by e.g. Strawson in *Individuals*, *see* (e.g.) pp. 15, 16, 29, 137 *et al.*

Chapter I

Sortals, Names and Predicables

A. Sortals: a Detailed Investigation

A.1. As I have said in my preliminary account of a 'sortal' given in my 'Introduction', by a 'sortal' I shall initially mean a symbol which furnishes us with a principle for distinguishing and counting particulars and which does so in its own right relying on no antecedent principle or method of so distinguishing and counting. Grammatically a sortal in this sense takes the form of a common noun which:

(i) takes the indefinite article in its own right;
(ii) takes the plural form in its own right.

This account of 'sortals' rules out as sortal expressions:

(a) Common nouns which are derived from other parts of speech (e.g. adjectives and verbs) via the criterion of 'taking the indefinite article and plural form in its own right';
(b) Common nouns such as 'gold', 'snow', 'water', namely what Strawson calls 'material names'[1] or words designating 'feature' universals[2] and what other writers have variously called 'mass' terms (Quine), 'matter' terms (Aristotle), 'substantival general' terms (Geach) via the same criterion and, by implication, it rules out such noun forms as being forms of symbol which furnish, in their own right, principles for distinguishing and counting particulars. That such nouns are so ruled out and that they may not, as such, furnish such a form of symbol, I shall presently seek to establish.

To consider (a) first. This may be illustrated as follows. (i) 'Author', e.g. in Russell's famous example 'The author of *Waverley* was Scotch', is a common noun but it is derived from the verb 'to write'; an author is one who writes or has written. 'Author' does not take the indefinite article in its own right since we may only speak of 'an author' in that we may speak of an *F* (where '*F*' is a sortal expression) who writes or who has written. Again we may only speak of 'authors' in that we speak of *F*'s who write or have written. 'Author' takes the indefinite article and hence the definite article by derivation from the common noun 'man' or '*F*'. We may only have the form of expression, form of words, 'The author of *Waverley*' in that we have the form of expression, form of words, 'The *person* (*man*, *F*) who wrote *Waverley*'.

Similarly with a verbal noun such as 'runner' or 'swimmer'. (ii) We may only count the number of authors in that we may count the number of persons who write or who have written. A person could not begin to comply with the instruction: 'Count the number of authors' unless he understood what was to count as one author and to understand what it is to count as one author he has to have the concept of 'one *F* who writes or has written'. Indeed the point that such nouns as 'author' may not be regarded as the grammatical form of symbols which supply, in their own right, principles for distinguishing and counting particulars accidentally comes out in Russell's 'informal' analysis of definite descriptions when he offers an analysis of 'The author of *Waverley* was Scotch' in terms of: 'At the least and at the most one person wrote *Waverley* and whoever wrote *Waverley* was Scotch',[3] but its impetus is not recognised by him since he reduces all sortals, (e.g. 'person') to indefinite descriptions and in due course to predicates.

The characteristic feature of sortals taking the form of common nouns exhibiting the features (i) and (ii) above have been well noted by Frege,[4] and by Geach in his comments on Frege,[5] but they were not thought by Frege to mark off a distinct category of symbol; rather for him, as we shall see, they mark off what he calls a 'concept', "properly so called". Again whilst Geach rightly draws attention to: (a) the focusing by Frege on such symbols; (b) the importance of such symbols, he likewise does not recognise that they form a distinct category.[6]

To consider (b). Such common nouns as 'gold', 'water' etc. are ruled out as grammatical forms of sortals in the sense I am considering. They do not, of themselves, take the indefinite article or the plural form. It is clearly nonsense to speak of '*a* gold' or 'gold*s*' or of '*a* water' or 'water*s*',[7] and in that it is nonsense such nouns may not be said to be the grammatical form of symbols which furnish, in their own right, principles of distinction and counting for particulars. In order to take the indefinite article or plural form they have to be precursed by a phrase such as 'bit of', 'piece of', 'parcel of', 'vein of', 'stretch of', 'area of'.[8] As Quine has aptly observed such 'mass terms' do not divide their reference of themselves.[9] By contrast, a sortal expression, common nouns such as 'cat' or 'apple', in no wise have to be precursed in order to furnish a principle of distinction and counting for particulars.

Here however it will be said that whilst it is a *sufficient* condition of X being a symbol which furnishes, in its own right, a principle of counting and distinction for particulars that X should have as its grammatical form a common noun exhibiting the features (i) and (ii) above, it is not a *necessary* condition. For whilst such common nouns as 'gold', 'water' may not, of themselves be the grammatical forms of such symbols, such nouns, when precursed by e.g. 'parcel (of)', 'stretch (of)' they *do* present us with such a grammatical form. In other words such locutions as 'piece of gold', 'stretch of water', *do* present us with principles of counting and distinction for particulars and do so in their own right. In general, such items as 'piece of *M*', 'parcel of *M*' (where '*M*' is a 'material name', 'matter' term) present us with such principles. That they do may be clearly illustrated. One can be asked to count and one can proceed to count the number of bits or pieces of gold at a certain location or stretches of water in a certain area without having to introduce any antecedent principles of counting. One cannot claim that a person can only count the number of pieces of gold or stretches of water in that he can first count the number of cups, coins, rings that are golden or the number of lakes, rivers etc., that are water, since it will be objected[10] that it is only *contingent* that pieces of gold or stretches of water are cups, coins, rivers, lakes and ponds. In a given situation bits of gold, stretches of water may be all that one is faced with but one can go ahead and count the number present. More generally, to invoke Strawson's vocabulary, 'feature' universals are not reducible to 'characterizing universals'.[11]

Now I agree that such phrases as 'bit of *M*', where '*M*' is a 'material name', may be a grammatical form of symbols which furnish, in their own right, principles of counting and distinction for particulars and that 'feature' universals are not reducible to 'characterizing universals', central to my argument that sortals are required for the introduction of names and predicables (predicates) is the thesis that sortals are required for the introduction of the notion of a *single object* and just *this* might well be denied in the light of the position that noun phrases such as 'bit of *M*', 'stretch of *W*' are grammatical expressions of symbols which supply, in their own right, principles of counting and distinction for material particulars. That is, it might be held that such a principle is *sufficient* for the introduction of the notion of a 'single object'; *all* we need for such a notion is such a principle. It is not necessary that we introduce sortals.

A.2. I shall first consider this thesis and raise objections which I regard as sufficient to reject it thus holding that it *is* necessary that we introduce sortals. I shall then consider the converse thesis viz., that all we need for the introduction of particulars (and hence the names of particulars) introduced by 'matter' terms precursed as above are *sortals* i.e. that sortals are *primary vis-à-vis* principles of counting and distinction for material particulars in the sense that they are sufficient for the introduction of such principles.

To turn then to the arguments in favour of the thesis that principles of counting and distinction furnished by 'matter' terms precursed or modified in a certain way[12] are the *primary* principles of counting and distinction for objects, primary in the sense of 'sufficient'. The thesis is:

(1) It is only necessary that we have such symbols, i.e. it is sufficient that we have 'matter' terms precursed in the way already illustrated for the introduction of 'single objects' and hence for the introduction of names and predicables (predicates) relating to single objects.

This minimal thesis is to be clearly distinguished from:

(2) It is necessary that we *only* have such symbols;
(3) It is necessary *only* that we have such symbols;

and (1) does not entail (2) or (3). (2) is a much stronger thesis; it would entail that sortals are eliminable in favour of, in some way reducible to 'matter' terms so precursed - that for the sortal 'man', say, we could write either: (i) 'man of ----', which is absurd or; (ii) 'animated parcel of flesh and bone', whereas the truth of the matter is the other way round. That we have one animated parcel of flesh and bone is determined by the fact that we have *one* man. It is not contingent that a man is an animated parcel of flesh and bone; in that we have a man we have one such parcel. It is however contingent that an animated parcel of flesh and bone is a man; more generally that an animated parcel of flesh and bone is an *F* (sortal). (3) is the strongest thesis of all. This would of necessity eliminate sortals as principles of counting and distinction for particulars altogether. It would of necessity reduce sortals to predicates; in Strawson's vocabulary it would of necessity reduce sortal universals to characterizing universals.

Having thus distinguished (1) from the stronger theses (2) and (3), let us now return to (1). This minimal thesis I contend is unsatisfactory. Having *only* such concepts as 'piece of gold' (in general 'piece of *M*', 'bit of *M*', etc.,) is *not* sufficient for the introduction of the concept of 'a single object' or the introduction of the proper names of a single object since:

(i) the concept of 'a piece of gold' (e.g.) is the concept of a relatively indeterminate parcel of matter; 'a coin', 'a ring', 'a cup' etc. are concepts of quite determinate parcels of matter. A coin, a ring, a cup etc., are quite determinate parcels of matter in that they are such objects i.e. we have one parcel of matter in that we have one coin, one ring etc. We cannot, I hold, from merely being in possession of a concept such as 'piece of gold' ('piece of *M*') claim we are in possession of a sortal such as 'cup', 'coin', 'ring'; more generally a 'material object' sortal. Here it might be objected that we can *derive* the concept of 'an *F*', where '*F*' is a material object sortal, from the concept of 'a piece (parcel) of *M*' (where '*M*' is a 'matter' term) together with the notion of 'shape' or 'form' in apparently Aristotelian fashion[13] but: (a) the thesis here being considered is that 'piece of *M*' itself is sufficient to yield a material object sortal; (b) not *every* object falling under a sortal has a distinctive shape or 'form' which can be specified independently of the sortal e.g. pipe, car, man, tree, tulip, glove. In such cases, and many more which one may produce, one has to introduce the sortal or at least have implicit reference to the sortal in order to introduce the appropriate 'shape of', 'form' which, together with the notion of a 'parcel of *M*', is supposed to be sufficient for the introduction of the sortal. There is *no* distinctive shape or 'form' of e.g. pipe, car, man, tree, tulip, or glove which can be specified independently of pipes, cars, men, trees etc.[14]

(ii) from the fact that Mr. A can count e.g. the number of pieces of gold at a certain location or the number of stretches of water in a given area it does not follow that he can count the number of coins, rings, or cups which are golden, of gold; or again, the number of lakes or ponds. Being able to count pieces of gold is not sufficient for being able to count *F*'s that are of gold. He can simply be said to count pieces of gold of some shape or 'form' or again the number parcels of water of some shape or 'form' but not necessarily pieces of gold or parcels (stretches) of water that are ponds, lakes.

A.3. To consider the converse case, namely, that sortals are *primary* viz., sufficient for the introduction of particulars introduced by 'matter' terms precursed in the way illustrated; in that we introduce names by reference to sortals we therein or thereby[15] have the means of introducing names introduced by reference to 'matter' terms so precursed.

Let me first draw the appropriate parallel with the minimal thesis advocated in (1) above. On the current thesis one would have to hold:

(i) that the notion of an '*F*', where '*F*' is a material object sortal, is *sufficient* to give us the notion of 'a piece' or 'a bit' ('a parcel') of matter. In that (e.g.) we have the concept of a 'lake' or 'pond' we therein have the concept of a 'stretch' or 'expanse of water'. In that we have e.g. the concept of a 'coin', 'cup', or 'ring' we therein have the concept of a 'quite determinate parcel of matter';

(ii) from the fact that Mr. A can count the number of rings, coins, cups etc., it follows that Mr. A can count the number of pieces of gold etc., or again, in that he can count the number of lakes, ponds, etc., it follows that he can count the number of stretches of water.

(i) and (ii) clearly will not do. It will be objected that in that one has e.g. the concept of a 'lake' or 'pond' all that one has is the concept of a 'stretch of water' which is a 'lake' or a 'pond'. One cannot be said to have the concept of a 'stretch of water' *simpliciter* since it is contingent that stretches of water are lakes or ponds. And this consideration prevents one from arguing as follows:

(a) In that we have the concept of a 'coin' or 'ring' or 'pond' we have the concept of a quite determinate parcel of matter; in general, in that we have the concept of an *F* which is material, we therein have the concept of a quite determinate parcel of matter but we *thereby* have the concept of a relatively indeterminate parcel of matter i.e. a parcel of matter which is *simply* a parcel of matter e.g. 'a bit or piece of gold or bronze'. In other words it prevents us from arguing that the notion of an '*F*' which is a quite determinate parcel of matter is *sufficient* to give us the notion of a relatively indeterminate parcel of matter; from arguing that in that we have the concept of the determinate we have the concept of the relatively indeterminate - that there is no extra concept which we need for this; the indeterminate rests on the determinate - it is derived from the determinate. To use Quinean language in that we have terms which divide there reference of themselves we *thereby* have terms ('mass' terms) which do not so divide their reference;

(b) In that Mr. A can count the number of lakes, ponds etc., he therein can count the number of parcels of matter, stretches of water, which are lakes, ponds etc., and he *thereby* can count the number of stretches of water on the same principle - ability to count the determinate entails ability to count the indeterminate.

And there is a further argument which prevents us from advocating such a thesis. The thesis that in that we have the concept of a material sortal (e.g. a 'cup', a 'ring') we *thereby* have the concept of a parcel of matter which is *simply* a piece or lump of matter also breaks down for the reason that it construes such a phrase as 'piece of gold' ('piece of *M*') as analysable as 'piece (of gold)' and not as a functional expression 'piece of ----', which, on investigation, it is.[16] That such a thesis relies on a non-functional account is made clear by the position that *coin* (e.g.) is held to be a quite determinate parcel of matter, whereas *piece* is held to be an indeterminate parcel of matter. The case for holding that e.g. 'piece of gold' 'stretch of water' are not to be analysed in such a way, but as functional expressions 'piece of ----'; 'stretch of ----' is: *necessarily* any piece, stretch, chunk etc., is a piece, stretch etc., *of* something or other and gains its identity from that of which it *is* a piece, stretch, chunk etc. That is, there are no such X's as pieces, stretches, chunks *simpliciter*: we have one piece only in that we have one *piece of M* (where '*M*' is a 'matter' term); we have the same piece only in that we have the same *piece of M*. Strawson comments[17] that 'feature' universals, (*snow*, *gold* etc.,) whilst not furnishing criteria of identity and distinction for particulars may be 'modified' to do so by such items as *lump*, *dump* etc.; I have spoken of 'matter' or 'mass' symbols (or 'terms') being precursed in a certain way i.e. by 'piece (of)', 'lump (of)', so as to yield principles of counting and distinction in their own right. This is strictly incorrect in the light of the above. Rather one should say that one may form a principle of counting and distinction (in its own right) for particulars by taking the functional expression 'bit of ----', 'piece of ----' etc., and completing it by a 'matter' term, or, in Strawson's vocabulary by an expression for a 'feature' universal. Alternatively, that one may form such a principle from a 'matter' term by enclosing it in the context of a functional expression 'piece of ----' etc.

It may be counter objected here that one may distinguish between: (i) how such a phrase as 'piece of gold', 'stretch of water' is formed or may be formed; (ii) how such a phrase is to be analysed. After all we *form* a subject-predicate sentence by writing the subject first and then the verb. This however is a mere accident of English and in any case what matters is how the phrase is to be analysed.

It will now be asked: 'Granted the analysis of such phrases as functional expressions, how can we make the original and surely correct claim that: (i) in that we can count the number of lakes in a certain place we can count the number of stretches of water which are lakes, ponds; (ii) that in that we have the concept of a 'lake' or a 'pond' we have the concept of a stretch of water which is a lake or pond? We can make such claims and grant such an analysis on the grounds that whilst *necessarily* any lake, pond, etc., is a stretch of water, it is only *contingent* that any stretch of water is a lake, is an *F*.

A.4. As I have rejected the thesis that principles of counting and distinction furnished from 'matter' terms precursed or 'modified' in the way illustrated are *sufficient* for the introduction of sortals, I equally reject the thesis that sortals are *sufficient* for the introduction of material particulars i.e. particulars introduced by reference to principles of counting and distinction formed from 'matter' terms. Neither principle is sufficient for the introduction of particulars introduced by the other. And neither principle *itself* is sufficient for the introduction of the other since both are principles of counting and distinction for *particulars*. In that I have shown that neither principle is sufficient for particulars introduced by reference to the other I have therein shown that neither principle itself is sufficient for the other. There is no other way in which one could show this. I am thus forced to the conclusions:

(1) that these two principles of counting and distinction for particulars are independent principles; neither is sufficient for the other; neither is *primary*;
(2) that we *do* need sortals in the sense I have already introduced for the introduction of the notion of a 'single object' in the sense that principles of counting and distinction formed from 'matter' terms are *insufficient* for the introduction of *this* notion;
(3) that we do need principles of counting and distinction furnished from 'matter' terms for the introduction of the notion of a 'material particular' in the sense of 'piece of *M*', 'stretch of *W*' etc.;
(4) under the heading of 'material particular' therefore we have *two* independent categories or types of particular: (i) sorts of material particular; (ii) kinds of material particular, each independent of the other. In more crude general terms, sorts of thing are not reducible to *kinds* of thing or *vice versa*.

It might here be interjected that all that has been established in the above arguments are the necessary (and trivial) truths that:

(a) material sortals are needed for the introduction of the notion of a 'single material object'; or more generally that sortals are needed for the introduction of the notion of a 'single object' and hence for the introduction of a name for such an object *vis-à-vis* other principles of counting and distinction for particulars;
(b) principles of counting and distinction formed from 'matter' terms are required for the introduction of the notion of a 'parcel of matter' *vis-à-vis* other principles of counting and distinction for material particulars.

However in so far as it is necessary that sortals are required for the introduction of the notion of a 'single material object' or a 'single object' it does not follow that either: (a) above is necessarily true or that it is a necessary truth in the sense: 'Necessarily it is true that ...'. All that follows from (a) is that it is true that it is necessary that material sortals or sortals in general are required, *not* that the proposition itself is necessarily true and from: 'It is true that it is necessary that *P*' it does not follow: 'Necessarily it is true that *p*'. But even if it did so follow such that we could hold (a) to be a necessary truth it is not thereby trivial, since not all, if indeed any, necessary truths are trivial. Consider: 'Nothing is both red and green all over'. Further, (a) *does* need to be argued for in the way I have attempted to do earlier since: (i) there is historical precedent for in effect holding that we may analyse out sortals in favour of 'matter' terms together with the notion of 'form' or 'shape'. Aristotle, for example, held that one could offer an analysis of 'house' (e.g.), in terms of bricks and mortar arranged in a certain way or order; the 'form' of a house was the arrangement of the bricks and mortar (or other building materials).[18] And this may well be regarded as expressing in a material mode what one expresses in a formal mode by saying that principles of counting and distinction for particulars furnished by 'matter' terms are sufficient if supplemented by the notion of an 'arrangement', 'form', for the notion of a principle of counting and distinction furnished by a sortal in the sense I have introduced. (ii) Geach has argued, contra Quine, that we may introduce quantifiers that bind variables other than proper name variables, but he only introduces 'substantival general terms' in this connection, not 'substantial ones'.[19] (b) needs to be argued for in the light of Aristotle's terrible struggles in the *Metaphysics* as to whether 'matter' or 'form' is prior.[20]

A.5. Having thus discussed the issue of the alleged primacy of principles of counting and distinction furnished from 'matter' terms for the introduction of the notion of a 'single object' and maintained the thesis that it is necessary that we have sortals for the introduction of the notion of a 'single object', I shall now return to the issue of what further classes of expression are ruled out as expressions for sortals via the criteria presented at the beginning of A.l.

Whilst failure to take the indefinite article and plural form in its own right is a *sufficient* condition for a common noun failing to be the grammatical form of a symbol which in its own right furnishes a principle of counting and distinction for particulars in the sense of 'objects', it is not a *necessary* condition of a word so failing that it should be a common noun. Adjectives and verbs which may and typically do perform a characterizing role in speech - what Strawson, speaking ontologically, designates as 'Characterizing universals' - may not be regarded as the grammatical form of such symbols, of sortals. In Strawson's terms 'Characterizing universals' do not furnish us with such principles. A 'Characterizing universal' I here speak of as a 'predicate symbol'. I now give my reasons for thus speaking. I use 'symbol' because I am, in the main, conducting my discussion in terms of categories of symbol for reasons given in my 'Introduction'; I use 'predicate symbol', as opposed to 'predicative symbol' since Strawson rejects the notion of 'predicative symbol' in that he rejects the notion of 'predicates' as *fundamentally incomplete* symbols[21] - that is he rejects the Frege/Geach account of predicates in terms of functions and ultimately treats predicates as a sort of name.[22] Embarrassment is caused by the locution 'Characterizing (universal)' in that Strawson confuses: (i) a category of symbol - in Geach's terms a category of expression (term);[23] (ii) the role or function of a symbol on an occasion of use. Strictly 'characterizing' is a role or a function a symbol may perform on an occasion in speech. A 'Characterizing universal', a 'predicate symbol' is clearly to be distinguished from a sortal. Such a universal, such a symbol, whilst supplying a principle for distinguishing and counting particulars supplies such a principle *only* for particulars already distinguished or distinguishable in accordance with some antecedent principle or method. 'Characterizing universals', predicates in this sense, typically take the form of a verb or adjective e.g. 'runs', 'red'. That 'Characterizing universals', predicates in this sense, *do* suppose an antecedent principle for distinguishing and counting has been well presented by Strawson[24] and may be illustrated as follows. Take the case of 'red'.[25] One may only count the number of red things in this room, say, in that one understands *what* one is to count. One cannot be asked simply to count the number of reds. One may only *start* counting the number of red things and only make an *end* to such counting in that one understands that one is to count the number of red tables, cups, saucers, bottles (the number of red *F*'s); or again that one understands that one is to count the number of red pieces of paper, lumps of clay, bits of stone.

A.6. The account of 'Sortal' which I have given thus far is clearly an adaptation of Strawson's introduction to the notion of 'Sortal universal' on p. 168 of *Individuals* tightened up by a more precise account of the kind of common noun implicit in the introduction of such universals. This is readily acknowledged but I wish to stress the point that it is an adaptation of Strawson's account, not a recitation. The ways in which my account differs from that account are crucially important. They are:

(1) I simply introduce the form: 'sortal symbol' or 'sortal' and not 'sortal universal' since I do not wish to be committed to Strawson's particular metaphysics. Indeed I see no good reason to commit myself to his doctrine of particular and universal *terms* and for him a sortal universal is a term.[26] Indeed I hold that there are good reasons for *not* being so committed.[27]

(2) Strawson writes (*ibid.*): 'A sortal universal supplies a principle for distinguishing and counting individuals which it *collects* ...'.[28] This supposes that we must introduce, by some antecedent means, the notion of 'an individual particular', a particular falling under which is then to be 'collected' by a sortal universal. My position is to be clearly distinguished from that put forward by Strawson for it is *my* thesis: (a) that in the light of the discussion in A.2. above it is plausible to hold that the notion of an 'individual particular', in the sense of an 'individual object', itself is sufficiently introduced via the notion of a

'sortal'; (b) that the notion of a 'sortal' itself is necessary is for the introduction of the notion of an 'individual particular' in the sense of an 'individual *object*'. Whilst a sortal *may* appear in the role of collecting particulars the appearance of a sortal in such a role is *secondary* to and rests on the prior introduction of some sortal which allows us to introduce the individual particulars which are to be collected. Thus 'animal' may serve to collect a number of individual men, horses and dogs, but in performing such a role it presupposes the introduction of the respective sortals 'man', 'horse' and 'dog', hence 'man', 'horse' and 'dog' cannot themselves be thought to be principles of collection relative to 'animal' unless we are able to produce some further 'lower order' sortal which itself specifies *which* individuals are to be collected by 'man', 'horse' or 'dog' as principles of collection. What is collectable must first be specifiable and countable and what is countable cannot be specified solely by reference to a principle of collection. This conclusion would seem to be denied by Strawson. In the light of what he later says,[29] he would seem to hold that at the 'lowest level' of sortal such sortals supply principles for distinguishing and counting individuals which they *in turn* collect.

At this point Strawson has either:

(i) to claim that what is collected by a sortal at the 'lowest level' is a *pure* particular;

or:

(ii) to claim that what is collected is a particular introduced by reference to a principle of counting and distinction formed from 'matter' terms as discussed earlier;

or:

(iii) to claim that what is so collected is a particular specified by reference to a 'lower order' ('lowest' order) sortal which is *itself* a principle of collection for particulars.

I shall later argue that it is (iii) that he is committed to.[30] I reject (iii) on the grounds that *what* is collected by a principle of collection cannot *itself* be specifiable and countable *only* by reference to such a principle of collection; yet this is what Strawson certainly seems to require. I shall develop this point later in one of my principal criticisms of Strawson - namely that he fails to realise that:

(1) *Necessarily*: to be a particular object is to be of a certain sort:

rather his thesis is:

(2) To be a particular object is *necessarily* to be of a certain sort.

More generally he fails to realise that:

(1)(a) *Necessarily:* to be a particular *is* to be of a certain sort or kind;

rather his thesis is:

(2)(a) to be a particular is *necessarily* to be of a certain sort or kind;

and concomitantly as regards proper names:

(1)(b) *Necessarily*: to be a particular proper name is to be the name of some object of a certain sort or some particular of a certain kind;

Strawson's thesis is rather:

(2)(b) to be a particular proper name is *necessarily* to be the name of some object of a certain sort or some particular of a certain kind;

It is important to note that whilst (2), (2)(a), (2)(b) entail (1), (1)(a) and (1)(b) respectively the converse is not the case. '*Necessarily* p is q' does *not* entail 'p is *necessarily* q'.[31]

Taking (2), (2)(a), (2)(b) permits Strawson to treat a sortal 'universal' as a *predicate* which I shall argue he in fact does,[32] and hence permits him to raise the question: 'What instances sortal universals?' and allow as *possible* answers:

(a) 'pure' particulars;
(b) particulars introduced by reference to 'principles of collection' furnished by 'matter' terms, for him 'feature' universals;[33]
(c) particulars which invoke for their specification sortals *themselves* of which they are said to be *instances*.

But this (c) is surely absurd. Just as what is collected cannot be specified *only* by reference to a principle of collection but must be specified *independently* of such a principle, otherwise how may we speak of what is collected? - equally what is an instance of *P*, where '*P*' is some predicate, must be specifiable *independently* of *P*, otherwise how may we speak of *one* or *many* instances of *P*? This point will be explicated and developed further when I set out the alternatives available to Strawson in trying to answer the question: 'What instances sortal universals?' (ch. IX). My present aim is simply to point out how my position differs from his. For me, sortals, sortal symbols, do not have the primary function of being 'principles of collection' but rather of specifying and counting what is to be collected by sortals in such a role; in crude terms sortals are required in order for us to introduce the notion of 'objects which are to be collected' by some 'principle of collection'. Further, for me sortals *themselves* may not be regarded as *having instances*, but rather: (i) are necessary in so far as we speak of 'objects'; are necessary for the very notion of 'an instance'; (ii) they themselves furnish the notion of 'an instance'. In other words they themselves are not 'predicates' which may *have* instances but underlie the introduction of that which may be said to have instances.

B. That Sortals are a Distinct Category of Symbol in that they are *not* Reducible to Names or Predicables/Predicates

Having thus set out what I initially mean by 'sortal symbol' or 'sortal' and distinguished my position from that advocated by Strawson, I now turn to my thesis that sortals *do* constitute a distinct category of symbol: that they are not reducible to either names or predicables (in more common parlance 'predicates'). I shall now seek to demonstrate this and its implications.

B.1. I shall first attempt to show that sortals may not be regarded as *singular names, proper names*. It might be thought that no one could seriously entertain such a view; but such a view might even be attributed to Strawson in his doctrine of 'sortal universals'[34] and to Plato in so far as in his earlier and classical doctrine of 'Forms', Man (e.g.) is said to be a Form.[35] My case for holding that sortals themselves may not be regarded as singular names is: (i) *it is a mark of a singular name that it may not take the indefinite article or plural form*. In that one might speak of 'Many Aristotles' or 'Many Johns' or 'an Aristotle' or 'a John', one speaks of 'Many men', 'many dogs', 'many *F*'s' called 'Aristotle', 'John'; 'a man (*F*) called "Aristotle"' or 'a man (*F*) having the peculiar gifts and abilities of the man (*F*) called "Aristotle"'. In general, in that one speaks of 'Many N's' or 'An N' one speaks of 'many *F*'s called N' or 'an *F* called N'.[36] (ii) *the definite form of a sortal is itself required for the introduction of singular names, proper names*. We need the singular form of the sortal 'The *F* who ...', 'the *F* which ...' to introduce the bearer of any proper name we introduce or have introduced since in that we introduce a proper name we introduce it as the proper name of some *one* object, some one individual and we need the concept of a 'sortal' to supply *that* notion. Again, in that we introduce a proper name we introduce it with a view to it being used to refer to one and the same individual, object, on a number of occasions and we need the notion of a 'sortal' in order to speak of referring to *one and the same object* on a number of occasions; or again, in order to speak of using the same proper name on more than one occasion. The notion of a 'proper name' which *could* only be used on one occasion is useless as regards communication.[37] That is, we need the general notion of a 'sortal' in order to determine that we are speaking of 'an individual object' and we need the definite form of a sortal to determine which individual object we are going to talk about or are talking about. In the case of proper names *currently* being introduced, we need to introduce the singular form of a sortal in order to produce the rules for the continued use of the names concerned e.g. by 'Mehitabel' I shall mean the cat, the *F*, which ...

Now whilst it may be agreed that as concerns any proper name we introduce, we introduce it as the proper name of some one object and again in that we introduce a proper name we introduce it with a view to it being used to refer to one and the same object on a number of occasions, the thesis that we need the singular form of a sortal for such introduction may well be queried. Why will it not suffice to say that all we need is either: (a) a 'matter' term precursed in the way already discussed above; (b) a predicate or set of predicates; (c) a definite description in Russell's sense? Why cannot a suitably prefixed 'matter' term, a predicate (or set of predicates), or a Russellian indefinite description furnish what Geach calls 'a criterion as to identity' for the proper names of individuals in the sense of 'objects'? To hold that all we need is (a) has already been argued against in A.2. above and so I now turn directly to (b) and (c).

One who advocates (b) regards a predicate as itself a category of symbol as opposed to, as Geach has put it, what a predicable becomes when completed by a name in subject position;[38] in my terms the function a predicable performs in a given proposition - that of being part of a predication, that of performing a predicative role. However a predicate conceived of as itself a category of symbol could not, even so, furnish us with such a criterion of identity for it is essentially an 'incomplete symbol', a 'propositional fragment'. Any attempt to introduce the bearer of a proper name N or produce a criterion of identity for the continued use of the name by such means would indirectly lead us back to invoking the definite singular form of some sortal viz., 'N' is the proper name of *the F which* ..., the predicates (predicables) e.g. '---- wrote *Waverley*', '---- died in misery' hold true: in general, of which the predicates (predicables) '---- is *P*' hold true. A singular term which was alleged to be either a predicate or held in some way to be reducible to a predicate or set of predicates, could never serve in such a role or roles since predicates are true or false *of* something or other - technically predicables are true or false of something or other;[39] and in order to introduce the very notion of 'that which is true or false of something or other' we have to have symbols which serve to introduce and furnish a criterion of identity for that *of which* the predicables may be true or false, or that *of which* the predicates are true or false - a function which cannot be performed by, carried out by, predicables themselves (or predicates themselves, conceived of as a category of symbol).

Further, in that predicates, conceived of as a category of symbol, are essentially propositional fragments they may not serve to introduce individuals or form a criterion of continued identity for individuals since they gain their identity from the propositions in which they occur, of which they are fragments; but symbols which may serve to introduce individuals and form a criterion of continued identity for individuals are required in order to determine the identity of propositions concerning individuals and hence the sense of predicates. Indeed in general it cannot be claimed that a type of symbol whose sense is given by reference to some set of propositions in which it occurs can furnish a criterion of identity for an individual since we should have no way of determining the identity of the individual on such a claim. This may be illustrated as follows. Let us consider the thesis that the sense of a proper name and hence the identity of a proper name is given by the set of true propositions in which the name occurs. Now in order to determine the *set* of true propositions which allegedly gives the sense and hence the criterion of continued identity for the name we have *already* had to decide that it is the *very same* name occurring in all such propositions which form the set. In other words we have to understand the sense of the name in order to determine the set. Yet on the thesis being advocated we may only determine that it is the very same name in that we understand some set of true propositions in which it occurs. This argument serves to establish that the sense and identity of names cannot be given in terms of some set of true propositions in which they occur[40] and *ergo* that it cannot be given by reference to a set of predicates which are propositional fragments. It also, I hold, establishes that what may serve as a criterion of identity cannot be reduced to or be thought of as a predicate or predicable.

With regard to (c), a definite description has the formal feature of 'being satisfiable'; as Anscombe has commented: 'A proper name will stand for its object because that object is called by that name; but a description, if it stands for its object, does so because the object satisfies it, which is clearly quite a different relation.'[41] Here then we have to have the notion of 'the object which satisfies a description', but the concept of 'an object which may satisfy a description (definite description)' may not itself be furnished by a definite description. We need the *prior* conception of 'an object' which is furnished by a sortal. For example, we may invoke a named individual as the satisfaction of a description, as in 'Scott' satisfies the definite description 'The author of *Waverley*'; 'Tony Blair' satisfies the definite description 'The current Prime Minister of Great Britain', but as I have already argued proper names such as 'Scott' or 'Tony Blair' rest on the introduction of sortals - in this case 'man'; ('person'). Again, we may directly invoke the definite singular form of a sortal e.g. 'The man who was taught by Socrates' satisfies the definite description: 'The philosopher who wrote the *Republic*'. Since then definite descriptions themselves rest on the introduction of sortals, in that they have a necessary reference to the notion of 'an object which satisfies a description', it cannot be held that we may introduce the bearers of proper names *solely* by reference to definite descriptions or produce a criterion of identity for the continued use of a proper name *solely* by reference to definite descriptions. At this point it will be protested that when one speaks of 'the object which satisfies the definite description', it is a misunderstanding to think that we need to invoke an independent notion of 'an object' or 'the object' furnished by some sortal since for there to be such an object *is* for the description to be satisfied. For the author of *Waverley* to exist, for example, *is* for the definite description to be satisfied. However, if for there to be such an object *is* for the description to be satisfied, how can one speak of 'the object satisfying it?' To claim that e.g. for the author of *Waverley* to exist is for the definite description to be satisfied, is to claim that there is at the least and at most one object which satisfies the description and we cannot claim that the account of the object satisfying the description is furnished by the description itself, otherwise how could we introduce the notion of 'the object satisfying the description'? Indeed a definite description, as conceived of by Russell, could never serve to introduce the bearers of proper names or produce criteria of identity for the continued use of such names; since a definite description in his sense is precisely an 'incomplete symbol' having sense only in the context of a proposition. It could in no wise feature in determining the identity of a proposition. Its sense is determined by the sense of some one or some set of propositions in which it occurs; not the other way round. In this way, as we shall see,[42] in crude terms, Russell holds that the sense of a definite description is given by a set of predicates but as I have already argued a set of predicates cannot serve to introduce or determine the bearer of a proper name or give us the criterion for the continued use of a name. Indeed as the locution 'definite description' itself suggests the most such a form of symbol could do is to describe just one particular or object; but as I shall argue in my next chapter, description itself rests on identification and from this perspective also we may contend that sortals in that their primary role in speech is to identify, to specify, may not be reduced to descriptions; or again that we may not identify the bearers of proper names primarily by reference to definite description in Russell's sense.

It may now be commented that although the above arguments support the thesis that we cannot introduce the bearer of a proper name by invoking a definite description in Russell's sense or by reference to some predicate or set of predicates, it still has not been shown that we need to introduce the definite form of some sortal to so introduce a proper name. It will be suggested that all we need to so introduce is *a* singular form of a sortal e.g. a sortal expression proceeded by a demonstrative 'That cat'; 'That house' together with a pointing gesture. In reply to this suggestion I shall be arguing later[43] that where such a singular form does serve to identify the bearer of a proper name, it is always, in theory, writable as the definite form of a sortal: 'That house' - 'The house which ...'; 'That cat' - 'The cat which ...' as Quine (e.g.) has urged and as has been argued by Robin Attfield and myself.[44]

The point that *all* we need to introduce the bearer (or indeed the reference) of a proper name is a singular form of a sortal, namely a sortal expression precursed by a demonstrative, may lead to a more radical attack on my thesis viz., that we can effect the introduction of proper names *solely* by reference to the use of demonstratives together with a pointing gesture e.g. 'That (pointing) is the bearer of the proper name "Felix"'. I shall be arguing in Chapter II that such a thesis presents a misunderstanding of the nature of demonstratives, as does the thesis that demonstratives themselves may be regarded as a type of name.

In the light of the above considerations I thus hold that neither sortals themselves, nor their singular definite form, may be regarded as singular, proper, names but rather such category of symbol is precisely required for the introduction and continued use of singular, proper names.

B.2. I now face the issue: 'May not sortals be regarded as *general* names - names that stand for a plurality of objects or at least more than one object?' My detailed answer to this will be presented in my discussion of Geach later.[45] For present purposes I give my answer in shortened form. The thesis that sortals are general names standing for a number of objects, that sortal expressions signify in that they stand for a number of objects, has itself to invoke the notion of a 'number of objects'. This notion itself invokes the notion of a sortal since to speak of a 'number of objects' is to speak of a number of *F*'s (where '*F*' is a sortal), subject to the reservations:

(i) that there are no such entities and can be no such entities as 'pure' objects, 'pure' particulars, objects *simpliciter* viz., that it is nonsense to assert that there are such X's or posit the existence of such X's or speak of 'the number of such X's';

(ii) that sortals themselves cannot be shown to be reducible to or rest on the introduction of some *other* and allegedly prior category of principle of counting and distinction i.e. principles formed from 'matter' terms, 'feature' universals.

That sortals cannot be shown to be so reducible or to so rest, I claim to have shown already in A.2. above: I thus turn to (i).

B.3. To say that there are objects in the above sense reveals a gross confusion. To so say necessitates combining the quantifier 'There is (are)' with what Wittgenstein calls a 'formal' concept. He writes at *Tractatus* 4.1272:

> So the variable name "*x*" is the proper sign of the pseudo-concept *object*.
>
> Wherever the word "object" ("thing", "entity", etc.) is rightly used, it is expressed in logical symbolism by the variable name.
>
> For example, in the proposition "there are two objects which . . .", by '$(\exists x, y)$ . . .'.
>
> Whenever it is used otherwise, *i.e.* as a proper concept word, there arise senseless pseudo propositions.
>
> So one cannot, *e.g.* say "There are objects" as one says "There are books". Nor "There are 100 objects" or "There are $\aleph_0$ objects".
>
> And it is senseless to speak of the *number of all objects*.
>
> The same holds of the words "Complex", "Fact", "Function", "Number", etc.
>
> They all signify formal concepts and are presented in logical symbolism by variables, not by functions or classes (as Frege and Russell thought).

Now Wittgenstein's comments reveal that it is nonsense to assert that *there are* 'pure' objects, 'absolutely simple particulars'. To so assert is to construe the formal concept 'object' as a proper concept. However, although Wittgenstein's illuminating comments show that it is a confusion to *assert the existence* of objects falling under a formal concept (for this is a 'pseudo' concept), it may be said that they do not show that it is a confusion to hold that there must, in some sense, *be* such objects in order for us to formulate propositions which *do assert* existence i.e. in order for there to be propositions of the form: '$(\exists x, y)$...'; '$(\exists x)Fx$'. It might be said that whilst such objects cannot be *asserted* to exist, they nevertheless have existence, indeed have necessary existence: their existence is necessary for the very formulations: '$(\exists x)Fx$'; '$(\exists x, y)$'.

This thesis also rests on a confusion and one which Wittgenstein recognises:

(a) the formulations '$(\exists x)Fx$'; '$(\exists x, y)$', the 'logical symbolism' is precisely a symbolism, a formula, a schema. We can conceive of asking what there must be in the sense of entities, in order for us to propound a particular proposition or even propositions of a certain kind, but we cannot intelligibly ask what there must be, in the sense of entities, in order for there to be propositions of a certain form i.e. a certain form of proposition. To so ask confuses the logical form of a proposition with a proposition itself. It confuses a formula with what is an interpretation of a formula. To adopt Wittgenstein: it confuses the formal with the 'proper'; it makes the logical form of a proposition itself a sort of proposition.

(b) in terms of the *Tractatus* distinction between *saying* and *showing*, it confuses *showing* with *saying*. '$(\exists x)Fx$'; '$(\exists x, y)$...': 'There exists an object such that that object is *F*', 'There are two objects which ...', do not *say* anything; they are not *themselves* propositions - rather they show what the forms of certain propositions are.

(c) to require that there must, in some sense, *be* absolutely simple particulars, objects *simpliciter* for the very formulations (e.g.) '$(\exists x)Fx$'; '$(\exists x, y)$...' requires that we may intelligibly ask for the reference (referent) of an individual variable, '*x*', which we may not so do. To ask for the referent of an individual variable '*x*' would be to construe an individual variable '*x*' as itself a value for that variable. We can ask of any value which we substitute for the variable what *it* stands for or refers to, but not of the variable itself without confusion and misunderstanding. Of necessity a variable is not *itself* a type of value.

Thus far then I have contended that either to assert the existence of 'pure particulars', 'objects simpliciter' or to claim that such 'objects' are needed for our formulations of assertions of existence rests on confusions brought out by Wittgenstein in the *Tractatus*. But those with an interest in 'pure particulars', 'objects *simpliciter*' and their 'existence' might press a further argument - namely, that although it is an error, based on the confusions mentioned, to introduce 'pure particulars' as the reference of individual variables, do we not nevertheless need to posit such particulars as the reference (referent) of values for an individual variable in the sense of the ultimate reference of such values, as the ultimate reference (referent) of proper names? It might indeed be argued that Wittgenstein's 'simple objects' of the *Tractatus* sought to answer the question: 'What are the *ultimate, final* referents of the values for an individual (proper name) variable?', as it is plausible to argue that his elementary propositions of the *Tractatus* gave the *ultimate*, *final* meaning of all non-elementary propositions which were simply truth functions of such elementary propositions.[46] Yet, as Wittgenstein himself later came to see,[47] it is a misunderstanding to ask for *the ultimate or final* meaning of a proposition; the *ultimate or final* interpretation of a sentence. There can be *no* such ultimate or final meaning or interpretation. Equally, we may press it is a misunderstanding to seek for an *ultimate*, final meaning for proper names and equally such a misunderstanding to seek for the ultimate, final references (referents) of values for an individual variable. There *can be no* such final or ultimate references (referents). We may substitute 'Michael' as a value for the variable '*x*' in '*x* is fat' and form the proposition 'Michael is fat'. In relation to the proposition so formed we can ask for the reference of 'Michael'. That is, we can ask for the reference of 'Michael' in the context in which such a proposition is formed or propounded: the context in which such a sentence is used. In one context the reference of 'Michael' may be a certain sheep dog; in another the husband of a person we know; in yet another the mathematics teacher in the local school. We can proliferate the examples. The reference of the proper name 'Michael' varies according to context, from which it follows that we cannot ask: 'What is *the* reference of "Michael"?' as opposed to (e.g.) 'What is the reference of "Michael"? in the proposition you are putting forward for consideration?' or 'What is the reference of "Michael" in the context of your sentence "Michael is fat" (or your use of the sentence; "Michael is fat")?'. And from this it follows that we cannot ask for *the* ultimate reference i.e. the necessarily 'context free' reference of 'Michael' - or indeed of any other proper name or value of an individual variable.

We have thus yet a further argument against the introduction of 'pure particulars' or 'objects *simpliciter*'; in all the contexts considered to introduce such particulars, such entities, is based on confusion and misunderstanding. At this point however it will be interjected that although I have produced arguments for showing that such entitles are illegitimately introduced, I have not shown that the notion of a 'pure particular', an 'object *simpliciter*', is itself an impossible notion. I do not however need to go thus far in order to make my point in B.2.(i) above. What an objector to my case needs to maintain is that in order to speak of the 'number of objects', it is not necessary to invoke the notion of, to speak of 'the number of *F*'s', where '*F*' is a sortal. All that it is necessary for him to maintain is that we can speak of the number of objects *simpliciter*, the number of 'pure particulars'. But, I hold, it *is* impossible that we *should* speak of 'the number of objects *simpliciter*' or of the number of 'pure particulars'. This Wittgenstein clearly brings out at *Tractatus* 4.1272 cited above: '... it is senseless to speak of the number of all objects'. It is senseless so to speak for to so do confuses 'formal' concepts with 'proper' ones. We may expand on this as follows. It would be impossible to count or even *speak of* 'the number of objects' in the sense of 'pure particulars', 'objects *simpliciter*' since *ex hypothesi* 'they' are particulars concerning which it is necessary that they fall under *no* principle of counting or distinction for particulars - otherwise 'they' would not be 'pure' or 'absolutely simple' or 'objects *simpliciter*'. Indeed in that it is necessary that they in principle are not subject to *any principle* of counting and distinction for particulars, then of necessity we cannot speak of 'them' or mention 'them'. The concept of a 'pure particular', 'absolutely simple object', 'object *simpliciter*' may be said to be an impossible notion, an impossible concept, in that it is not possible for us to even introduce the idea of objects falling under it, for we could not, in principle, speak of either 'one' or 'many' such objects. But to so comment renders such a concept impossible only if it is (mistakenly) treated as a 'proper' concept which those who posit 'pure particulars', 'absolutely simple objects' take it to be. It may still retain its status as a 'formal' concept.

I thus claim that I have now produced sufficient reason for rejecting the reservations to the thesis put forward in B.2. above and hence that one may hold that sortals may not be regarded as *general* names for the reason there offered and to be developed in more detail later as there mentioned.

C. Considerations on a General Thesis which may arise from the claim that Sortals are *not* Reducible to Predicables/Predicates

My thesis that sortals may not be reduced to predicates or 'predicables' has, in effect, been argued for earlier. I shall re-state it in brief here and then go on to consider an interesting more general theoretical thesis which might be thought to arise from the arguments advocated in support of my case that sortals are not reducible to predicates or predicables.

C.1. Sortals may not be reduced to predicates or predicables, I have claimed, since a sortal in its definite singular form is required in order to introduce the very notion of 'a particular individual' in the sense of 'a particular object', which is to be the subject of a predication. It might be suggested that this does not of itself show that the indefinite form of a sortal (a so-and-so, an *F*) is not reducible to a predicate or predicable. This, in turn, however can be easily shown; indeed is implicit in the point already made. In that it is acknowledged that the definite form of a sortal is required for the very introduction of the notion of 'a particular object' which is the potential subject of a predication, it is therein acknowledged that the indefinite form of a sortal is required. We could not consistently acknowledge that the definite form of a sortal is required, but that the indefinite form is not inherently required, for we may only introduce the form: 'The *F*...' in that we acknowledge the form: 'An *F* ...'. One may generalise on this point as follows. The indefinite form of a sortal may not itself be regarded as a predicate or predicable since it itself is required in order to introduce individuals (objects) which themselves may form the subjects of predication.

C.2. Since sortals are thus required for the introduction of particulars (objects) which are the subjects of predication and predicables are *of their essence* incomplete symbols, one may be led to suggest a more general theoretical thesis – namely, the thesis that at the most general level of logical theory, sortals may not be reduced to predicables (predicative functions) since they are required for the introduction of the notion of an 'individual variable' in the sense that they are required in order for us to determine the range of values which may coherently be substituted for the variable in a predicative function and so yield a proposition which is either true or false. Whilst I shall contend that sortals are required for the introduction of the notion of 'an individual variable', I shall reject the above mentioned thesis. I shall claim that this thesis is based on a misunderstanding.

I shall first of all set out an argument in support of this thesis and consider an 'internal' objection to it. I shall attempt to reply to such an objection and then consider a counter thesis holding that the counter thesis fails. I shall then produce a fundamental objection to the original thesis - the objection that it is based on a misunderstanding. However whilst I hold that the original thesis is based on a misunderstanding, its rejection does not entail that sortals are not in any sense required for the introduction of the notion of 'an individual variable "*x*"', such that one may, without any reservation whatever embrace Russell's doctrine of 'an essentially undetermined variable "*x*"'.

C.3. What considerations then may be adduced in support of the suggested thesis? We may offer the following. If, for example, we are given the function '*x* is wise' it may be claimed that we have to know what range of values may coherently be substituted for '*x*' such as to yield a true or false proposition, for it is not the case that the proper name of any object whatsoever may be so substituted. We may not, it may be said, substitute the proper name of a number as a value for '*x*' in '*x* is wise'. 'Five is wise' is not simply false; it is absurd. What can clearly and coherently be substituted is the proper name of, for example, a *person* and 'person' is a 'higher order' sortal, a 'formal', 'categorical' concept.[48] Again, take the function: '*x* is heavy'. What can clearly and coherently be substituted for the '*x*' in the function is e.g. the proper name of a material object and the 'formal concept' of 'material object' is simply the formal expression of the 'proper' concepts furnished by material object sortals as I have previously argued.[49] Sortals then, in the sense of 'proper' concepts, it may be suggested, *determine*, at least indirectly, what range of values may be substituted for '*x*' to yield a true or false proposition. They determine that we can say that e.g. '*x*' ranges over the proper names of persons, the proper names of material objects, the proper names of colours, in a given 'universe of discourse'. In that sortals are thus implicitly required at this general theoretical level to determine the range of values that may coherently satisfy a predicative function they themselves may *not* be regarded as predicables or as reducible to such.

Such a general thesis will be objected to on 'internal' grounds viz. in terms of the way in which it is expressed. It will be said that the most that might be argued for is that sortals are required in order to specify or indicate the range of values i.e. the proper names of which entities may coherently be substituted as values for the variable in a function such as '*x* is wise'. In this case '*x*' ranges over the proper names of e.g. persons, but in that sortals are so required to specify or indicate[50] then they cannot be said to *determine* what range of values may coherently be substituted in the function. To perform a specificatory function is not to perform a determinative one and indeed necessarily excludes it. This objection may be answered, as I have indirectly indicated above, namely, by distinguishing between different levels or orders of sortals such that at one level sortals may be said to specify or indicate, whilst at another level they may be said to determine. For example, as regards the function '*x* is white' or '*x* is in pain', the 'highest' order sortals (e.g. 'spatial object'; 'sentient creature') - that is sortals in the sense of the *categories* formed from 'lower' order sortals - serve to *specify* what range of proper names may coherently stand for the variable in the function; the 'lower level' or 'lower order' sortals (e.g. 'desk', 'table', 'house') *determine* in that such 'higher' or 'highest' order, 'categorical' sortals are simply a function of such 'lower' order sortals, the formal expression of such 'lower' order or 'lower' level sortals as I have previously tried to explain.

The thesis that sortals are required to determine and specify what range of values may coherently stand in the argument place in a function will now be objected to on different grounds namely on the grounds of the thesis that what may so stand is determined by *the function itself*, not by anything extraneous to the function: a function *itself* determines it own values, its own arguments. This is the 'counter thesis' I alluded to earlier.

Such a thesis is, I think, open to quite devastating objections. First, we may ask: 'How can e.g. "---- is white", "---- is *P*" itself determine what may coherently stand in the argument place?' How may one determine what may stand in an argument place simply by considering the function itself? That one *may*, it can be held, is based on a misunderstanding of the notion of a 'function'. Such a thesis appears to construe a function not as an *essentially* incomplete symbol, but rather as a symbol which only *per accidens* contains an argument place such that what can stand as an argument is determined by e.g. 'is white'; 'is *P*'. What can stand as an argument are the names of (say) coloured objects determined by 'is white', or, more generally, the names of '*P*-ish' objects determined by 'is *P*'. Such an account *does* precisely show a misunderstanding of the notion of 'a function'; it goes right against the doctrine of functions as *fundamentally* incomplete symbols. It would deny Frege's doctrine that 'concepts' are *essentially* predicative, 'unsaturated'.[51] In that a predicative function is of its essence incomplete we are back to the question of how such a function may determine its own values. One cannot determine what may stand as a argument *simply* by introducing the function for the function, of its essence, contains an 'empty place'. Indeed, in general, an explanation of X, or whatever determines X, cannot be given by reference to X itself; nothing can be its own explanation or its own determinant.

Is there *any* way in which we might hold that a function, conceived of as an essentially incomplete symbol, may be held to determine its own values? The following way might be put forward. What can replace '*x*' in the function '*x* is *P*' is determined by the set of propositions (or set of sentences) from which the function is derived. On such an account we are immediately faced with the difficulties:

(a) the set may be infinite - in which case how may we form the function and determine the arguments?

(b) in any case, even to determine that the set *is* infinite we have to have some principle by which we may decide that the set *is* infinite. In other words we are faced with the question: 'How do we or may we determine the set of propositions (sentences) from which the function is formed?'

Here it certainly seems that we are in an impasse. One may draw a parallel with an earlier argument which I advanced concerning the sense of proper names. I there[52] considered and rejected the thesis that the sense of a proper name is given by the set of true propositions or sentences in which the name occurs. I argued that in order to determine *the* set of true propositions (sentences) in which the name occurs, one would already have had to determine that it was the same proper name occurring in all the members of the set and hence that one could not determine the sense of a proper name by reference to the set of true propositions (sentences) in which *it* occurs. In parallel fashion here we cannot say that what arguments can stand in the argument place in a function is determined by the set of propositions, sentences,[53] from which the function is derived or formed since lacking 'external' criteria we can only determine the set of propositions (sentences) from which the function is derived or formed (and hence the set of arguments which may stand in the argument place) in that the set of propositions (sentences) are all instances of the same function. In other words we have to have the notion of 'the same function' in order to determine the appropriate, relevant, set of propositions (sentences) from which the function is derived, formed. The identity of the function therefore cannot be given by reference to the set of propositions (sentences) from which it is formed and hence neither can the sense and identity of the argument place, the variable, in the function. Again, we may not hold that the sense of an argument place in a function, the sense of a variable in a function and hence what may stand as arguments in a function, values of the variable, is determined by the set of functions in which *it* occurs since in order to determine the members of the set, and hence the set, we *already* have to understand '*x*' as the *same* variable.

We should further note here that from the fact that a predicable, a predicative function, of its *essence* contains an 'empty place', a variable, it does not follow that the notion of an 'argument place', a 'variable', of its essence is *part* of a predicable, *part* of a function. Indeed this is necessarily not the case otherwise we would never be able to write that a predicable is a fundamentally incomplete symbol completed by an argument. We would never be able to say that it of its essence contains an empty place if the notion of an 'empty place',[54] 'argument place', itself of its essence were part of the whole '---- is *P*', since a part gains its identity from the whole of which it is a part.

I conclude therefore that on the thesis that a predicable, a predicative function, may determine its own values we are faced with severe difficulties - difficulties sufficient for us to reject the thesis and indeed we are faced with a dilemma both horns of which are equally unacceptable. We either have to treat such functions as not of their essence incomplete symbols or, treating them as such we are faced with an impasse, a vicious circularity.

C.4. I now come to the charge that the thesis originally under consideration is based on a misunderstanding. The charge of misunderstanding takes the form that to raise the question: 'What range of values may coherently be substituted for the variable "*x*" in a predicative function "*x* is *P*"?' is a *mistake*. *Any* proper name whatsoever may be so substituted where '*x*' is a 'proper name variable', and to think that we may raise the question of what proper names may coherently be substituted is to make the mistake of thinking that there may be different *types* of proper name; whereas '*a*' is either a proper name or it is not. Admittedly to hold that any proper name whatsoever may be substituted for the variable in a predicative function initially does give rise to certain well known difficulties. If there is no restriction on the range of proper names which may replace the variable '*x*', say in the function '*x* is white' or in the function '*x* is in pain', then it may be contended that we may not distinguish between *falsity* and nonsense. On such a thesis we should be committed to the position that e.g. 'Five is white' (where 'Five' is the proper name of a number), 'Saturday is in pain' (where 'Saturday' is the proper name of a day of the week) are simply *false* propositions whereas they are clearly pieces of nonsense manifesting a 'type' or 'category' error.[55] However it will rightly be counter objected that it is not true that on such a thesis we may not distinguish between falsity and nonsense. Such a thesis would commit one to saying that such propositions as 'Saturday is in pain', 'Five is white' were species of false proposition, but that we may clearly distinguish such false propositions from: (a) propositions which are contingently false; (b) propositions which are necessarily false (contradictions). (i) Such propositions are clearly not merely contingently false propositions such as 'Vienna is the capital of France'; 'Lady Thatcher lives in Scotland' and (ii) they may be distinguished from contradictions as follows: 'Five is white'; 'Saturday is in pain' are *absolutely* false, viz., *necessarily* they are false, as opposed to '*a* is both *P* and not-*P*' or 'A bachelor is not an unmarried man' which are *necessarily* false. Or, to put the matter the other way round: *It is not possible that* 'Five is white', 'Saturday is in pain' is true, as opposed to: '*a* is both *P* and not-*P*' or 'A bachelor is not an unmarried man' *is not possibly true*. And 'It is not possible that *a* is *P*' does not entail '*a* is not possibly *P*' any more than 'Necessarily A is false' entails 'A is necessarily false'. On such an account of absurdity one is able to maintain the intuitively true proposition that every proposition is either true or false.[56]

On these grounds I accept the charge that the question: 'What range of values may coherently be substituted for the individual variable in a predicative function?' is a mistaken question. But from this it does not follow: (i) that the need for sortals is eradicated; (ii) that one may, without any reservation whatever, embrace Russell's doctrine of 'an essentially undetermined variable "*x*"'. Although we may not say that sortals are required in order to determine and specify what may coherently replace the variable in a predicative function, sortals are *still* required in that we can say that any proper name *whatsoever* may replace the variable in the function, for any proper name whatsoever is *still* the proper name of *some F*, where '*F*' is a sortal, and it is not contingent that a predicative function contains an empty place to be filled by a proper name. It is in *this* sense that sortals are implicitly required for the introduction of predicative functions (predicables) that contain proper name variables.

We may not, without reservation, embrace Russell's doctrine for the following reasons. He is *right* in rejecting the thesis that every variable is *essentially* determined in that this thesis commits one to the view that for any proper name '*a*', '*a*' is *necessarily* the name it is and could have been no other; more generally, that any *F* is the *F* it is and could have been no other; but *wrong* in rejecting the thesis that *it is essential* that every variable is determined. That it *is* essential that every variable is determined is made clear in the statement that what may stand as a value for a proper name variable in a predicative function is the proper name of *any F whatsoever.* It is essential that every variable is determined in that *necessarily* there is a reference to some sortal or other but from this it does not follow that every variable is *essentially* determined such that for any (proper name) variable only a certain specific range of values may coherently be substituted viz., that there is essential reference, not to some sortal or other, but to some one sortal or range of sortals. 'It is essential that *a* is *P*' does not entail '*a* is essentially *P*' and '*Necessarily* there is reference to some sortal or other' does not entail either: (i) there is reference to some one sortal; or (ii) 'There is a necessary reference to some *one* sortal'. Russell's doctrine of 'an essentially undetermined variable "*x*"' is, I shall claim, to be rejected on the grounds that in his use one is committed to an unintelligible doctrine of 'logically proper names'. Such a doctrine, I shall claim, arises out of his failure to recognise the category of the sortal[57] together with that of holding that: 'It is essential that every variable is determined' entails 'Every variable is essentially determined'.

D. Summary of Chapter

In this chapter then I have attempted to present a case for holding that sortal symbols:

(1) May not be reduced to proper names or general names;

(2) May not be reduced to predicables or in more common parlance, 'predicates';

(3) underlie the introduction and continued use of proper names and underlie the introduction of predicables (or 'predicates') in the sense explained.

I have contended that sortals, in the sense introduced, present us with a category of symbol which furnishes a principle of counting and distinction for particulars in the sense of 'objects' and that they do so in their own right presupposing no other category such as the category of symbol formed from what I have spoken of as 'matter' terms in the way illustrated earlier. Sortals permit us to form the notion of 'a particular' in the sense of 'an object' and likewise 'one'; 'many'; 'several' particulars (objects). In this sense I claim sortals present us with a quite distinct category of symbol which needs to be recognised as such.

Notes

1. In 'Particular and General', repr. in *Logico-Linguistic Papers*, *op. cit.*, pp. 36-39.
2. In *Individuals*, ch. 6, sec. 6.
3. *See Introduction to Mathematical Philosophy*, ch. XVI and *also* my ch. VI.
4. *See* my discussion in ch. V.
5. In *Logic Matters* (Oxford: Basil Blackwell, 1972), ch. VI; and *also* in his paper on Frege in *Three Philosophers*.
6. *See* my discussion in ch. VII.
7. It may be objected that it is not clearly nonsense to speak of 'waters'. Consider: 'By the waters of Babylon we sat down and wept'; but the author of Psalm 137 is here speaking of the *rivers* of Babylon. What about someone who in the eighteenth century went to Bath to 'take the waters'? We may contend that such a person went to Bath to take the waters of the *springs* which are present in that city; or again that he went to take the *types* of water that are so present.
8. cf. Strawson: 'Particular and General', *op. cit.* For a full discussion of such phrases as 'bit of gold' etc., cf. below and in ch. X.
9. In *Word and Object*, paperback edn. (Cambridge, Mass.: The M.I.T. Press, 1964), p. 91.
10. As, by implication, Geach does in *Reference and Generality*, ch. 2, pp. 39-40.
11. For 'characterizing universals' cf. *Individuals*, pp. 167ff. and pp. 20-21 of this chapter.
12. 'Modified' is Strawson's term; cf. *Individuals*, ch. VI, p. 202.
13. I shall be returning to this matter anon. cf. note 18 below.
14. *See* below.
15. *Vide* my discussion in ch. X.
16. *Ibid.*
17. It is to be noted that Strawson includes principles of counting formed from 'matter' terms as 'sortals'. I shall be discussing this in ch. X. I shall speak of such principles as 'sortals' in a 'secondary' or 'extended' sense.

18. cf. W. Charlton in his translation and notes on Aristotle's *Physics* I and II (Oxford: Clarendon Press, 1970), pp. 70ff., esp. pp. 71, 76. This however will not suffice since in order to determine the appropriate arrangement one first has to have the concept of 'a house'.
19. Geach, *op. cit.*, ch. VI.
20. Especially in *Metaphysics* H.
21. *Individuals*, p. 153 where a 'predicate' expression is given as e.g. 'is wise', *not* as '---- is wise'.
22. cf. my argument in ch. VIII.
23. cf. Geach's review of Strawson's *Subject and Predicate in Logic and Grammar*, *see* my 'Introduction', note 43.
24. *Individuals*, p. 168.
25. cf. Frege's discussion of this very case in ch. V of this book.
26. That this is so derives from p. 146 of *Individuals* and elsewhere; cf. ch. VIII.
27. Again, *see* ch. VIII.
28. My italics.
29. *Individuals*, pp. 169ff.
30. in ch. IX.
31. *See* Michael Dummett, *Frege: Philosophy of Language* (London: Duckworth, 1973), pp. 110ff., esp. pp. 112, 124-126. In '*Necessarily: p is q*', '*necessarily*' operates on the whole proposition; in '*p is necessarily q*', '*necessarily*' relates to the predicate '*q*'; '*Concerning p, it* is necessarily *q*'. We may only derive '*p is necessarily q*' from '*Necessarily p is q*' by an illegitimate shift of the modal operator '*necessarily*'. One crucial consequence of this 'operator shift' may be illustrated by reference to theses in epistemology. Some philosophers have argued from: '*Necessarily* what is known is true' (which is itself true) to: 'What is known is *necessarily* true' (which is false), thus holding that the only propositions which can be known are necessary truths.
32. In chs. VIII and IX.
33. *Individuals*, ch. VI, p. 202.
34. cf. ch. VIII.
35. *See* my paper 'Plato, the "Third Man" and the Nature of the Forms', in *Southern Journal of Philosophy*, Vol. XVII, No. 3, 1979, pp. 287-304.
36. cf. my discussion of Frege on this point in ch. V.
37. For this reason I reject Russellian 'proper names'; cf. ch. VI.
38. *Reference and Generality*, p. 25.
39. Geach, *ibid.*, ch. 2, p. 32.
40. I thus reject Wittgenstein's view that a name only has sense in the context of a proposition, as advocated in the *Tractatus*. *See* also my paper: 'On the Thesis that a Name has Sense only in the Context of a Proposition; A Problem', in *Cogito*, Vol. XVII, No. 1, 1993, pp. 47-49.
41. In her *An Introduction to Wittgenstein's Tractatus*, (London: Hutchinson University Library, 1959), p. 41.
42. In ch. VI.
43. In ch. II, sec. G, esp. p. 55.
44. In 'The Irreducibility of "Meaning"', in *Nous*, Vol. VII, No. 3, 1973, pp. 282-298.
45. Ch. VII.

46. *See* Peter Winch 'The Unity of Wittgenstein's Philosophy', in Peter Winch (ed.), *Studies in the Philosophy of Wittgenstein* (London: Routledge and Kegan Paul, 1969), pp. 1-19.
47. *Ibid.*
48. For 'higher order', 'categorical' concept, *see* below. That 'person' is such a 'higher order', 'categorical' concept I argued for some years ago in *Theology and Intelligibility* (London: Routledge and Kegan Paul, 1973). I now have some reservations on this matter but as I currently am unable to produce decisive objections to this thesis I shall, for present purposes, which are purely illustrative, let it stand.
49. In that they yield categories; *see also* next paragraph.
50. cf. Quine's 'Range Indicators' in *Mathematical Logic*, rev. edn. (Cambridge, Mass.: Harvard University Press, 1961), pp. 68-69.
51. Frege, 'On Concept and Object', repr. in Peter Geach and Max Black (eds.), *Translations from the Philosophical Writings of Gottlob Frege* (Oxford: Basil Blackwell, 1952), pp. 42-55.
52. pp. 25-26.
53. I give both 'propositions' and 'sentences' here since both forms are used in the literature.
54. Geach speaks of 'empty place'; *Reference and Generality*, ch. 2.
55. Type error in Russell's terminology; 'Category' error in Ryle's terminology.
56. cf. Quine on 'categories' in *Word and Object*, pp. 228-229.
57. cf. ch. VI.

Chapter II

Sortals and Identification

Part I: The Priority of Sortal Identification over both Naming/Referring and Predication

A. Of the Priority of Sortal Identifying over Naming/Referring

I have argued that sortals are not reducible to names or predicables (predicates) from the point of view of language viewed as a symbolic system, but are to be recognised as a distinct category of symbol logically prior to the introduction of names and predicables (predicates). In corresponding fashion I now want to argue that, from the point of view, aspect, of language viewed as the use of symbols, viz. from the aspect of speech or thought,[1] the distinctive role which sortals play, namely that of identifying particulars in the sense of: (i) serving or being used to pick out or discriminate particulars; (ii) serving or being used to specify what they are, is not reducible to naming/referring on the one hand or predicating on the other, but is to be recognised as a function, a mode of speech activity which is prior to both operations.

A.1. This thesis may be objected to as follows. It might be said that I can *allocate* a name to a particular without being able to identify it in any way, let alone identify it in the sense of picking it out by reference to some sortal. I can allocate a name to a particular by simply pointing and saying, e.g. 'I name *this* Mehitabel'. However the use of a demonstrative together with a pointing gesture of itself in no way enables us to determine *what* one is pointing at: the use of a demonstrative *per se* in no way determines reference or identification. The only way in which a demonstrative, together with a pointing gesture, can succeed in indicating or picking out some particular for us is if we have already understood some further expression which specifies *what* is indicated or picked out by the demonstrative in question: if we understand *what* is being pointed at.[2] Even so, an objector will press, it does not follow that in order to understand *what* is being pointed to - in order to understand the reference of a demonstrative in such a context - we have to understand some *sortal*. We may understand what particular is being indicated by reference to either: (i) some non-sortal according to the criteria I have introduced; (ii) a definite description; (iii) a vague general symbol such as a 'thing' together with a set of predicates (descriptions) which, for present purposes in the immediate context, determine the reference. Let us now consider this case.

First, the case of some non-sortal according to the criteria I have offered earlier. It might be held that I can determine the reference of the demonstrative 'this' if I understand that what is being pointed at is, e.g. a piece of stone or an expanse of snow. 'A piece of stone' or 'an expanse of snow' can answer the question 'What is it that is being pointed at?' with perfect adequacy, yet 'piece of stone' and 'expanse of snow' are not sortals according to the criteria offered. Yet they do, as I have already acknowledged, furnish principles of counting and distinction as such, in their own right. I thus am not denying that symbols, other than sortals, may serve to identify particulars but in that the alleged or putative reference of the demonstrative is an *object*, as opposed to a parcel of matter, then it is necessary that some sortal be introduced, as I have already argued,[3] and have identification by reference to some sortal is equally necessary. And we primarily speak in the context of objects, not in the context of parcels of matter.[4]

Next the case of 'definite descriptions' which will, *ex hypothesi*, contain no sortal. For example, by 'Mehitabel' I shall mean 'the winner of the 3.30 at Doncaster'; by 'Mehitabel' I shall mean 'the swimmer who is the first to reach the French coast'; or, by 'Mehitabel' I shall mean 'the inhabitant of this cave' (pointing). In all these cases some sortal is clearly implied as I argued earlier.[5] In all these cases specifying the reference by such descriptions carries with it an implicit reference to some sortal - the *F* (sortal) who/which won the 3.30 at Doncaster; the *F* (sortal) who is swimming towards France and who reaches the coast first; the *F* (sortal) who/which inhabits this cave, i.e. the cave which ...[6]

Thirdly, consider such cases as 'I name *this* viz., the rough, round, red thing right at the end of my finger "Mehitabel"'; or by '"Mehitabel" I mean (or shall mean) that dark, streaky object on the horizon which is moving between those two poplar trees (pointing)'. Now whilst 'the rough, round, red thing at the end of my finger' may indeed serve to pick out an object, may in a context serve to pick out a reference for and future bearer of the name 'Mehitabel', it may do so only upon our prior understanding of some sortal. That is, '*thing* which is rough, round and red' may only serve to pick out in that '*F* (sortal) which is rough, round and red' may also serve to pick out. The use of an indeterminate form (of expression/words) such as 'thing which is φ' to pick out an individual or allegedly specify an individual rests upon the use of a quite determinate form (of words or expression) '*F* which is φ' to pick out an individual. We could not, as a matter of logic, pick out *things* which are rough, round and red if we could not pick out *F*'s (where '*F*' is a sortal) which are rough, round and red since to be able to pick out things which are rough, round and red we have to be able to *count* them and we can only *count* the number of things which are rough, round and red in that we are able to count the number of *F*'s which are rough, round and red.[7] We can only carry out the instruction: 'Count the number of rough, round, red things in this room' if we know the *F*'s we are supposed to be counting. Thus whilst it might be thought to be too extreme to hold the thesis that for there *to be* things which are rough, red and round is for there *to be* *F*'s which are rough, red and round, the *lesser* thesis that for us to introduce 'thing which is rough, red and round' presupposes that we have already introduced '*F*' which is rough, red, and round (where '*F*' is a sortal) is plausible on the grounds offered[8] and is all that is required in order to sustain my thesis. Similar comments apply to the second case given above. Whilst 'dark, streaky object on the horizon which is moving etc.', may serve to pick out an object it may only do so on the presupposition that there are sortals (expressions) which may serve to pick out objects. That is, logically we can only pick out dark streaky objects on the horizon in that we can pick out dark streaky *F*'s on the horizon, where '*F*' is a sortal. We can only use an indeterminate form of words/expression - 'that dark streaky object ...' to pick out an individual in that we can understand the use of a quite determinate form of words/expression: 'that dark streaky *F*...' on the horizon. Indeed it is to be noted that we use such indeterminate forms of words/expression to pick out an object precisely when we don't know what the thing in question is. If we were to hold that we could use, e.g. 'that dark streaky object on the horizon' to pick out an object for us without understanding what it is to pick out a dark streaky *F*, then we should be committed to the thesis that we could count the number of, e.g. dark streaky objects on the horizon without being able to count the number of *F*'s on the horizon, i.e. that 'dark streaky object' could be a principle of counting in its own right. On the contrary, however, we can only count the number of dark streaky objects if we have some concept of what it is to count *one* object, i.e. if we have the concept of an *F*, where '*F*' is a sortal.[9]

A.2. Thus far I have dealt with objections concerning *allocating* a name to a particular. We must now turn to a second objection, namely that I can *use* a proper name to refer to a particular without being able to pick it out, identify it in any way let alone having to understand the singular form of a sortal (expression), i.e. I can use the name 'Mehitabel' to refer to an individual particular without being able to identify that particular in *any* way. Such a contention, I hold, is most certainly absurd. On such a contention I should be quite unable to say what the name 'Mehitabel' refers to and hence to speak of using it to refer *at all*. It won't do here to say that on such a contention I could speak of 'Mehitabel' having an indeterminate reference, as opposed to a determinate one, since the contention advocates supposed reference to a particular without the user of the name being able to pick out the particular, or specify the particular in any way. If this contention is modified to read that I can use a proper name to refer to a particular without being able to identify it in a quite determinate way then: (i) its initial force against my thesis fails; and (ii) we are faced with the problem of how the contender can claim that he can use, e.g. 'Mehitabel' to refer to *just that* particular.

Once again it will be pressed that whilst it cannot be held that I may use a proper name to refer to a particular which I cannot in a least *some* way pick out or specify, it does not follow from this that the picking out or specification *has* to be done by reference to a definite singular form of a sortal. It will again be held that such picking out can be done by means of 'matter' terms - suitably precursed, or descriptions, or phrases of the form 'the object which ...' as enlisted above. I have already acknowledged the situation as regards the first, and explained my position; in as much as the case for the other two fails above it fails here too.

A third objection may well be introduced here, but it meets with the same difficulties as the former two objections. This objection is that I can refer to a particular by the means of a demonstrative without being able to identify it in the sense of 'picking it out'. The answer to this objection has already been provided under A.1. above. Or again, it might be said that I can refer to a particular by the means of a definite description not formed from a sortal without being able to pick it out by reference to a sortal. This too, I hold, has been answered under A.1. above.

B. A Case in which Referring *may* be taken as prior to Identifying

It may now be pressed: 'Surely, in spite of all the above arguments, there is a sense in which in order to identify a particular I must be able to *refer* to it and thus there is a sense in which referring is prior to identifying.' Now one sense in which referring *may* be taken as prior to identifying concerns that sense of identifying in which we say, more precisely, *what* some particular is. For example, '*That* is a pineapple', '*That* is an ostrich'. But in such cases referring is not *absolutely*, but only *relatively*, prior to identifying. *Relative* to identifying the individual in the sense of saying more precisely what it is, we have to be able to refer to the individual, but our referring to the individual by means of the demonstrative presupposes our being able to identify it in a wider sense. We have, in accordance with arguments advanced earlier, to be able to specify the reference of the demonstrative, e.g. 'That', in the case of '*That* is a pineapple' refers to a sort of fruit or an individual (piece of) fruit, so that for 'That is a pineapple' we have to be able to write 'That fruit (piece of fruit) is a pineapple'. Similarly for '*That* is an ostrich' - 'That bird is an ostrich'.

Not all cases however in which referring may initially be thought to be prior to identifying can be explicated in this way, i.e. by showing that referring is only relatively prior. I shall consider one more specific case at this point and then turn to consider a general attack on my thesis which I have alluded to earlier.[10] The specific case I have in mind at this point is such sentences as: 'Socrates is a man'; 'Fido is a dog'. In the case of the use of such sentences, can it be claimed that referring is only relatively prior? Initially, there are two possible interpretations of such sentences. On the first interpretation, 'man' and 'dog' may be construed as serving to specify more precisely or specifically what is referred to by 'Socrates' and 'Fido' respectively, and on this interpretation 'Socrates' and 'Fido' will already have had their reference identified by reference to a more general or wider sortal, e.g. 'that rational animal who ...'; 'that creature which ...', and so on this interpretation identifying in the sense of a specification by reference to some more general or wider sortal is prior to referring; that is, as with the earlier case, referring is only *relatively* prior to identifying. On the second initial interpretation which is possible here, each of our sentences will be construed as giving us a general rule for the use of the names 'Socrates' and 'Fido' respectively. On this interpretation 'man', 'dog' serve to introduce the bearers of the respective names and as this is the case the question of reference or of reference being prior does not arise. On such an interpretation the names 'Socrates', 'Fido' are introduced or mentioned, not used; on such an interpretation we cannot ask for the reference of 'Socrates' or 'Fido'. To revert to the language of reference, on this interpretation, 'Socrates is a man' serves to specify the reference of 'Socrates' and similarly with 'Fido is a dog'. 'Socrates' and 'Fido' in such a context, on such an interpretation, cannot be spoken of as 'referring'.

There is however a third interpretation which must be considered. The construction of 'Socrates is a man' and 'Fido is a dog' as a general rule for the use of the names 'Socrates' and 'Fido', may be objected to on the grounds that on this construction 'Socrates is a man', 'Fido is a dog' are writable as '"Socrates" is a man', '"Fido" is a dog' and this is implausible, since in the original there are no inverted commas round 'Fido' and 'Socrates' to indicate that these names are being mentioned as opposed to used. In that we *take* Fido and Socrates to be the subjects of the sentences, the original sentences serve to specify what Fido, Socrates *are*, not merely, if at all, how the names are to be used; these sentences convey not the 'nominal' essence of 'Fido' and 'Socrates', but the *real* essence of Fido and Socrates. They tell us what it is for Fido and Socrates to *exist*, not merely something about their *names*. Now it might be argued that to give the nominal essence of *N*, the name, is simply to express in the 'Formal mode' what is expressed in the 'material mode' by the real essence of *N* the individual; to say what it is for *N* (Socrates, Fido) to exist, is to express in the 'material mode' what sense is to be given to the name: to give us a general rule for the name. Leaving aside such an argument for the moment in fairness to the initial plausibility of the objection, is it not the case that referring may be thought of, indeed must be thought of as prior to identifying? The answer to this, I hold, is *no*. My explanation of this is as follows. The use of a name in a referring role presupposes that the object putatively referred to by the name exists or has existed, or will exist;[11] it presupposes the truth of an existential proposition to the effect that there is, was, or soon will be just one *F* such that ..., but the possibility of forming such an existential rests on our being able to say *what* the individual is whose existence is presupposed by the referring use of the name, otherwise we cannot articulate the appropriate existential.

C. Consideration of a general attack on my thesis that Sortal Identification is prior to Naming/Referring

The general attack I have alluded to earlier may be expressed as follows: 'Surely I *can* succeed in referring to something *without* being able to identify it in the sense of being able to say *what* it is. You must allow that there are many things which I myself and others can successfully refer to without knowing what they are, i.e. *without being able to introduce some sortal.* Further, not everything is able to be specified by reference to some sortal; there are things which can only fall under some 'substantival general' term in Geach's sense.[12] How then can your thesis that referring logically presupposes identifying by reference to a sortal be upheld?

First, once again I readily acknowledge that not everything, in the sense of not every particular, is able to be specified by reference to some sortal: particulars introduced by 'substantival general' terms may not be able to be so specified; my concern is not with such particulars but with objects, particulars in the sense of objects, as I have explained. Hence I shall not in future consider objections based upon such cases since as I have already argued one may not derive the notion of an object from such substantival general terms.[13]

Now it is indeed true and it is not part of my thesis to deny that one may successfully refer to a particular, without being able to identify it in the sense of being able to say specifically or precisely what it is. I further have to admit that it is *possible* to refer to a particular, to an object, without in any clear sense being able to identify it; without in any clear sense being able to say what it is: but to admit of such possibilities does not destroy my thesis that identifying by means of sortal is prior to referring. The basis of my thesis is that the very introduction of the notion of a particular, an object, (as opposed to a mere piece of matter) presupposes the presence of a principle of counting and distinction, in its own right for particulars, in other words a sortal; and identifying in the sense of specifying what something is, is primarily and basically the function of sortal at the level of speech. They naturally answer the 'What is it?' question; Aristotle's 'τί ἐστι;' question[14] precisely because they are symbols which furnish such principles. In that they furnish such principles they perform, we use them to perform, the role in speech of specifying what something is; for it is not one thing for *F* to furnish a principle of counting and distinction for particulars and another distinct thing for *F* to serve to specify what something is; for *F* to have this role or use. In that *F* furnishes a principle of distinction for particulars, *F* does furnish a specification of what something is: *F* does furnish specification of things; *F* has this role at the level or aspect of use in speech.

My thesis that sortal identification is prior to referring is not contravened by the existence of a type of case in which it may be held that a person may successfully refer to a particular object even though he/she cannot in any clear sense say what it is - or indeed say what it is at all. Let us pursue an example. First, as I have admitted one may successfully refer to a particular without knowing *precisely* what it is, e.g. one may successfully refer to the piano in the corner of the room by the use of 'That musical instrument which ...' (completed by a set of uniquely distinguishing predicates). Or again one may successfully refer to the dog in the corner by the use of 'that animal which ...' (completed by a set of uniquely distinguishing predicates). However an objector will press that successful reference may be made without a person being able to say what the *x* in question is at all. He may press that one may successfully refer to the piano in the corner by the means of such an expression 'that object in the corner which is ...' (completed by a set of uniquely distinguishing predicates); or again refer to the dog by means of such an expression as 'that sentient being which ...' (completed by a set of uniquely distinguishing predicates). Now it might by replied that in these cases *too* the speaker has to be able to specify *what* is being referred to and he has done so in this one case by the use of the word 'object' - or the fuller version 'spatio-temporal material object': in the other case by the use of 'sentient being'. However to claim that 'object' or 'spatio-temporal material object' can tell us *what* something is, is open to objection and equally to claim that 'sentient being' can perform this function. Such words introduce categories, not classes of objects. As I have earlier claimed, 'spatio-temporal object' (and also 'sentient being') are 'formal' concepts as opposed to 'proper' concepts.[15] 'Formal' concepts cannot *say* anything only 'proper' concepts can: in terms of the *Tractatus* only proper concepts can ('represent') be part of saying: only 'proper' concepts could specify what something is. Further, even if 'object' or 'spatio-temporal material object' ('being' or 'sentient being') *could*, *per impossible*, in some very general sense tell us *what* something is, it cannot *then* be claimed that in order to identify a particular in the sense of saying what it is we *have* to invoke a sortal, since neither 'object' or even 'spatio-temporal material object' can be held to be a sortal. One cannot simply count the number of objects or even the number of spatio-temporally determined material objects (or 'beings' or 'sentient beings').[16]

We have here then a type of case in which it cannot be claimed that in order to successfully refer to a particular I have to be able to identify it in the sense of being able to say what it is by the introduction of some sortal. How then can one admit this type of case and retain the general thesis that referring presupposes identifying by the means of a sortal?[17] One can explain this as follows. The *possibility* of this type of case rests on the existence of *other* types of case - types of case in which successful reference *does* depend on one being able to specify what one is referring to. That is, such a phrase as 'That object which ...' may, in general, only succeed in referring to a particular if we *already* have in our possession the use of a phrase of the form: 'The *F* which ...' where '*F*' serves to specify what the object is. The use of such a phrase as 'That object which ...' to refer to a particular rests on, and necessarily rests on, the back of the use of other phrases which refer to particulars by reference to specifying what they are. A condition of our being able to use phrases of the form 'That object which ...' to refer, where we are unable to specify the reference, is our ability to use phrases of the form 'That *F* which ...', where '*F*' is a sortal and hence where we *are* able to specify the reference.[18]

It will now be asked: 'How is it shown or to be shown that the type of case just considered is, as it were, the "odd man out"; that it does so rests for its possibility upon the existence of cases in which a sortal is present and we are able to specify reference?'

One may start off by saying that specification of the reference is primarily required in that if we are unable to specify the reference we are quite unable to determine the reference. To so say, however, is clearly question begging in relation to the type of case under consideration; for in this case we are unable to specify the reference and yet it is plausible to hold that we can determine the reference.

Let us therefore approach the matter in another way. To *deny* that this type of case rests for its possibility on cases in which a sortal is involved and we are able to specify the reference entails denying that sortals are principles of counting in their own right. It entails the thesis that far from it being the case that we can only count the number of objects in that we can count the number of *F*'s, rather we can only count the number of *F*'s in that we can count the number of objects which *are F*, i.e. which exhibit the *characteristic* or *feature* of being *F*. *F* (a sortal) thus becomes a feature or characteristic of objects. In Strawson's language this is to reduce a 'sortal universal' to a 'characterizing universal'; but as he has argued and I have equally defended, 'sortal universals' are required as antecedent principles of counting and distinction for 'characterizing' ones, and common nouns which exhibit the features I drew attention to earlier[19] are precisely paradigm cases of sortal expressions and are introduced by Strawson as such.[20] We are thus faced with the situation of *either* (i) denying the necessity for such principles of counting in their own right *or* (ii) denying that the paradigm cases of sortals are symbols the grammatical form of which is a common noun exhibiting the features mentioned earlier and which serve to specify what something is, and implicitly holding that 'object' or some variant on 'object', e.g. 'spatio-temporal object' or 'spatio-temporal material object' is an expression, form of words, which supplies a principle of counting and distinction in its own right and can serve to specify what something is. I have already advanced arguments against the thesis that such a phrase as 'spatio-temporal object' or 'object' can serve to specify *what* something is, and to claim that 'object' or some variant of 'object' can supply a principle of counting and distinction for particulars in its own right lands one in the confusion introduced earlier, namely, construing a 'formal' concept as a 'proper' one. It lands one in the position of having to use such impossible formulations as: 'there are objects'; 'there are spatio-temporal objects' and of being prepared to affirm the existence of absolutely *simple* spatio-temporal particulars or spatio-temporal material particulars.[21] To hold (ii), I thus claim, lands one in impossibilities.

As regards (i), to deny the necessity for both principles of counting and distinction in their own right, to deny the necessity of sortals, is to claim that we may introduce the notion of 'a particular' ('an object') without reference to that of a sortal; all we need is a principle of counting and distinction formed from a 'matter' term or 'feature' universal, but this thesis I have already rejected.

It is my claim that whilst one may successfully refer to an object simply by the use of the phrase 'That object which ...', 'That (material) object which ...', this necessarily rests on the back of sortal identification. But now we must consider the type of case in which it is claimed that one may successfully refer to an object yet only be able to identify what one is referring to by invoking Geachian 'substantival general' terms or Strawsonian 'feature' universals (or parallel 'terms'). In such cases I hold that such identification is secondary, resting too on sortal identification in the following way. Let us consider, for the purpose of illustration one of the examples given earlier - viz. the piano in the corner of the room. I there acknowledged that: (a) a person does not have to know precisely what the *x* in question was in order to successfully refer to it; (b) he could have made successful reference to it by the use of 'the object which ...' plus a set of uniquely distinguishing predicates; but I did not consider the point that a person might successfully refer to the piano simply by the use of either (a) 'that piece of *furniture* in the corner which ...' or (b) 'that parcel of wood in the corner which ...'[22] or even (c) 'that wooden object in the corner which ...'. As regards (c), one may only count and discriminate the number of wooden *objects* in that one first has the notion of an *F* (sortal) which is wooden. In other words, (c) is covered by what I said earlier concerning 'object'. As regards (b), my case is that a person may only successfully identify *an object*, e.g. a piano *as* a parcel of wood in that he first has the concept of *an object*, which is to have the concept of a sortal. In the situation envisaged we are, after all, speaking of the conditions necessary for a person to successfully refer to an object, or a piano, and in this situation the person considers himself to be making reference to an object which he identifies *as* a parcel of wood. Unless he *first* has the concept of an object, an *F*, he could not be in a position to identify an *F as,* for example, 'a parcel of wood'. As regards (a), one may present the case that a person may only identify an object as a piece of furniture in that he may *first* identify an *F* which is a piece of furniture, since whilst there may be and indeed, it seems, actually are pieces of furniture which are not of any sort (e.g. Victorian 'what nots'), these cases are primarily secondary, derivative. Whilst one cannot go as far as to say that it is a necessary truth that all pieces of furniture are *F*'s, we may go as far as saying that it is part of the *ratio* of a piece of furniture that it is an *F* and thus unless we have the concept of a piece of furniture that was an *F* we could not have the concept of, simply, 'a piece of furniture'. This is reflected in the fact that we talk of pieces of furniture precisely and only in that we are unable to say what they are: when we have produced no classification for them. To maintain this position does not entail the *general* thesis that identification by reference to principles of counting and distinction formed from 'matter' ('pieces of *M*') terms is *secondary* to identification by reference to sortals. 'Pieces of furniture' is a rather special case, to be contrasted with cases such as 'piece of gold'. But in any case the thesis that sortal identification is prior to identification by reference to principles of counting and distinction formed from 'matter' terms or from Geachian 'substantival general' terms, as regards the identification of *objects*, is established by reference to the argument offered in (b). In the above example the person makes successful reference to the object *as* a piece of furniture.

Having presented the case for holding that sortal identification is prior to naming/referring, I now turn to my thesis that sortal identification is prior to *describing* or *predicating*.

D. Some general arguments in favour of the thesis that Sortal Identification is prior to Describing or Predicating

The thesis is that at the level of *role* in speech or thought, identifying by means of sortals is prior to describing, or predicating. I shall be tackling this question in more depth in Part II of this chapter with particular reference to Gareth Evans, but certain general arguments may be put forward in advance of this more detailed discussion. I have already argued that at the level of symbols sortals are not reducible to descriptions (analysed out as predicates), to predicates themselves or to predicative functions and hence to predicables; but rather that the notion of a description and a predicate/predicable rests on that of a sortal, for descriptions are satisfiable by something or other and predicables are *true* or *false* of something or other. At the level of symbols sortals are prior to the introduction of descriptions and predicables, primarily in that we have to introduce the notion of 'something or other', 'some individual or other' which necessitates the introduction of a principle of counting for individuals, which as I have argued, is furnished by a sortal. At the level of speech or thought, the role of a sortal in identifying, in either the sense of 'picking out an individual' or in the sense of specifying what it is, is prior to that of (A) describing, (B) predicating, in that (a) unless one can pick out an object *independently* of describing it, one could not determine that it was *that* object which satisfied the description; unless one could specify what it was that satisfied or purportedly satisfied the description, one could not determine whether the description was satisfied or not; (b) unless one could pick out the object *independently* of predicating something of it, then one could never determine the subject of one's predication and hence one could never determine whether the predication was true or false; unless one could specify what the object was *independently* of predicating something of it, then equally one could never determine the subject of one's predication and never determine whether the predication were true or false. This is *not* to maintain the thesis that *correct* sortal identification, *correct* sortal specification, is logically prior to describing or predicating or indeed to referring or naming such that unless a person can *correctly* specify what the subject of his purported predication is, then we can never determine the subject of the predication and determine whether it is true or false. This clearly won't do; a person may not be able to correctly specify what he is talking about, yet it be clear from the context what he is talking about. For example, a person may incorrectly identify an emu as an ostrich or a dromedary as a camel and yet we can determine the subject of his predication (it is an emu, a dromedary) and we can determine whether his predication is true or false. Indeed it is odd here to speak of correct sortal identification as being logically prior (or not) to describing or predicating: it is odd primarily because in so talking we are engaging in a 'category mistake'. *Correct* sortal identification, *correct* sortal specification, are not *modes* of identification, specification any more that successful referring is *mode* of referring.[23] We cannot apply 'correct' or 'incorrect' to a *mode* of identification itself but only to something falling under that mode, within that mode. *Within* the mode of sortal identification, A may correctly or incorrectly identify: correctly or incorrectly specify. The thesis, therefore, that A on an occasion may succeed in making a predication and yet may not be able correctly to identify what the subject of his predication is, is therefore compatible with my thesis and to bring it up as a critical case is to engage in a confusion of levels. My thesis is a thesis concerning a mode of identification being prior, not a thesis about what is prior within that *mode*. *Within* that mode it may be argued that identification, in the sense of picking out an individual, in general implies/supposes being able to say what it is viz., that within that mode being able to correctly specify it prior to 'picking out'. To hold such a thesis is not to say that a person can never pick out an object unless he is able to correctly say what it is, as witness the implication of the above examples. There are indeed cases of misidentifying and it may well be pressed that such cases are only able to be considered in that they *are* cases of misidentifying, mispecifying; cases of misidentification cannot be regarded as the norm, as witness the locution '*mis*identification'. The very possibility of these cases rests, it may be urged, on cases in which being able to pick something out demands that one can first correctly specify, identify it in this sense. Conversely, *within* that mode there may be cases in which a person can say what something is in the sense of giving a perfectly correct definition of that thing, yet when faced with an actual situation be quite unable to pick out an individual of that sort. One can imagine a person who prides himself on being able to give a correct definition of a gnu but be quite unable to pick out the gnu on a visit to the zoo. But in such cases we should have to say that he was merely possessed of a *verbal* definition; in that he was unable to pick out a gnu when actually faced with one, we should say that he has not the concept of a gnu, but merely a verbal definition. In other words speaking *within* the mode of sortal identification, being able to give a correct specification of something in general implies being able to pick it out; being able to pick something out in general implies being able to correctly say what it is. But to make these points is to digress from my thesis and to discuss what might reasonably be held to be the case *internal* to a mode of identification.

To return to my thesis and to further arguments that might be advanced against it. It may be objected that a person may in fact be able to describe something and succeed in pointing it out by reference to a description and yet *not* be able to identify it, in the sense of specify it or pick it out by reference to some sortal. This of course has to be acknowledged but without detriment to my thesis. The typical type of case in which a person is able to describe something and succeed in picking it out without being able to identify it by reference to some sortal, is one in which he or she uses such a form or words as 'That thing over there which is *Y*-ish' or 'that *Y*-ish (circular, red, roundish) thing in the bushes (pointing)'. However as I have argued earlier, one could only use such a form of words to pick out something if one could first use such a form of words as 'That *F* (sortal) which is *Y*-ish' or 'That *Y*-ish *F*'. In other words, picking out an object by the means of a description in this sense logically rests on being able to pick out an object by means of a sortal and we may note here that, e.g. 'reddish', roundish, roughish object' does not *itself* introduce a *sort* of object, i.e. 'that reddish, roundish, roughish object' is *itself* a sortal.

A parallel argument may be advanced against the argument that logically, predicating or the enlisting of a set of predicates is prior to identifying by reference to a sortal. We may indeed in fact pick out an individual by reference to a set of predicates - 'its that *thing* over there which is black, shapeless and slowly moving towards us' - and we may be in no doubt as to what we are speaking about. Once again however we have introduced the form of words: 'that *thing* which ...'; with all that the intelligible use of such a form of words supposes.

It might now be objected that the notions of 'logical priority' and 'rests on' as introduced in my discussion above are not sufficiently clarified. When it is claimed that sortal identification is *prior to* naming/referring on the one hand and describing/predicating on the other, is the claim that sortal identification, as a mode of identification, is *presupposed* by naming/referring or *implied* by naming/referring - and similarly in relation to describing/predicating? Both terms it seems may be used, and indeed have been used. It is not clear however, in *this* context, namely the context of the relation between various modes of activity or operations in speech or thought, how one can employ and why one should be prevailed upon to employ the strict logical distinction between presupposition and implication which concerns relations between *propositions*. If I were however to be pressed into the straight jacket of having to say by analogy, whether by 'logical priority', 'rests on', is meant presupposition or implication, I would come down on the side of *presupposition* as opposed to implication. But as I have said I see no clear reason why I should be forced into this straight jacket: sometimes it seems more natural to speak in terms of 'implication'; others more natural to speak in terms of 'presupposition' and indeed not only sometimes more natural to speak in terms of 'implication' as opposed to 'presupposition'; but sometimes necessary so to do if one is to keep within the framework of a given text - as for example in my forthcoming discussion of Gareth Evans on the issue of the relation between Demonstrative Identification and Descriptive Identification. Here, if pressed into the straight jacket, then I would be inclined to say that Demonstrative Identification presupposes, as opposed to implies, Descriptive Identification but as we shall see the framework of Evans' text is in terms of that of implication and in the light of the consideration that it is not clear how one is obliged to enter the straight jacket of '*either* implication or presupposition', I reserve the right to speak in both ways as context dictates or suggests.

I provisionally conclude that identification by reference to sortals in the sense of picking out, discriminating by reference to sortals and specifying by reference to sortals is logically prior, as a speech activity, to naming/referring on the one hand and describing/predicating on the other. I say 'provisionally conclude' here since before any final conclusion may be advocated it is necessary to consider some writings of Gareth Evans;[24] in particular his concept of 'Demonstrative Identification', which he claims is *sui generis*, and if such a claim can be substantiated it is in conflict with my claim concerning sortal identification. The question naturally arises: 'How is Demonstrative Identification related to sortal identification and subsequently to naming/referring one the one hand and describing/predicating on the other?' I shall devote Part II of this chapter to a discussion of Evans' position and its apparent conflict with my own.

Part II: The Contribution of Gareth Evans

E. An introduction to Evans: His concept of 'Demonstrative Identification'

Evans advocates a concept of 'Demonstrative Identification' which he claims is *sui generis* and by implication has no reference, even implicit, to what he calls 'Descriptive Identification', which on one interpretation is what I call 'Sortal Identification'. I shall claim that his concept of 'Demonstrative Identification' is minimally based on a misunderstanding of the role or use of demonstratives in relation to the function of sortals; is maximally based on a misconstruction of demonstratives in the context of demonstrative sortal phrases. As a background to this claim I shall first: (i) re-state my case that demonstrative sortal identification, i.e. the identification of particulars by reference to phrases of the form 'this *F*', 'that *F*', where '*F*' is a sortal, is logically prior to the identification of particulars, or alleged identification of particulars, by the means of a demonstrative *only* together with a pointing gesture; (ii) re-state my case that 'Demonstrative Identification' by means of a demonstrative alone, together with a pointing gesture is strictly *impossible*; (iii) argue that such a concept of 'Demonstrative Identification' is based on either a misconstruction of demonstratives in the context of demonstrative sortal phrases, or upon a misunderstanding of the role or use of demonstratives in relation to the function or use of sortals and correlatively a misunderstanding of the concept of 'spatial object' or 'spatially located material object'. I shall then turn to my claim in regard to Evans. Having attempted to establish this claim, I shall then maintain, contrary to him, that 'Demonstrative Identification' in his sense *does* rest on 'Descriptive Identification' in his sense in that for his 'Descriptive Identification' we may write 'Sortal Identification'. It rests on 'Descriptive Identification' (Sortal Identification) in the sense of what I termed 'sortal specification'.[25] Further, in that 'Demonstrative Identification' is to be a mode of *identification*, as opposed to a mode of *reference*, I shall claim that it is reducible to sortal identification in the sense of sortal 'picking out', sortal 'discrimination'.[26]

I shall then consider two further possible interpretations of 'Descriptive Identification' in Evans' sense, claiming that in the one case 'Descriptive Identification' actually reduces to sortal identification in the sense of sortal 'picking out', sortal 'discrimination': in the other case that 'Descriptive Identification' can be categorised as '*quasi* sortal identification'.

Finally, in the last part of this chapter I shall consider Evans' case that identification in *thought* is prior to identification in *speech*. Since I shall be charged with making identification in speech prior, I shall hold that whilst I find no absolutely overriding reason for accepting his thesis, even if such an overriding reason could be found, my case could be re-cast in terms which would acknowledge such priority. However, in the light of the fact that I can find no such overriding reason and for certain other reasons to be given, I shall carry out my programme, at least by implication, in terms of identification in speech.

F. That Sortal Identification is logically prior to Demonstrative Identification

My case for this is that we logically could not identify a particular, i.e. an object, or pick out or discriminate a particular, simply by the use of a demonstrative together with a pointing gesture unless we are able to identify a particular by reference to a demonstrative together with a sortal phrase: i.e., the use of a demonstrative together with a pointing gesture to identify a particular, to pick out, discriminate a particular, rests on the use of a demonstrative together with a sortal accompanied by a pointing gesture; precisely since the use of a demonstrative itself, even with a pointing gesture, cannot determine *what* is being discriminated or picked out. The use of a demonstrative, even with a pointing gesture, has no discriminating or discriminatory power, and indeed no referential power. Cases where we may claim that the use of a demonstrative on its own together with an 'appropriate' pointing gesture, may serve to identify, to discriminate an individual, do exist - as in '*That's* the one', '*That's* it', but on investigation we are able to see that all such cases are writable as 'That *F* (where "*F*" is a sortal) is the *F* I mean':[27] 'that *F* is the *F*'. In short, in order for 'that' together with a pointing gesture to identify, i.e. to discriminate, pick out an individual, we logically have to be able to specify *what* it is we are pointing at, without being able to do this no discrimination can be intelligibly claimed.[28]

G. The Case against Demonstrative Identification (1): the *Misconstruction* of Demonstrative Sortal Phrases

Demonstrative identification, in the sense of picking out, discriminating an object by the use of a demonstrative alone, is strictly impossible because the use of a demonstrative alone together with a pointing gesture has, as I have just argued, no discriminatory power; it cannot distinguish what is to be discriminated and the pointing gesture cannot imbue the demonstrative with the power, since it itself cannot distinguish what is to be discriminated.

My claim that to hold that demonstrative identification in this sense is possible is based on a *misconstruction* of demonstrative sortal phrases is as follows. It construes such phrases as 'That car' ('This tree' ['This *F*']) as a logically complex phrase, formed from two elements 'that' and 'car' ('that' and '*F*') of which the demonstrative is a *name*, or at least a sort of *name*, in the sense that (a) it has an independent sense (b) is a complete symbol, indeed a 'simple' symbol in Russell's sense.[29] Now to this it may be objected (i) that a demonstrative cannot be said to have a sense at all, let alone a complete sense. That is, it cannot be said to have a sense in Frege's sense of 'sense', in which case it could not be said to be able to perform a discriminating role or be used to perform such a role. However, it might now be counter objected that if we construe 'sense' in terms of meaning and meaning in terms of use, a demonstrative can certainly be said to have a sense in that it can be said to have a use. But the claim would then be that a demonstrative can be said to be a sort of name in that it has an *independent* sense or use. We thus must offer an account of what this independent sense or use is. The answer, on such a construction, would readily be on hand - namely it has the sense or use of serving to locate some definite spatial location and in that it has the use of serving to locate it, pick out, discriminate some definite spatial location it may serve and does serve to pick out, discriminate, a spatial object. My point however is that 'this' of itself ('that' of itself) even with a pointing gesture *cannot* be said to have an independent sense, in that it can, of itself (even with a pointing gesture), be used to discriminate any definite spatial position. We may only determine whether 'this' ('that') has such a role in that we understand that 'that *F*' signifies, serves to specify some *spatial object*. Without *that* understanding we cannot determine the role, the sense of the demonstrative. 'That' together with a pointing gesture, cannot determine that what is being discriminated, picked out, identified, is *even* a definite spatial location. In this regard therefore, even if we construe 'sense' as above, a demonstrative cannot be said to have an *independent* sense: it may only be said to have a sense in the context of a sortal which signifies, specifies a spatial object.

At this juncture however an objection will be raised, namely, that all we need to understand the 'sense' of a demonstrative, is not the concept of some sortal, but solely the concept of 'spatial object'. 'That' may perform the role of serving to locate some definite spatial location, provided we know that what is to be located is (simply) some spatial object; we do *not* have to understand some sortal. This line of thought however is itself based upon a misunderstanding, since for us to have the concept of a 'spatial object' (or 'spatial material object') is not independent of our having the concept of a sortal. Indeed we may only have the concept of *a*, *one*, spatial object (or indeed spatial material object) in that we *do* have the concept of a sortal. We may only pick out, determine the number of spatial objects in a given area of three dimensional space in that we understand what is to count as *one* such object and the answer to this question is furnished by a sortal. One cannot begin to obey the instruction: 'Pick out the number of spatial objects in this room (say), unless we know what is to count as *one* such object. To think that one *can* is once again to confuse what Wittgenstein calls 'formal' concepts with 'proper' concepts.[30]

Indeed if it be said that all that we need for a demonstrative to have such a 'sense' or 'role' ('use') is that we understand, here, the notion of a 'spatial object' and that we are able to have the notion of 'a spatial object' independently of any sortal, then we are left with a doctrine of 'simple' spatial objects which can be identified, 'picked out', discriminated, by demonstratives and this naturally leads to our construing a sortal as a *predicate* viz., 'that', 'this', have sense in that they serve to locate a spatial object which *is F* or an *F*: in such a proposition as 'This is white', we have a subject predicate-proposition in which 'this' serves to locate a (simple) spatial object *of which* it is true or false that *it* is white.

I thus hold that a demonstrative may not be said to have a sense independently of our understanding some sortal; and to claim that it may be said to have a sense independently of some sortal, but not independently of the notion of a spatial object, is a mistake based upon a misunderstanding of the nature of the concept of 'spatial object'. One who holds however that a demonstrative is a sort of name holds that a demonstrative has a sense, can have the use of serving to discriminate a spatial object, independently of the notion of even 'a simple spatial object'. One who so holds moves *from* simply saying that a demonstrative has sense in that it can serve, is used, to locate the spatial position *of some spatial object to* saying that it has sense in that it can serve, is used to, can be used, simply to *locate*, or *locate some spatial position.* This however, for reasons already offered is absurd, since (A) there is no such activity and can be no such activity as simply *locating* - one seeks to locate something or other; (B) 'this' may only serve to locate in that we understand what it is that is purportedly located.

As regards my implication that a demonstrative may not be said to be a complete symbol, some explanation is required. First, it cannot be claimed that a demonstrative is an *incomplete* symbol in the sense that a predicate is: such a claim is obviously absurd. Yet it cannot be said to be a *complete* symbol in that to understand its 'sense', its potential function, we do need to understand *something else.* That is, for reasons offered, a demonstrative may only perform its function in that it is accompanied by, or understood to be accompanied by a sortal. It may only perform its function of serving to discriminate an individual in the *context* of a sortal and therefore, I hold it more perspicacious to construe a demonstrative, as of its *essence* an element in a complete phrase of the form 'this *F*', 'that *F*', where '*F*' is a sortal, and *not* as a sort of name. On my view therefore a demonstrative has sense only in the context of, or implied context of a sortal: whereas I have already argued that it cannot be held that a name has a sense only in the context of a proposition. To construe a demonstrative, as of its *essence* an *element* in a complete phrase of the form 'this *F*', 'that *F*' seems to be the best way of putting the matter. To speak in terms of a demonstrative being, of its essence, *part* of a complete phrase 'this *F*', 'that *F*' leads to the difficulty that a part is defined by reference to a whole and thus strictly we could not even understand the alleged independent use of demonstratives: to speak in terms of a demonstrative being an incomplete symbol, essentially of the form 'this ---' brings forth the difficulties (a) that we could never intelligibly speak of 'this'; (b) that there is a severe problem in saying exactly how a demonstrative differs from a predicable as regards the sense of completion.

H. The Case against Demonstrative Identification (2): Consideration of a stronger and weaker thesis as regards its base

I shall argue that 'Demonstrative Identification', in the sense advocated by Gareth Evans, is at the very least based on a misunderstanding and a correlative mistake; at the most based upon a misconstruction of phrases of the form 'this *F*', 'that *F*'; where '*F*' is a sortal. It is based on a misunderstanding, in that while it concedes that a demonstrative is not a sort of name, having an independent sense, nevertheless it sometimes subscribes to the view that whilst a demonstrative has a dependent sense, its sense is determined, not by reference to some sortal, but by reference simply to the concept 'spatial object' viz. that a demonstrative may serve to locate some definite spatial location provided it is understood that what is to be located is (simply) some spatial object (or some spatial material object) which is *F*. Thus, on this view a demonstrative has sense, is able to perform the function of uniquely discriminating an object in that it primarily operates in the context of the notion of 'spatial object', or 'spatial material object' and not, as I have argued in the context of a sortal.[31] The correlative mistake is that this view commits one to the thesis that 'spatial object' or 'spatial material object' is a 'proper', as opposed to a 'formal' concept and that indeed there are such *x*'s as (simple) 'spatial objects' (or '[simple] spatial material objects'). At the most it is based on a misconstruction of such demonstrative phrases, treating demonstratives as themselves a kind of *name*, and the whole of such phrases as logically complex in the manner described at the beginning of Section G.[32]

a. In consideration of the strongest thesis that Evans is committed to construing demonstratives as a kind of name and to construing sortal phrases, 'that *F*', as a complex containing the demonstrative as a *naming* element, one may, *prima facie*, advance the following.

a.(i). *The general context of Russell*

Granted Evans' background to his doctrine of Demonstrative Identification, namely the background in Russell's doctrine of naming, this is precisely what one would expect, as Russell does regard demonstratives as a type of name; indeed[33] cases *par excellence* of a 'proper name' in the sense of 'logically proper names'. Evans writes: 'Russell introduced us to the idea that demonstrative identification is a mode of identification quite unlike descriptive identification, and to the idea that it is apt to underlie the use of Russellian singular terms.'[34] To expand: genuine Russellian singular terms are logically proper names and what underlies their use is the fact that there is present to the observer (sensory agent) an object of immediate perception, namely a sense datum. Unless there is such an object immediately present to an observer (sensory agent), then the (logically) proper name is no name and nothing is expressed by such a sentence (atomic sentence) as 'This is white'. A condition of their use then is that there is some sense datum present which they immediately name. Evans rejects that part of Russell's thesis which would seem to commit him to a doctrine of private language, but accepts the general thesis that (1) demonstrative identification is to be radically distinguished from descriptive identification: (2) that in demonstrative identification there must be present to the subject just one object with which he is in 'informational contact' in order for a singular term (demonstrative singular term) to have a genuine use. [In order for it to perform its identifying role, in order for it to be a genuine singular term]. If there is no such object then no thought is expressed, no proposition, true or false, is expressed by the sentence containing such a singular term. In Evans' own terms, which places a premium on understanding demonstrative identification and upon demonstrative thoughts as opposed to making demonstrative identification and demonstrative statements, we have:

> ... a demonstrative Idea of an object is not reducible to any other sort of Idea, and in particular cannot be regarded as a species of descriptive identification. One has an adequate Idea in virtue of the existence of an information-link between oneself and the object, which enables one to locate that object in egocentric space .... Consequently, demonstrative thoughts about objects, like 'here'-thoughts, are Russellian. If there is no one object with which the subject is in fact in informational 'contact'... then he has no Idea-of-a-particular-object, and hence no thought. His demonstrative thought about a particular object relies upon the *fact* of an informational connection of a certain kind, not upon the thought or idea of that connection; hence it is unconstruable, if there is no object with which he is thus connected.[35]

It will however be protested that the fact that Evans' doctrine has its background in Russell does not entail that Evans takes over Russell's doctrine of demonstratives. Indeed, it will be said, his actual examples of demonstrative identification do not involve such Russellian atomic sentences as 'This is white' but rather, e.g. 'This *G*', 'That person', 'That player',[36] as in e.g. 'That person is *F*': 'This *G* is *F*': 'That player is good looking'. But whilst we may cite such examples, we may equally cite others where Evans does *not* speak of 'This *G*' or 'That *G*', but rather of 'This is *F*', 'This was *F*', 'This is *F* now'.[37] What one has to establish is that it is essential to Evans' theory that he regards demonstratives in the way that the strong thesis depicts. That it is so essential may be claimed as follows.

a.(ii). *Direct evidence*

His very raising and answer to what is for him a central question, namely: 'What makes demonstrative identification of spatially located material objects possible?'

He writes:

> Given that the existence of an information-link between subject and object is not by itself sufficient for identification,[38] what makes it possible to have, in the standard case of demonstrative identification, a mode of identification that is free of the conceptual element we have been considering? The answer is that

> in standard cases, not only is there an information-link, but also the subject can, on the basis of that link, *locate the object in space*.[39]

And again, in direct answer to the question: 'What makes demonstrative identification of spatially located material objects possible?', he writes:

> In the ordinary perceptual situation, not only will there be an information-link between subject and object, but also the subject will know, or will be able to discover, upon the basis of that link, where the object is.[40]

Thus, it seems on Evans' thesis, what makes demonstrative identification of spatially located material objects possible, that is to say identification of such objects by the means of such phrases a 'That *F*', 'This *F*' (where '*F*' is a sortal) possible in standard cases, is the ability of the subject, on the basis of the information-link, to locate the object in space.

This, I contend is wholly misleading, and more directly does show that Evans assumes that a demonstrative itself is a complete symbol, a sort of name, in that it may of itself serve to uniquely discriminate a spatially located object. That Evans is committed to *this* thesis, as opposed to the *lesser* thesis that a demonstrative may only *have sense* in the context of the notion of a 'spatial object' or 'spatially located material object', can be seen from the fact that for him a central question is: 'What makes demonstratives identification of spatially located material objects possible?' He could not even *raise* this question if he were to hold the lesser thesis. To raise this very question presupposes or assumes that there can be demonstrative identification of spatially located material objects even independent of the notion of 'spatial object'. To answer this question in terms of the subject, on the basis of an information-link, being able to locate the object in space, supposes that the subject is so able to locate independently of the nature of the object, independently even of knowing that the object to be located is a spatial one, since this is primarily an answer to the question 'What make demonstrative identification of spatially located material objects *possible*?' In that it is an answer to *this question*, then it necessarily cannot be answered that the subject understands that the nature of the object to be located is *spatial*.

Now *in that* Evans cannot even allow that a demonstrative only has sense in the context of the notion of 'a spatial object', he certainly cannot allow that a demonstrative only has sense in the context of a sortal. In the case of e.g. 'that cat wants some milk', that he would have to say that it is simply the spatial position located by 'that' which may serve to uniquely discriminate or individuate some one cat from any other cat or object. That we can understand the function of 'that' in 'that cat' independently of understanding that it is a cat that is to be uniquely discriminated and necessarily so. This might be questioned in the light of his doctrine of the information-link between subject and object. It might be said that the information-link provides the subject with the information that there is e.g. a cat present to his senses and on the basis of this information the subject can locate the very cat in space which is the subject of his demonstrative thought. That is, that after all, Evans is not holding that the subject can locate the object in space independently of understanding the nature of the object; independently of knowing what it is that is to be located in space. He knows, on the basis of the information-link, that it is a *cat* (say) which is to be located. But that we simply know, or understand, on the basis of an information-link that it is a *cat* which is to be located, does not entail that for him a subject must understand what it is that is to be located, in order for him to be able to discriminate the object concerned. It is merely *contingent* to this activity that the subject understands what it is that is to be located since he is primarily concerned with an answer to the question 'What makes the identification of spatially located material objects *possible*?'.

a.(iii). *Indirect evidence*

Spelling out in answer to the question 'What makes demonstrative identification of spatially located material objects possible?', Evans writes:

> Given the subject's general knowledge of what makes propositions of the form $\ulcorner \pi = p \urcorner$ true, for arbitrary π, when p is an Idea of a position in his egocentric space, and given that he has located, or is able to locate, the object in his egocentric space, he can thus be said to know what it is for $\ulcorner$This = the object of π now$\urcorner$ to be true (for arbitrary π). Hence he can be said to have an adequate Idea of the object.[41]

Here 'π' is a position in public space, where the object is, so Evans' position is that granted the subject knows what it is for propositions of identity between items of public space and egocentric space to be true and given that a subject is able to locate the object in his egocentric space, he can then be said to know what it is for a proposition 'This is identical with the object at π now' to be true. Note that Evans writes *not* 'This *object* = the object at π now', but simply 'This = the object at π now'. Thus he seems to be committing himself to the view that 'this' has an independent sense and can serve on its own, together with a pointing gesture to indicate some object. And it is no accident that Evans writes in this way; it is no slip of the pen. It is, on the contrary, *crucial* that he write $\ulcorner$This = the object at π now$\urcorner$ since what it is to be locatable in egocentric space involves no reference to any description, or in Evans' terms, no reference to any 'descriptive thoughts'.[42] It is only on the thesis that location in egocentric space *does* involve reference to descriptions or descriptive thoughts that Evans *could* write $\ulcorner$This *object* = the object at π now$\urcorner$, but Evans explicitly repudiates any such thesis. It is essential to his doctrine of egocentric space that 'here', 'there', directly designate spatial positions relative to the subject conceived of as the centre of his space; and in parallel fashion 'this' can so directly designate without reference to any 'description' and hence is not of its nature or essence an element in a phrase of the form 'this *F*' ('this object'). It is central to my position that Evans' doctrine of egocentric space be refuted in the sense that it be shown, in his terms, that location of objects in egocentric space does, of its essence, involve a reference to descriptions, to descriptive thoughts. I shall engage in this exercise later, namely in my case that Demonstrative Identification in Evans' sense at least rests upon Descriptive Identification in his sense: in *my* terms that Demonstrative Identification as least rests on sortal identification, though it is not to be assumed, for reasons later to be offered, that what I mean by 'sortal identification' is what Evans means by 'Descriptive Identification'. Although I shall argue that in terms of Evans' distinctions Demonstrative Identification does at least rest on Descriptive Identification, I shall then reject the notion of 'Descriptive Identification' in Evans' sense in that it accepts Russell's doctrine of descriptions and reduces descriptions to predicates. I shall in effect claim, as I have already done, that all 'descriptive' identification, properly understood, is reducible to sortal identification.

Thus far therefore the purported evidence is in favour of the strongest thesis advocated above.

b. Can this alleged evidence be challenged such as to show that Evans is only committed to the weaker thesis, namely that a demonstrative does *not* have independent sense; it has *dependent* sense in that it primarily operates in the context of the notion of 'spatial object' or 'spatially located material object'? To decide this issue it will be necessary to reassess the alleged evidence in section a.(ii) and a.(iii) above; the evidence for a.(i), I hold was not of itself decisive.

First a.(ii). It will be objected that I have given too wide or too general interpretation to the question 'What makes demonstrative identification of spatially located material objects possible?'. Evans is not as such raising the question 'What makes demonstrative identification of spatially located material objects possible?', such that he cannot invoke the notion of 'spatial object' or 'spatially located material object' in his answer, *rather* he is raising the question; 'What make the demonstrative identification of spatially located *F*s (where '*F*' is a sortal) possible - viz. what makes the demonstrative identification of *F*s, cats, trees, houses possible?', and the answer is that the information-link provides the subject with the information that there is an *F* present to his senses and on the basis of this information the subject can locate the very *F*, the very cat (say), which is the subject of his demonstrative thought. So what makes Demonstrative Identification possible is the information that there is some *F*, some cat (say) present to the subject's senses and that the subject is able to locate *it*.

This answer, however, while it does not commit Evans to counting 'that' as having an independent sense, as a name, nevertheless commits him to the view that 'that' in 'that *F*', or 'that cat', can serve to locate the cat concerned, the *F* concerned, independently of our understanding that it is an *F*, a cat, which is to be identified (picked out, discriminated). Now it might be said here that, 'that' in 'that cat', can serve to locate the cat concerned, independently of our understanding that it is a cat which is to be identified (picked out, discriminated): viz. independently of our understanding the *particular* sortal - it could serve to pick out, discriminate the cat concerned even if we don't know that it is a cat that is to be identified, but only know that it is an *animal*. But on this account Evans' thesis goes further, viz. it commits him to the view that 'that' in 'that cat' can serve to locate the cat concerned, not necessarily independently of our understanding the particular sortal, but independently of our understanding *any* sortal. More generally that 'that' in 'that *F*' can serve to locate the *F* concerned independently of our understanding *any* sortal, otherwise he would not be raising the general question 'What makes the identification of *F*'s, cats, trees, houses, etc., possible?'. All that is required in order for 'that' (a demonstrative) to perform its function of serving to locate the very *F* which is the subject of his demonstrative thought, is that it should have sense in the context of the notion of 'spatially located material object' or 'spatial object'. But to hold that 'that' in 'that *F*' can serve to locate the *F* concerned, independently of any sortal, merely by reference to the notion of 'spatially located material object' or 'spatial object', is a claim based on the mistake of construing a 'formal' concept 'spatially located material object' as a 'proper' concept. We may only have the concept 'spatially located material object' in that we have an appropriate proper concept, namely a sortal, for we may only speak of *one* spatially located material object in that we may first speak of one *F*, as I have already argued earlier.

I thus concede that a.(ii) gives a too wide or too general interpretation of the question: 'What makes demonstrative identification of spatially located material objects possible?' and I therefore further concede that the alleged direct evidence discussed is only direct evidence for the lesser thesis and thus that Evans, on this count, is only open to the charge of misunderstanding and being mistaken.

Let us now turn to a.(iii) - the alleged indirect evidence. It will be objected here that from the fact that Evans holds that Demonstrative Identification is *sui generis* and does not reduce to or rest upon 'Descriptive Identification', it does not follow that for 'that' he cannot write 'that object'. It would only so follow if he held that one could only have the notion of 'spatially located material object' in that we first had the notion of some sortal: but this he clearly does not hold. That is, he could still *hold* the view that 'that' only has sense in the context of 'that object' and *deny* that he was indirectly invoking a description and hence that his doctrine as to the *sui generis* nature of Demonstrative Identification was seriously challenged. But in that he wishes to hold that 'that' only has sense in the context of 'that object', he would have to treat 'object' in 'spatially located material object', as *itself* a proper concept; in his terms, following Russell, a *description*. And note he can hardly avoid treating 'the object at π now' in ⌜this = the object at π now⌝ as a definite description. Hence the attempt to claim that Evans is only committed to the *lesser* thesis in a.(iii) is incompatible with his doctrine of the *sui generis* nature of Demonstrative Identification. Only the doctrine that 'this' in ⌜this = the object at π now⌝, is a name is compatible with his doctrine of Demonstrative Identification, and only this doctrine coheres with his thesis that in egocentric space, in our thoughts about egocentric space, 'here', 'there', directly designate a spatial position relative to the subject.

In the light of the above discussion I conclude that Evans, in his doctrine of Demonstrative Identification, is at the very least committed to a misunderstanding and a mistake and at the most committed to a misconstruction of demonstrative phrases, treating demonstratives as names. On either count we have good reason to reject his doctrine of Demonstrative Identification, but in the light of (a) the clear importance of this doctrine, (b) its antecedents in Russell, I shall now proceed to substantiate my thesis that Demonstrative Identification is not *sui generis* in that it in one sense rests on (presupposes) 'Descriptive Identification', and in another sense can be said to be reducible to it in that by 'Descriptive Identification' Evans means or includes 'Descriptive picking out', 'Descriptive discrimination'.

I. That Demonstrative Identification (in Evans' sense) rests on, presupposes, Descriptive Identification in his sense and indeed on Sortal Identification

As we have seen from the passage cited from p. 173 of *The Varieties of Reference*,[43] Evans claims that Demonstrative Identification is unique in that the speaker, in order to have a demonstrative Idea of an object and hence express a demonstrative thought concerning an object, must be in 'informational contact' with that object and have, by implication, no 'descriptive thought'. No descriptive thought need be ascribed to such a speaker.

Earlier[44] he forcefully argues against the view that some clear descriptive thought can be ascribed to a subject who essays a demonstrative identification but is not perceiving anything. He is in favour of the view that the production of a clear descriptive thought in demonstrative identification can only be ascribed to a subject in that that subject perceives the object and further claims that it is a *misconception* to think that a clear descriptive thought can be ascribed to a subject who essays a demonstrative identification but is not perceiving anything.

I wish to argue for the truth of the following theses:

(1) that Evans does not convincingly establish the stronger thesis viz. the 'misconception' thesis just mentioned;

(2) that an argument can be produced in favour of the thesis that *some* clear descriptive thought in the sense of a uniquely distinguishing descriptive thought can and indeed must be ascribed to a subject who essays a demonstrative identification but is not perceiving anything;

(3) that whilst Evans produces some cases for the thesis that no clear descriptive thought, in the sense of no uniquely distinguishing descriptive thought, can be so ascribed: (a) this is not conclusive in virtue of the argument advocated in support of (2) above; (b) that he has no sufficient case for the wider thesis that no clear descriptive thought in the sense of a descriptive thought which is relatively specific

can be so ascribed; (c) no case at all for the widest thesis that no descriptive thought at all can be so ascribed to a subject.

In that I can establish the truth of these theses, I can claim that Demonstrative Identification in his sense, rests upon, implies, Descriptive Identification in his sense: and rests upon, implies 'sortal identification', in the sense of sortal specification in that we can construe what Evans means by 'Descriptive Identification' as identification by means of 'descriptions' which contain or are formed from sortals.

At this point it might be commented that even if (b) and (c) as just set out are established, this is compatible with Evans' thesis since what he is principally holding is that in demonstrative identification no clear descriptive thought in the sense of no uniquely distinguishing descriptive thought can be ascribed to a subject who does not perceive anything. Although this is admittedly his central thesis, he must also hold the theses mentioned in (b) and (c) above in order to retain his thesis that demonstrative identification is not in principle reducible to descriptive identification; that it is *sui generis*. To limit his case to the point that in demonstrative identification no clear descriptive thought, in the sense of no uniquely distinguishing thought, can be ascribed to a subject who does not perceive anything, is not sufficient to establish that demonstrative identification is not a species of descriptive identification; not sufficient to establish the *sui generis* nature of demonstrative identification and the crucial importance he wishes to attach to it.

First then, (1) above - the 'misconception' thesis. Evans' case for this strongest thesis is that a subject can demonstratively identify and think about one object in an enormous array of closely packed and indistinguishable objects viz., in a situation where the subject could provide no precise egocentric spatial description. In such a situation a subject can so demonstratively identify 'provided, as we say, he keeps his eye upon it'.[45] He produces the case of a table covered with indistinguishable coloured pills. In relation to such a case the point he wishes to make is that 'the Idea of a point *p* in egocentric space precise enough to be adequate to individuate the pill exists only because there is something at *p* - the pill - for the subject's perception to latch on to'; the Idea of *p* depends upon the perception of the pill and not *vice versa*. It is for this reason that it is a misconception to think that some clear descriptive thought can be ascribed to a subject who essays a demonstrative identification but is not perceiving anything.

I shall now argue that Evans' 'pill' case is not so convincing as may first appear and that indeed in holding such a thesis he is committed to some implausible positions and even apparent absurdity.

(a) To hold that the Idea of a point in egocentric space precise enough to be adequate to individuate the pill exists only because there is something at that point - the pill - commits one to the implausible position that the existence of the pill is necessary to the identification of the point. This is implausible because it is only contingent that the pill is at that place; the place (point) remains the same independently of the existence of the pill and one *could* specify the point (place) and hence have the Idea of point (place) independently of the pill. Theoretically one must be able to do this and in practice one could viz., in terms of three dimensional position in egocentric space relative to other pills in relation to the table and oneself as subject.

(b) If one holds that we only have the Idea of a point *p* in egocentric space in that there is actually something *at p*, then how could we ever be in a position to formulate such a hypothesis? In order to determine that there is something *at* position *p* or point *p*, we have to be able to specify position *p* independently of any object existing, yet this is precisely what he denies. In so denying, I submit Evans has denuded the notion (Idea) of an object being *at* a position or at a point *p* in egocentric space of any clear sense.

(c) To claim that the 'Idea of *p*' is equivalent to the Idea 'where the pill is' presents us with the further difficulty that equally were one to *remove* the pill from its place one could not have the Idea of the spatial position formerly occupied by the pill independently of the pill; but surely one could - e.g. one could specify the spatial position formerly occupied by reference to an area of three dimensional space relative to other pills and in relation to the table and oneself as subject as envisaged in (a) above.

(d) Strictly speaking on Evans' thesis one could not speak of 'removing the pill from *its* place'; but even in this interesting and problematic case one can still speak of 'removing the pill from its place - removing the pill from the *place it now occupies*'. Strictly speaking Evans would have to deny that we can so talk. If the Idea of position *p* in egocentric space is equivalent to the Idea 'where the pill is' then for 'removing the pill from its place, from the place it now occupies', one could, and indeed one would have to write 'removing the pill from where the pill is' but 'where the pill is' can only serve to identify some place if we *can* independently specify the place. It is for this reason that it strikes one as absurd to hold that for 'removing the pill from its place, the place it now occupies' we can write 'removing the pill from the place *where the pill is*'.

In the light of these implausibilities I hold that Evans does not conclusively establish his strongest thesis concerning misconception. I now turn to (2) above.

Evans considers the position of Schiffer.[46] This writer, in considering a case rather like the 'pill' case produced by Evans, suggests that the object would have to be identified by the means of some such description as 'the one I have fixed my gaze upon'. Evans contends that this gets things the wrong way round; 'it is the fact that I have my gaze fixed on something, not the Idea that I have my gaze fixed on something that determines which object is the object of my thought'. However, unless I can provide *some* description; unless I can, in

Evans' terms, essay *some* descriptive thought I cannot even begin to specify *which* fact it is which determines the object of my thought. Further, Evans cannot hold that it is *simply* the fact that I have my gaze fixed on *something* which determines the object of my thought for this is not sufficient to determine *the* object of my thought. Rather we must hold that, if anything, it is the fact that I have my gaze fixed on some *one* thing, i.e. just one thing which determines the object of my gaze; but unless I can provide, not merely some description, but some uniquely individuating description - in Evans' terms essay some uniquely descriptive thought - then I cannot uniquely determine *which* supposed fact it is that allegedly determines which object is the object of my thought. Even granted that Evans is correct in objecting to Schiffer, the advocacy of his own position supposes that we have to entertain, essay, some uniquely descriptive thought concerning the object in order to determine the identity of the *fact*.

I now turn to (3) above, specifically (3), (b) and (c). Evans writes:

> The importance which being able to locate the object has may encourage the idea that the demonstrative identification of spatially located things can be *reduced* to the identification of positions in egocentric space, so that 'this *G*' would be equivalent to 'the *G* (now) at *p*'.[47]

He holds that this is not in general possible. I am not concerned here with the 'reduction' thesis as such but to establish (3)(b) and (c); hence my interest in this case is the arguments the author appeals to, to support his case.

Evans comments that in a great many cases a subject may make a demonstrative identification of an object without actually knowing where it is, i.e. the information-link with the object may enable the subject to locate the object without providing very specific information about its location - 'for example, when one is able to home in upon the beetle eating away in a beam'.[48] Here it may be agreed that we might not be able to produce any *quite* specific location of the beetle *vis-à-vis* the beam but from this it does not follow that no spatial description is necessary, no 'descriptive thought' is necessary, for the demonstrative identification of the beetle in particular or objects in general. In this case in order to make demonstrative identification of the beetle we have at least to say that the beetle is *in the beam*, which entails locating the beetle in public space and, more precisely we have to be able to say it is at some position in the beam which is within the direct focus of the subject's eyes or direct focus of the beam of his torch. In terms of the subject's system of egocentric space it is the beetle upon which he is now focusing his direct attention - 'the one right before my very eyes'. Unless the subject is able to produce: (a) some more general spatial identification (specification) then his reference (demonstrative identification) is quite indeterminate and Evans cannot even introduce the beetle example; (b) some more specific spatial identification (specification) either (i) within his own egocentric space; (ii) within public space; (iii) within a combination of the two, then the reference and hence the alleged demonstrative identification is relatively indeterminate and hence problematic as an identification.

Evans however would argue that in such a case as the 'beetle' case the information-link places the subject in a position rather like that of a man who feels something tugging at the end of a fishing line. In such cases we are placed in a position in which we have the *practical ability* to locate the object and it is not necessary to construct some concept ('the one at the end of my line') in order to allow the subject's thought to reach out to its object when he can effectively do so himself.[49] The difficulty here for Evans is that it is certainly arguable that the subject *can* only have the practical ability to make the demonstrative identification (to locate *the* object) in that he *does* at least have the concept 'the one at the end of the line' viz. in that he can produce some relatively specific description (specification) otherwise his identification is not determinate. 'The one at the end of the line' is *not* some extra 'concept' which is constructed *post eventum* by a philosopher, but a concept, a description, which it is necessary that the subject have in order to make the relatively precise demonstrative identification he claims to so make. The man in the fishing line example is alleged to have the practical ability to locate the object but in order to do this he must at least have the relatively specific concept 'the one at the end of the line'; 'the thing at the end of the line' to make the relatively precise demonstrative identification he claims to make. Evans is only able to claim that such a concept is not necessary by conflating 'something' with 'some one thing', i.e. 'just one thing', for he moves without apology from speaking of a man who feels *something* tugging at the end of a fishing line to having the practical ability to locate *the* object. From the fact that he feels something tugging at the end of his line it does not follow that there is just one thing so tugging so that he can move without apologia to speaking of locating *the* object.

I thus hold that Evans has to concede that some relatively specific descriptive thought is necessary for demonstrative identification to be relatively determinate. And, one might add, he has to concede the more general point that *some* descriptive identification, some descriptive thought is necessary for a subject who essays a demonstrative identification in that the very introduction of the analogy itself demands it - 'rather like that of a man who feels something tugging at *the end of a fishing line*'.[50]

The further contention[51] that it would be wrong to define 'this *G*' in terms of 'the *G* at *p* now' since we are certainly able to identify an object (demonstratively) on the basis of perception even when it is moving too rapidly for there to be any question of assigning to it a position in egocentric space, certainly holds as regards locating such an object at a *definite* place at a given time. But this does not militate against the truth of the contention that we have to be able to provide some *relatively* specific spatial description in order for demonstrative identification to be determinate viz., some such description as 'The object moving from left to right across my immediate visual field now' or even, in the case where there are a number of fast moving objects involved, 'The first (second, etc.) of the objects moving from left to right across my visual field now'. If no such relatively specific description is furnished then no determinate identification can be made; we should have no clear idea as to whether such demonstrative identification was successful.

In that I have argued that in order for demonstrative identification to take place some relatively specific description has to be given or understood; some relatively specific descriptive spatial thought has to be essayed or understood and hence have established (3)(b) above, I have also established (3)(c). Indeed that what I have designated as the 'widest' thesis cannot be held by Evans without inconsistency has: (i) already been independently indicated; (ii) can be further and importantly shown by reference to his own writing on the matter of identification of places in egocentric space which is central to his thesis concerning the *sui generis* nature of demonstrative identification.

I have previously introduced in another context,[52] the 'formal' statement of Evans' answer to the question 'What makes demonstrative identification of spatially located material objects possible?' and I have commented that I would be returning to the matter of his doctrine of location in egocentric space, with a view to showing that a reference to description, descriptive thoughts is supposed. To aid the reader in my present discussion I shall again cite Evans' 'formal' answer; repetition is sometimes helpful, even if in general annoying.

In the 'formal statement' of his answer to the question 'What makes demonstrative identification of spatially located objects possible?' he writes:

> Given the subject's general knowledge of what makes propositions of the form ⌜$\pi = p$⌝ true, for arbitrary π, when p is an Idea of position in his egocentric space, and given that he has located, or is able to locate, the object in his egocentric space, he can then be said to know what it is for ⌜This = the object at π now⌝ to be true (for arbitrary π). Hence he can be said to have an adequate Idea of the object.

One may initially comment here that since, on this view, in order to have an adequate Idea of the object a subject has to know what it is for a proposition of the form:

⌜This = the object at π now⌝

to be true he has to know the truth of some proposition of identity which itself invokes the description 'The object at π now', Evans can hardly deny that descriptive identification is implicit in the demonstrative identification. It is, even so, reasonable to suppose that he would so deny. His line of argument I conceive to be as follows. The subject can know what it is for ⌜This = the object at π now⌝ to be true independently of understanding any 'descriptive identification' since: (a) he is able to know what makes propositions of the form '$\pi = p$' true; (b) his is able to locate the object in egocentric space; and these two are conjointly sufficient conditions for knowing what it is for ⌜This = the object at π now⌝ to be true. Neither of these conditions which are conjointly sufficient for the truth of ⌜This = the object at π now⌝ implies descriptive identification or any descriptive thought.

However: (a) it is certainly *arguable* that to know what makes propositions of the form '$\pi = p$' true does imply some descriptive thought; correlating egocentric space and public space implies a reference to at least some descriptive thought;[53] (b) it is plausible to hold that a subject's ability to locate the object in his *own* egocentric space has a necessary back reference to some descriptive identification, to some descriptive thought and hence can be said to rest on, imply descriptive identification, descriptive thought. It is only necessary for me to argue for the truth of one of these theses in order to refute Evans' claim here and I shall argue for the truth of (b) in the light of the importance he attaches to the priority of egocentric space and the role this is alleged to play in demonstrative identification.

It is argued[54] that the way in which a subject identifies a place in egocentric space cannot be regarded as a species of descriptive identification. Thoughts of this kind rest upon dispositions to react to certain ways in and to objects in one's immediate environment but that these dispositions cannot be guaranteed by the apprehension of any thought invoking a descriptive mode of identification. However, even granted that these dispositions cannot be *guaranteed* by any thoughts invoking a descriptive mode of identification it does not follow that thoughts invoking such a mode are not necessary to their identification. I shall contrariwise argue that they *are*. Let me now present such an argument.

Evans argues against the thesis that in our 'here'-thoughts we identify the place by descriptions.[55] We have a special way of gaining knowledge in virtue of occupying a place which is relevant to our 'here'-thoughts and in these 'here'-thoughts we do not identify the place 'here' as 'the place I occupy' such that a descriptive thought is implied. Now whilst we may acknowledge the point that 'here' in a system of public space to which systems of egocentric space are to be correlated cannot be identified as 'the place where I am' for the good reason that no place gains its identity from the person or *x* who (which) occupies it - otherwise how could we speak of 'it' without vicious circularity - even so it does not follow that 'here'-thoughts in relation to public space are not to be identified by description or have no necessary reference to descriptive identification. For 'here' in its typical use to express where I am, as in 'I am here' one can always write e.g. 'in the garden', 'in the dining room', 'in the bathroom' etc., i.e. 'here' in relation to public space typically goes proxy for a definite description of a place. Indeed for me to understand my use of 'here' in 'I am here' and for you to understand my use, both the subject (speaker) and the hearer must be able to produce *some* description, for the question 'Where is "here"?' or 'What do you mean by "here"?' is always pertinent.[56]

To this Evans would reply that whilst identification of places in public space may have a reference to descriptive 'thoughts', descriptive identification, *his* point relates to *egocentric* space and thoughts within a system of egocentric space. In our 'here'-thoughts in egocentric space we do *not* identify the place by description; in such thoughts we do not and indeed cannot identify 'here' as the place I occupy.[57] Now whilst Evans does indeed produce a convincing argument in favour of *this* thesis, even so it does not follow that identification of places in egocentric space does not ultimately rest on descriptive identification, descriptive thoughts. He is surely right to press that to see how 'here'-thoughts work in egocentric space we must realise that they belong to a system of thought about places, e.g. 'its *F over there*'; but even in such cases the problem arises as to how the subject can understand the use of 'over there' (e.g.) in his system of egocentric space if he can furnish no description whatsoever, if he essays no descriptive thought whatsoever. If no descriptive thought at all is involved how can he understand his own different uses of 'over there'? To raise a problem however is not to present a counter case. What does present a counter case is the following consideration taken from Evans' own writing. He argues[58] that 'I' does not take primacy over 'here' such that 'here' can be identified as the 'place I occupy' but he *has* to say that in the egocentric mode of thought the subject conceives himself to be *the point of origin, the centre of space.*[59] Thus the egocentric mode *itself* rests upon identification by description, essays 'descriptive thoughts' in that there is a necessary reference to: (i) the point of origin; (ii) the centre of space. Thus whilst he can with some justification point out that we may think egocentrically about space and introduce a system of spatial relations egocentrically based and that *within* such a system we make spatial identification which does not explicitly invoke descriptive identification in terms external to the system, even so descriptive identification, 'descriptive thoughts', are involved in giving an account of the framework of the system and further sortal identification is indirectly invoked here in that for the subject to conceive of himself to be the point of origin, the centre of space, he conceives himself to be an *F*, a person who is the point of origin, centre of space, even though 'here' cannot itself be identified as the place I occupy.

In that my advocacy of theses (1)-(3) as introduced at the beginning of this section has been successful, I maintain that Evans cannot claim that demonstrative identification is *sui generis* in the way he demands and that, as a positive thesis, I have indirectly established that demonstrative identification rests on descriptive identification. In Evans' mode of speaking that a subject may only be said to essay a demonstrative identification and indeed may only be said to perceive an object, in that *some* descriptive thought may be ascribed to him. The ascription of some descriptive thought is necessary for us to understand the alleged demonstrative identification and indeed (indirectly) for the subject himself to understand it within his own egocentric space.

J. In that Demonstrative Identification is to be regarded as a mode of *Identification*, as opposed to a mode of Reference, it is reducible to Sortal Identification, in the sense of 'Sortal "Picking-Out"', 'Sortal "Discrimination"'

I have already argued: (a) that demonstrative sortal identification is logically prior to demonstrative identification by the means of a demonstrative alone; demonstrative identification in Evans' sense; (b) that demonstrative identification in Evans' sense is strictly impossible; (c) that demonstrative identification is fundamentally or basically 'demonstrative sortal identification'. My case for the above thesis is that if Evans, or anyone who construes a demonstrative as a *name* or 'simple' symbol in Russell's sense, then it could never serve in an identifying role, never be used by a speaker to pick out, discriminate, an object but at the very most only be used to refer to an object, precisely because in that it is thus construed it has no sense. The only way in which a demonstrative may serve to identify, pick out an object is when it is located in part of a complete phrase formed from a sortal; when it is used essentially as an element within a complete phrase incorporating a sortal as explained above. But that a sortal phrase[60] of the form 'This *F*', 'That *F*' may only serve in the role of identifying, picking out an object in that it is writable in the form of a definite sortal phrase 'The *F* which ...' is clear. For in that a person seeks to pick out, discriminate an object by the means of such a sortal phrase as 'This *F*' or 'That *F*', 'This cat' or 'That cat', as opposed to refer to an object by the means of such a phrase, the question: 'What do you *mean*?' or similarly 'Which one do you *mean*?', as opposed to the question: 'Which one are you referring to?', 'What are you referring to?', is always raisable, even though it may, in the case of 'This cat', where 'This cat' is used to pick out the cat before your very eyes, be absolutely obvious which cat is meant. In that this question 'What do you mean?': 'Which one do you mean?' is a request for the speaker to give the meaning of 'This *F*', 'That *F*', it is a request for the speaker to spell out his meaning (*the* meaning here) in the form of a definite sortal phrase 'The *F* which ...'; in these circumstances therefore 'This *F*' is equivalent in meaning to 'The *F* which ...' (some uniquely individuating 'description'). And 'This *F*' performs its identifying function, discriminating function or is used by a speaker to perform such a function in that it is equivalent in meaning to a definite sortal phrase 'The *F* which ...', which serves to identify, in the sense of pick out, uniquely discriminate. In this way then 'demonstrative' identification is reducible to sortal identification, in the sense of sortal 'picking out', 'discrimination'. Crudely, 'that', 'this' ('That *F*', 'This *F*') only serve in an identifying (picking out) role, or may only be used in such a role, in that they are equivalent in meaning to 'descriptions' ('descriptions' containing sortals), which perform such a role or may be used to perform such a role.

Indeed we have here a plausible criterion for determining whether 'That *F*' is serving in an identifying role, is being used to identify some *F*, as opposed to serving in a referring role or is being used simply to refer to some *F*, namely the criterion of whether for 'That *F*', we can substitute, without change of sense, a sortal phrase 'The *F* which ...'; in more traditional language whether for 'That *F*' we may substitute, without change of sense, a 'definite description', 'The *F* which ...'. Where we cannot make such a substitution, then 'That *F*' serves in a referring role, is being used to refer as opposed to identify. I shall return to this matter again[61] when I discuss the question of how one determines whether such a phrase as 'That *F*' is occurring in an identifying or referring role, or being used to identify or refer and what the relation(s) is/are between referring on the one hand, and identifying on the other and existence. For the present however I am concerned with the 'reduction' thesis as set out in the heading of this section.[62]

Evans of course would deny any such reduction thesis. This, I suggest, is primarily because he confuses *demonstrative identification* with *demonstrative reference*. He would deny that for (e.g.) 'This cat wants some milk' one can write *salva veritate* 'the object (cat) with which I am now in direct observational contact wants some milk'. But it is plausible to argue that this may *not* be claimed in that one is concerned with identification as opposed to reference, since what is to be identified by 'This cat' is one and the same *F* (cat) as 'The object (cat) with which I am in immediate informational contact'. Evans would deny equivalence on the grounds that whilst the subject can have the descriptive thought 'The object (the *F*, the cat) with which I am in direct informational contact wants some milk' and yet there *be* no object, no *F*, no cat, the subject can't have the 'Idea-of-a-particular-object' hence the demonstrative thought expressed by 'That cat wants some milk', if there be no object, no cat. However, one can plausibly hold that the subject *can* have the demonstrative Idea of an object, in that he has the 'Idea-of-an-object' - 'The object, *F*, cat, with which I am in direct informational contact' and he can have this 'Idea of an object', in that 'The object (*F*, cat) with which I am in direct informational contact' can be substituted *salva veritate* for 'this cat'. We may only *deny* that 'The object (*F*, cat) with which I am in direct informational contact' may be substituted *salva veritate* for 'This cat' and that a subject cannot be said to have the 'Idea-of-a-particular-object' in that he has the Idea 'The object (*F*, cat) with which I am in direct informational contact', in that we hold that the 'this' in 'This (cat)' is a *name* and *not*, to use the traditional language, a 'veiled description', in my language a 'veiled sortal'. That Evans requires that 'this' is a name is a thesis which I have already argued for above: in that he allows that 'This cat' ('This *G*') may be used by a person to demonstratively identify an object, he *also* must regard 'This cat' as a name, is clear from his requirement that in order to have the 'Idea-of-a-particular-object', it is necessary that that object exists. This is a plausible contention in the case of names, but not, and Evans would agree, in the case of descriptions. One may indeed hold with Russell (and others), that in the case of a *name* for which there is no object, a name which stands for no object, then a sentence containing that alleged name (an empty name) can express no thought. It is because Evans treats all demonstratives as names, and by derivation such phrases a 'That cat' ('That *F*') as a sort of name, that he demands that a subject cannot have the 'Idea-of-a-particular-object', and hence the demonstrative thought if there is no object. Yet in that he does treat 'this', 'that' as names, and in parallel fashion 'This *F*', 'That *F*', as a sort of name, then he cannot speak of 'demonstrative identification', as opposed to 'demonstrative reference'. This might well be challenged. It might be said that 'This cat' e.g. might well be used to identify some cat in the sense of uniquely discriminating some item of my immediate sense experience, but I contend that it may only do so in that it is writable as a 'definite description'. In that it is not so writable, not so equivalent, and on Evans' view it is *essentially* not so writable, not so equivalent, it can only serve to *refer* to some item of my immediate (current) experience which has been uniquely discriminated by reference to a description, or is so uniquely discriminated. To hold that 'This *F*', 'This cat' may be used to uniquely discriminate some item of my current sense experience, some item furnished by my 'informational channel', and is *not* writable in the form of a definite description is tantamount to holding that the demonstratives 'this' and 'that' may be used to discriminate, a position which I have argued against at length above and which is central I claim to Evans' position.

K. Of the relationship of 'Descriptive Identification' to 'Sortal Identification' and 'Predicate Identification'

K.1. It may be said that what Evans means by 'Descriptive Identification' cannot be restricted to descriptive identification by means of a sortal: 'descriptive thoughts' may be expressed by sentences containing definite descriptive phrases which do not contain sortals such as: 'The author of *Waverley*'; 'The fastest runner in the UK'; 'The minister responsible for administrative affairs'. How are such cases of 'descriptive identification' related to sortal identification? I have already argued[63] that in such cases some sortal is clearly implied. To take the matter up in the context, not of indicating a reference but in the context of 'descriptive identification' in the sense of 'picking out'; 'discriminating an object'. 'Descriptive Identification' by means of such phrases at least implies sortal identification in the sense of 'sortal picking out', discrimination. That is, such 'definite descriptions' as 'The author of *Waverley*', 'The fastest runner in the UK' only may serve to pick out an object, discriminate an object, in that 'The *F* who wrote *Waverley*', 'The *F* who runs the fastest in the UK', so serve or may be used to discriminate or pick out. 'The author of *Waverley*', 'The fastest runner in the UK' etc., may only serve to pick out, discriminate in that 'The *F* who φ's' may so serve; since what is purportedly being picked out or discriminated is an object, yet one can only have the concept of 'an object' in that we already have the concept of a sortal. As we only may count the number of *F*'s who write or run, equally we only may pick out an author or runner in that we may pick out one (an *F*) who writes or runs. The proposition:

> 'The author of *Waverley* died in misery'

is true *only if*:

> 'The man who wrote *Waverley* died in misery'

is true. In short, a proposition containing a 'definite description' - a proposition of the form: 'The *A* ... is φ (φ's)', where '*A*' is a non-sortal, grammatically a noun formed from a verb - is true *only* if some proposition containing the definite form of a sortal, some proposition of the form:

'The *F* who ... is φ (φ's)' is true.

The truth of some proposition of the form 'The *A* ... is φ (φ's)', where '*A*' is a non-sortal, implies the truth of a proposition of the form:

'The *F* who ... is φ (φ's)'

where '*F*' is a sortal.

One may further argue, however, that descriptive identification in the sense of picking out, discriminating, by the means of such phrases as I am considering not only *implies* sortal identification, sortal 'picking out', but is actually *reducible* to it, since not only is 'The *F* who ... φ's' serving to pick out a necessary condition of 'The author of *Waverley*' etc., serving to pick out, but is also a *sufficient* condition of the 'The author of *Waverley*'; 'The fastest runner' etc., serving to so pick out, discriminate. In that one picks out, discriminates the *F* who wrote *Waverley* (e.g.) one has picked out the author of *Waverley*. In this sense, a non-sortal phrase such as 'The author of *Waverley*', 'The fastest runner' is reducible, it seems, to a sortal phrase of the form 'The *F* who ... φ's'. The truth of some proposition of the form 'The *F* who ... is φ (φ's)' is a necessary and sufficient condition of the truth of some proposition of the form 'The *A* ... is φ (φ's)'.

Whereas the truth of some proposition of the above form is a necessary and sufficient condition for the truth of some proposition of the form 'The *A* ... is φ', it might also be suggested that the converse is also the case, viz. that sortal identification reduces to non-sortal identification, in the sense under discussion. This however certainly will not do as a general thesis since it would imply that all sortals are reducible to the relevant non-sortal form, in grammatical terms that all non-verbal nouns are reducible to verbal nouns, nouns formed from verbs, as 'runner', 'author' etc. are. And indeed one may only introduce verbal nouns upon the introduction of non-verbal ones, as witness the fact that we have to introduce the creation 'verbal noun'. Further we cannot plausibly hold that sortal identification, in the sense of sortal 'picking out', 'discrimination' reduces to non-sortal identification in the sense under discussion since what is purportedly being picked out, discriminated, is an object, and we cannot hold, for reasons given above, that what may primarily serve to pick out an object is a non-sortal (verbal noun) 'description'. Finally, whilst it is reasonable to hold that a non-sortal 'definite description' such as 'The fastest runner in the UK' etc., analyses out in sense as 'The *F* who runs the fastest in the UK' etc., it is *not* reasonable to hold that 'The *F* who runs fastest in the UK' etc., analyses out in sense as 'The fastest runner in the UK', since we only have the concept of *a* runner, *an* author, etc., in that we have the concept of 'one (an *F*) who runs', 'one (an *F*) who writes', etc. And one may observe: we may naturally ask for the meaning of 'an author', 'a runner', and be given an informative answer in terms of 'one who writes', 'one who runs', but (a) one does not naturally ask for the meaning of 'one who writes'; 'one who runs'; (b) even if one could conceivably raise this question the answer 'an author', 'a runner', conveys no additional information, indeed no information. Indeed, if one were to raise the question 'What is meant by "one who writes", "one who runs"?' one would simply be asking, not for an account of the meaning, sense, but for what one who writes or one who runs, is *called.*

K.2. One further interpretation of Evans' concept of 'Descriptive Identification' must now be considered, namely that what is intended by *descriptive identification* is identification by means of such phrases as 'The *F* who ...' where '*F*' is a sortal, *or* by 'The *A*', where '*A*' is a non-sortal, by 'definite descriptions' in *Russell's* sense, i.e. by the means of such phrases construed as 'incomplete symbols', analysed out in terms of predicates, so that in effect what is meant by 'descriptive identification' is *predicate* identification. A person may pick out an object, uniquely discriminate an object by reference to one or more of a set of predicates that are true of it. For example, the person I mean (pointing) is the one who is wearing a striped jacket and a boater; or, again: the elephant I mean (pointing) is the one waving his trunk up and down in an agitated manner; the chair I mean (pointing) is the one which is black, circular, with spikes on it. One must note here however that what serves to pick out, together with a pointing gesture, are not the predicates - 'is wearing a striped jacket and a boater' or 'is waving his trunk up and down', etc., such forms of words together with a pointing gesture, can serve to pick out nothing, for of themselves they lack a complete sense; they only have a complete sense and together with a pointing gesture may serve to pick out an object, in that we may supply an appropriate sortal, as in the examples viz. 'The one (the *F*) who is wearing the striped jacket and boater': 'The one (the *F*) who/which is raising his trunk up and down in an agitated manner'. That is to say a predicate may serve to identify, pick out, discriminate an object, only in that we may supply a sortal in completion. This is not to indirectly advocate a thesis to the effect that predicates are reducible to sortals: such a thesis is clearly absurd and would turn all descriptive adjectival terms into attributive ones.

Neither can we say that 'predicate identification' *reduces* to sortal identification, as was the case with identification by means of 'definite descriptions' not containing sortals as discussed above. To hold that 'predicate identification' reduces to sortal identification would entail holding that e.g., 'is wearing a striped jacket', 'is green' etc. - in general 'is *P*', could, of themselves, serve or be used to pick out an object. Whereas this is impossible. Here we must contrast so called 'predicate identification' with identification by the means of 'definite descriptions' not containing sortals, as discussed above. Whereas the meaning of e.g. the author of *Waverley* can be given as the man who wrote *Waverley*, the *F* who wrote *Waverley*, the meaning of 'is wearing a striped jacket' or 'is bald' cannot be given as e.g. 'The *F* who is wearing the striped jacket' or 'The *F* who is bald' since: (a) as Geach has pointed out,[64] a predicate does not show what the predication is about; (b) precisely because 'is wearing a striped jacket', 'is bald' has an incomplete sense we cannot speak of *it* being equivalent in meaning to 'The *F* who ...'.

In the light of this consideration, together with its complement that in order for a predicate to serve or be used to identify it has to completed by a sortal, predicate identification may not unmisleadingly be said to be *quasi* sortal identification. I shall thus call it *quasi* sortal identification. It is worth observing here that in that Russell reduces definite descriptions, which contain both sortals and non-sortals to predicates,[65] and in so far as anyone holds that on this analysis they may perform an identifying role, as one may hold Evans does, then I claim such people are mistaken. It is only in so far as they are *not* so reducible that they may perform, or be used to perform such a role.

L. Summary of Conclusions from Part I and Part II

It will be helpful at this point to summarise my conclusions thus far reached in Parts I and II of this chapter.

From Part I:

Identifying by the means of sortals both in the sense of (i) picking out, discriminating; (ii) specifying, is prior to naming/referring on the one hand and predicating (describing) on the other in that one is speaking in relation to *objects*: in relation to *particulars* in the prime sense and 'ordinary philosophical sense', of objects.

From Part II:

(1) Demonstrative Identification in Evans' sense is strictly impossible;
(2) In so far as it might be thought to be possible it rests on what Evans terms 'descriptive identification' which is construable as 'sortal identification' in the sense of 'sortal specification';
(3) In so far as Demonstrative Identification is to be regarded as a mode of *identification*, a mode of picking out or discriminating objects it is reducible to sortal identification in the sense of 'sortal picking out', 'sortal discrimination';
(4) 'Descriptive Identification' construed as identification by non-sortals is reducible to Sortal Identification;
(5) 'Descriptive Identification' construed as identification by 'definite description' in Russell's sense is regardable as 'quasi-sortal' identification.

In sum, therefore, at the level or aspect of the role or use of symbols in speech or thought, sortal identification is prior to other forms of speech or thought activity or other such forms of speech or thought activity are either reducible to it or may be regarded as a quasi-form of it.

My next and final task in this chapter is to face a further challenge presented by Evans and others, namely that I am at least implicitly assuming that identification in speech is prior to identification in thought; indeed that in my examples extra to my explicit discussion of Evans I have been concerned, either explicitly or at least implicitly, with identification in speech; whereas identification in thought is *prior*.

Part III: Thought and Speech

M. Of Evans' general thesis that Identification in Thought is *prior* to Identification in Speech

Evans advances several arguments in favour of his general thesis that identification in thought is prior to identification in speech. I shall now enumerate and discuss them.

M.1. *Arguments from exploiting a device of reference*

We have the particular argument that the central concept is not that of *making* a reference, but *understanding* a reference or identification.[66]

M.1.(a). Evans argues from the fact that one may exploit a device of reference in order to be understood in a certain way, when one is not in a position to understand one's words in that way. That is, since we may use a device of reference in order to manifest the intention to be understood in a certain way when we are not in a position to understand our words in that way, understanding is prior to making a reference. Evans uses the example of a blindfolded man. A blindfolded man may make a demonstrative reference by pointing and saying 'That person is *F*', but he is not fully in a position to understand his words as manifesting the intention to be understood in that way. He exploits an established device of reference in order to manifest the intention to be understood as referring to a certain person, but he is not fully in a position to understand his own words as making a demonstrative reference or identification.

Evans' thesis here, however, relies on the truth of the supposition that the blindfolded man is not in a position or at least not fully in a position, to understand his words as manifesting the intention to be understood in a certain way viz. as making demonstrative reference or identification. And he is allegedly not in such a position to so understand precisely because he is *blindfolded*. From this therefore it transpires that Evans holds that it is a necessary condition of a person understanding his words as manifesting an intention to be making a demonstrative reference or identification is, that that person be immediately acquainted with the object or at least the object be present to his senses, that the object be presented by the information-link. But whereas it might plausibly be held that it is a necessary condition of a person understanding his words as manifesting an intention to be making a demonstrative *reference*, that that person be immediately acquainted with the object or at least the object be present to his senses, that the object be presented by the information-link; it cannot be so plausibly be held in the case of a person understanding his words as manifesting an intention to be making a demonstrative *identification* since, as I have argued above demonstrative *identification* is reducible to descriptive identification. Yet in order for, e.g. 'That person is *F*' - 'The person who ... is *F*' - to manifest an intention to be making a descriptive identification, it is not necessary that there be any object present to his senses.

Evans further argues that[67] if we make *demonstrative reference* or identification prior in order to explain the blindfolded man case viz. how he can manifest his intention to be understood as making a demonstrative reference or identification, then we should not be able to discover any *essential connection* between demonstrative identification and information-links. But this argument in favour of the priority of understanding over making rests on the assumption that there *is* an essential connection between demonstrative identification and information-links viz. that it is essential to demonstrative identification that there be an object present to the senses, that the information-link furnishes an object present to the senses. Whilst, as I have admitted, this is a plausible hypothesis as regards demonstrative reference, it is not plausible as regards demonstrative identification. This further argument however draws our attention to the importance for Evans' thesis that *understanding* a reference or identification of a certain kind viz. demonstrative reference/identification, is prior to *making* such a reference or identification; of the thesis that it is *essential* to demonstrative reference/identification that the object referred to or identified in demonstrative reference/identification be present to the senses - be presented by the appropriate information channel - a thesis which I deny in the case of demonstrative identification as opposed to the case of demonstrative reference. And the same kind of argument holds against Evans' point that if we focus on what is involved in *understanding* a demonstrative reference or identification, then the connection between demonstrative identification and information-links holds. By focusing on understanding what is involved in demonstrative identification (demonstrative reference) we are able to retain the link. This point too, rests on the thesis that we must retain the connection between demonstrative identification (demonstrative referring) and information-links. It is my case that it is only necessary to maintain such a link in the case of demonstrative reference: not in the case of demonstrative identification.

M.1.(b). Let us now turn to a similar argument that Evans gives in a footnote.[68] The argument here once again is that: (1) there must be a close connection between demonstration and perception; (2) that we can see what the connection is if we can ask what is required in order to *understand* an utterance accompanied by one of these pointing gestures (e.g. 'That person is fat'); (3) in order to *understand* such an utterance (such a demonstrative identification) it is necessary to perceive the object pointed out (to make it out) and to have the thought about the object thus distinguished that the speaker is saying of it that it is thus and so. Thus the thesis that understanding a demonstrative identification is prior to *making* such an identification rests on the thesis that: (1) there is a close connection between demonstration and perception; (2) that in order to *understand* a demonstrative identification it is necessary to *perceive the object pointed out* (or at least to make it out). I raise the same objection to this as to the earlier discussed case and hence contend that thus far Evans has not established his thesis. Though there is the close connection Evans advocates between *demonstration, in the sense of demonstrative reference*, and *perception*, there is no such close connection between demonstrative identification and perception in that whilst it is plausible to hold that in order to *understand* a demonstrative reference it is necessary to perceive the object pointed out (referred to), this is not necessary in the case of *demonstrative identification*.

M.1.(c). Let us turn to another of Evans' arguments.[69] Here Evans holds that it is in general a necessary condition of understanding an utterance of a sentence containing a Russellian referring expression '*a* is *F*' that one have a thought or makes a judgement about the referent to the effect that it is being said to be *F*.

This is not, however, a necessary condition of making such an utterance in such a way as to say of the referent that it is *F*. This divergence arises because of the possibility that a subject may exploit a linguistic device which he does not properly understand. Given the divergence between the requirements for understanding and the requirements for saying, it would be absurd to deny that our primary interest ought to be in the more explicit conditions which are required for understanding.

However Evans' general argument here in favour of the thesis that understanding is prior to saying refers us back to the possibility that a subject may exploit a linguistic device which he does not properly understand and I have already criticised the tenets of that argument earlier under M.1.(a). The most Evans may claim here is that understanding is prior to saying in the case of Russellian referring expressions.

M.2. *Evans' other arguments in favour of his thesis*

Evans, however, produces a whole range of other arguments in favour of his thesis.

M.2.(a). *From his discussion of the 'composite' theory of proper names*[70]

According to this theory the only genuine way to describe what is required for understanding a sentence '*a* is *F*' is this. One must think of the referent, in whatever way one is accustomed to think of it and take the remark to be true just in case that object is *F*. But Evans contends the only principle capable of determining *what does go* ('What counts as understanding') must make essential reference to a particular object. The only thing that unifies the disparate states of mind is that they all embody ways of thinking *about one particular* object. If there is no one object to which appeal can be made there is no state of understanding, no saying.

So Evans' case here for the thesis that understanding is prior to saying is that understanding a sentence '*a* is *F*', where '*a*' is a proper name, is that there is just one object named by *a* which is presented by an information-link. However, one may plausibly argue that it is not a necessary condition of understanding a sentence '*a* is *F*', where '*a*' is a proper name that speaker and the hearer should be immediately acquainted with the object or that there should actually be an object so named. All one needs in order to understand a sentence containing a proper name is that one should have the *concept* of the one object which unifies all the disparate states of mind: but having the concept of the one object which unites all the disparate states of mind does not entail that that object exists. Indeed, not even having the concept of that object for which the proper name putatively stands entails that that object exists. Whereas understanding a sentence '*a* is *F*', where '*a*' is a proper name, does not entail that there *is* one object to which appeal can be made (one can understand fictitious names); for a sentence containing a proper name to *say* something (to express a proposition (either true or false)), it is required that that name has a reference, unless a proper name can be said to be a veiled definite description.

I thus contend that Evans' argument here in favour of the thesis that understanding is prior to saying breaks down, since his argument seems to be that *there being* just one object named by the name is a condition of understanding a sentence containing the name and hence a condition of the sentence saying something.

Evans, however, presents further arguments.

M.2.(b). *A linguistic distinction founded on a distinction of thought*[71]

He holds that to say that nothing has been *said* in a particular utterance is to say that nothing constitutes *understanding* that utterance, thus understanding is prior to saying. This thesis may certainly be challenged.

(I) Even granted the thesis that nothing has been *said* in '*a* is *F*', where '*a*' is a proper name which has no reference, it does not follow that nothing constitutes understanding that utterance (e.g. 'Cephalus guards the gates of Hades'). It only follows, on the case Evans has produced, that nothing can constitute understanding such an utterance if it is a necessary condition of understanding '*a* is *F*' is that there *is* something named by *a*; but I have argued that this is not a necessary condition.

(II) In general it does not follow that if nothing has been *said* in '*a* is *F*', then nothing constitutes understanding that utterance. One can understand the utterance 'Cephalus guards the gates of Hades' even though it is plausible to argue that nothing has been said. So it is not the case that to say that nothing has been *said* in a particular utterance is to say that nothing constitutes understanding that utterance and hence that saying has a necessary back reference to understanding.

M.2.(c). *The argument from linguistic counters - demonstrative singular terms*[72]

Evans writes that English contains a category of linguistic counters, genuine demonstrative singular terms, such that if anyone uses such a counter as e.g. 'that cup' with the intention of using a genuine demonstrative singular term, then it is necessary for understanding what is *said* that one has a demonstrative thought about the reference, which crucially depends upon the subject correctly perceiving that the object. Thus it is claimed one simply will not understand a normal use of the sentence: 'That cup is *F*' *unless*: (i) one can perceive the cup; (ii) one thinks in a way that depends on that perception.

This argument of itself does not present any fresh case for the thesis that understanding is prior to saying and indeed we may object to the thesis that a necessary condition of understanding what is *said* by a sentence '*a* is *F*', where '*a*' is a genuine demonstrative singular term, is that one has a demonstrative thought about the object which depends upon the subject actually perceiving the object, as I have indeed done earlier.[73] One can understand what is said by such a sentence containing such a singular term, in that that term performs an identifying role, by reference to 'descriptive thoughts' only,[74] which do not depend upon perception of the object. Once again one may make a point parallel to that made in M.2.(b) above. Even granted the thesis that nothing has been *said* in 'This cup is *F*', where there is no object, it does not follow that nothing constitutes understanding the counter as a genuine singular term. It only follows that nothing can so constitute understanding if it is a necessary condition of understanding 'That cup is *F*' is that one is currently perceiving the cup. But one *can* understand the use of 'that cup' as a genuine singular term even if one is not currently perceiving the cup. One can claim to understand 'that cup' as a genuine singular term even in the absence of the cup in that one understands that the speaker is making the claim that there is just one object present to his senses, just one object signified by 'that cup'. It is not necessary that one actually be perceiving the cup to understand this. So once again from the fact that there is no saying it does not follow that there is no understanding or that understanding is prior to saying.

I commented above that it is not a necessary condition of understanding what is said by a sentence 'That cup is *F*', where 'that cup' is a genuine singular term in an identifying role, that one should have a demonstrative thought about the object which depends on currently perceiving the object. One can understand what is said by reference to descriptive thoughts only. On this point it might in turn be commented that there I have implicitly made a concession to Evans to the effect that understanding is prior to saying in that I have introduced the notion of a 'descriptive thought'. Even so one could equally put the point by saying all that is necessary for understanding a singular term in an identifying role, a speaker's use of it in such a role, is that one understands what it is for some form of words to be a description. One understands definite descriptions in *speech*, in *discourse*. Hence there is no need to introduce 'descriptive thoughts'; only, simply, *descriptions*. And, in the light of the fact that to date Evans has produced no valid detailed argument in favour of the thesis that if there is no saying there is no understanding, which has a necessary reference to thoughts, yet there must be understanding, hence thought is prior to speech, I, to date, see no good reason to subscribe to Evans' view that thought is prior to speech. It is not necessary in order that one should understand a sentence containing a genuine singular term such as 'that cup', that one should have a demonstrative thought or descriptive thought. And again, to revert back to proper names, as the other type of linguistic counters Evans considers in these pages, it is not necessary in order that one should understand an utterance, a sentence, containing a proper name (*a* is *P*) that one should have a demonstrative thought which necessitates perception of the object putatively referred to by the name, or indeed, it seems, any thought. All one needs in order to understand a given speaker's use of a proper name *a*, it is initially plausible to hold following Wiggins,[75] is that one understands what particular '*a*' stands for, and this can be given by invoking a 'definite description' (for me a sortal 'description') e.g. by "Socrates" I mean the man whom everyone calls 'Socrates'.

Thus, so far I see no good argument for holding that:

(1) understanding is prior to speaking (speech);
(2) thought is prior to speech;
(3) we need to introduce thought in order to explain how it is possible for there to be understanding of proper names or singular terms.

It might now be commented that either I have misunderstood Evans or that Evans has got things the wrong way round. With respect I suggest the latter is the case. Evans clearly holds[76] that to say that nothing has been said in a particular utterance is to say that nothing constitutes understanding that utterance with the implication that understanding is prior to saying, whereas to claim that understanding is prior to saying he *should* argue that if nothing constitutes understanding a particular utterance, then nothing can be said. This, if it could be shown to be true, would establish the thesis that understanding is prior to saying. But this is not what Evans shows to be true or attempts to show to be true.

M.2.(d). *A more general argument: language as an intrinsically social phenomenon; the gaps between thought and speech*

Evans rightly points out[77] that Russell's sole interest in natural language was as a guide to the structures of human thought; for him language was not an intrinsically social phenomenon. Evans himself subscribes to the view, brought about by contemporary work in philosophy of language, that language is intrinsically a social phenomenon and one's interest therefore is in the phenomenon of language itself. Once one has this in focus, one is concerned with the way in which it functions as a means of communicating within a community. Hence one will be concerned with the utterances of individual speakers, but with the utterances of individual speakers as exploitations of a linguistic system which exists independently of any given speaker's exploitations of it.[78] Granted this view of language, Evans continues, there

> immediately opens up the possibility of a gap between what a speaker means to say by uttering certain words - what thought he wishes to express - on the one hand, and what he strictly and literally says, according to the conventional meanings of the words he utters, on the other.
>
> It is easiest to see the gap which arises by reflecting on the fact that uttering certain words with the intention of expressing the thought that *p*, is not sufficient for saying that *p*. For a person to say that *p*, it is at least required that the thought that *p* is one of the things which the words he utters may, in the circumstances of use, be conventionally used to express.[79]

Now whilst we may agree that on the view of language presented and subscribed to by Evans, there is indeed the possibility of a gap between:

(1) what the speaker means to say (the thought expressed by the speaker);
(2) what he strictly (actually) says;

and further:

(3) uttering certain words with the intention of expressing the thought that *p* is not sufficient for saying that *p*;

why does it follow from these considerations that the thought is prior to the saying? Why should it follow that a necessary condition of a person saying that *p*, is that there be the thought that *p* which can be expressed by the words he utters? We may indeed argue that the thought that the speaker wishes to express, what he means to say, is a *public saying* and it only is a possible thought in that it is a possible saying. That is, we may argue that what *is said* by the words he utters is prior to what he *means* to say: or at least what is potentially *sayable* by the words he utters, is prior to what he means to say by the uttering (use) of these words. Unless we had the notion of what is *said* or is sayable by the words he utters (uses), we could not have the notion of what he *means* to say; for he cannot *mean* to say *what* cannot be *said*. What is *said* or *sayable* by the words he uses is the norm, is *basic*. What he *means* to say is the *derivative*, necessarily *secondary* case. In that what he *means* to say is something different from what is said, it is *this* that requires explanation; one can only have the thought that *p*, intend to say that *p*, in that that *p* is a possible saying. Hence one can plausibly argue *contra* Evans that the expression of thoughts is secondary to saying. There is however a further point to which Evans draws our attention here,[80] namely, that natural language is full of ambiguities and that when one is faced with the utterance of an ambiguous sentence, the only way to solve the ambiguity ('disambiguate the utterance') is to refer to the speaker's intentions. He gives the example: 'This ship is veering to port'.

Now it can hardly be denied that the intentions of the speaker 'are uniquely relevant to disambiguating' sentences like 'This ship is veering to port'. In order to know *what* the speaker is saying we have to know in which sense the speaker is using the word 'port'. So Evans concludes, using the terminology of 'linguistic counters', that the content of what the speaker says would be determined by the meaning with which that counter is endorsed. Hence one might hold that it may generally be concluded that thought is prior to saying in that one has to understand what a speaker means by his words, which linguistic counter he is using, in order to understand and hence determine what is being said. Now whilst in *some* cases, i.e. plausibly in the cases of ambiguous words, it can be held that it is necessary to understand the speaker's intention in order to understand and determine what is being said, this is not true in all cases, for it is not the case and indeed could not be the case that natural language is all that full of ambiguities. Ambiguous words are precisely not the norm or standard even in natural language, otherwise there would be no need to point them out. That we need to point them out as a special category itself shows that they are not the norm. This case then for the priority of thought over speech will not suffice.

Here however it may be pressed that whilst Evans' case from the ambiguity of natural language does not establish that thought is prior to speech, it surely nevertheless *is* the case that in order to understand what is said, what the speaker is saying, in general even in non-ambiguous cases, we do have to understand the speaker's intentions. That is, there can be no understanding or determination of what is said independently of the speaker's intentions. We have to know in what sense he is using the words he uses: in which way his counters are to be taken. But even *granted this*, one has not thereby established that thought is prior to speech since it is perfectly plausible to hold, that *p* is a possible thought only in that *p* is a possible saying. And it is a least problematic that we may even grant *this* since:

(1) It may be pointed out that we only seek for a speaker's intention in his use of a sign, a token, when the context of its use to be not clear or determinate. Yet for the context of use to be not clear or determinate cannot be the norm. To even *raise* the question of in what sense he is using the words he uses implies a non-normal, non-standard, situation.

(2) We may only raise the question of what a speaker *means*, in Evans' case what thought the speaker intends to express, when *the* meaning is unclear. Yet unclarity of meaning cannot be the norm or standard - as witness the fact that it is precisely in case of unclarity of meaning that we ask for a clarification or explanation. That is, we only ask what thought the speaker intends to express by his use of the words, when what he intends to *say* is unclear.

(3) Agreed, we have to understand which linguistic counter a speaker is using; in which way the linguistic counter is to be taken, in the case of, say, proper names. Even so we may often do this without reference to the speaker's intentions. Context may well determine which linguistic

counter the speaker is using and it is in cases of unclarity that we resort to a speaker's intentions.

M.3. Thus I hold that one is not forced by the above arguments into Evans' position that:

(a) in order to understand what is *said* one *has* to understand a speaker's intentions;
(b) in order to determine the content of what is said we have to so understand;
(c) thought is prior to speech in the sense that we have to understand what a speaker intends or means by his words in order to understand what is said. On the contrary a speaker can only mean to say that *p* in that *p* is sayable. A speaker can only have the thought that *p* in that *p* is a saying or a potential saying.

However, even if some overriding argument could be produced in favour of the priority of thought over speech, this would not crucially affect my thesis, for my thesis could be replaced in terms which would acknowledge that priority - e.g. my thesis as to the priority of sortal identification in speech could be appropriately re-written in terms of the priority of 'sortal thought' *vis-à-vis* 'demonstrative' and 'descriptive thoughts'. In short, I could rephrase my thesis in terms appropriate to the categories Evans introduces.

However, since I see no overriding argument for adopting Evans' position plus the considerations:

(i) that I wish to discuss the writings of some logicians (including Strawson) who either explicitly or implicitly adopt the view that reference to the intentions of speakers (subjects) are secondary: that is, who do not explicitly adopt the view that language is an intrinsically social phenomenon in that a study of language *essentially* involves a reference to a speaker's meaning and hence to the introduction of *thoughts* which have a necessary reference to what a speaker means to say;[81]
(ii) that the relevant aspects of the works of authors I wish to discuss employ the categories of *language* (Frege, Russell, Geach, Quine) *not* conceived of as above, and *speech* (Strawson),

I have found and shall find it both relevant and convenient to write in terms of: (a) language viewed as a system of symbols pertinent to the expression of human thought; (b) language viewed as speech, i.e. language viewed as its use, as a form of human activity and hence in terms of identification in speech (at least implicitly), whilst allowing myself the latitude of sometimes writing in terms of 'speech or thought'[82] on the grounds that: (i) they are not mutually exclusive; (ii) a possible thought (as argued above) is a possible saying; (iii) to concede some degree of fairness to Evans; (iv) to underline my point that my thesis could be rephrased in terms of priority in relation to thoughts.[83]

Notes

1. *See* my discussion of Evans in pt. II, sec. M.
2. ch. I, p. 27 and as will be further argued in pt. II of this chapter.
3. cf. ch. I, sec. A.2.
4. Strawson, it is to be noted, does not include 'parcels of matter' in his list of what in familiar philosophical use comes under the heading of 'a particular' (*see Individuals*, p. 15) and he contends that material bodies (objects) and persons occupy a central position in our conceptual scheme (*ibid.*, p. 11).
5. ch. I, pp. 13-14.
6. *See* my later discussion of 'Descriptive Identification' in relation to Evans, pp. 72-74.
7. For a challenge to this thesis and a defence of it *see* pp. 45-48. I do not offer the challenge or defence at this point since the challenge and defence is more effectively to be given in a context in which a very strong general objection to my whole position is offered. Further, *see* my discussion of Evans later, pp. 63ff.
8. *See also* p. 41 and my discussion of Evans on 'Descriptive Identification' later, p. 61.
9. Again, I refer the reader to pp. 46-48.
10. *See* above, note 7.
11. One can refer to my as yet unborn child. [For problems and discussion of this issue, *See* G.E.L. Owen, 'Aristotle on the Snares of Ontology', in Renford Bambrough (ed.), *New Essays on Plato and Aristotle* (London: Routledge and Kegan Paul, 1965), pp. 69-95; repr. in G.E.L. Owen, *Logic, Science and Dialectic* (Ithaca, New York: Cornell University Press, 1986), pp. 259-278; Robin Attfield, 'How Things Exist: A Difficulty', in *Analysis*, Vol. XXXIII, No. 4, March 1973, pp. 141-144.]
12. cf. *Reference and Generality*, p. 39.
13. *See* ch. I, sec. A.2.
14. *Metaphysics* 1017a 25; 1028a 12, 14; 1028a 38; *Topica* 103b 23; *Categories* 2b 10.
15. *Tractatus* 4.1272; *see also* ch. I, pp. 28-31 and my later discussion of Evans *passim*.
16. cf. *Tractatus ibid.*
17. cf. p. 45 above.
18. 'Formal' concepts are manifested in 'proper' concepts; 'formal' concepts only come about as reflecting our use of 'proper' concepts.
19. Introduction, pp. 5-6; ch. I, p. 13.
20. *Individuals* p. 168.
21. *See also* my later discussion of Evans.
22. I would hold that successful reference here could not be brought about simply by the use of, e.g. 'That piece of wood'; a piano is not just one piece of wood but at least an object formed from several pieces of wood (and having a certain function).
23. I have spoken of 'successful reference' earlier (pp. 47-48); but that speaking in no wise commits me to this thesis.
24. Gareth Evans, *The Varieties of Reference*, ed. John McDowell (Oxford: Clarendon Press, 1982).

25. One might substitute 'sortal *description*' here provided that by 'description' is *not* meant that which is true of an object or satisfiable by an object. To avoid this I stick to 'specification', which in any case is more accurate.
26. There is no inconsistency here. Demonstrative identification rests on sortal identification in the sense of sortal specification: sortal specification is prior to such demonstrative identification, but demonstrative identification reduces to sortal identification in the sense of 'sortal picking out', 'sortal discrimination'.
27. cf. pp. 46-47, 50-51 of this chapter and my later discussion of Evans.
28. cf. *also* Wittgenstein on 'ostensive definition', *Philosophical Investigations* paras 28-30.
29. cf. Russell, *Introduction to Mathematical Philosophy*, ch. 16, p. 173.
30. *Tractatus* 4.1272.
31. cf. sec. G above, esp. p. 55.
32. It is to be noted that in the light of what I have said at the end of my previous section, I myself am *not* holding that 'this *F*' is a logically *simple* phrase, formed from 'this ----'; I also want to hold that 'this *F*' is a logically complex phrase, but *not* a logically complex phrase in which 'this' is a *name*, but a logically complex phrase in which 'this' is an element yet its role 'sense' being determined by the *other* elements in the complex phrase viz. the sortal.
33. in his 'Lectures on the Philosophy of Logical Atomism', *op. cit.*
34. Evans, *The Varieties of Reference*, p. 143.
35. *Ibid.*, p. 173.
36. cf. *ibid.*, p. 148 and p. 171.
37. cf. *ibid.*, p. 171 and p. 174.
38. cf. Evans' earlier discussion *ibid.*, 6.2, pp. 145ff.
39. *Ibid.*, p. 150.
40. *Ibid.*, p. 170.
41. *Ibid.*, pp. 170-171.
42. *See ibid.*, pp. 170-171.
43. Sec. H above, p. 57.
44. Evans, *op. cit.*, pp. 171ff.
45. *Ibid.*, p. 172.
46. Stephen Schiffer, 'The Basis of Reference', in *Erkenntis*, Vol. XIII, 1978, pp. 195-196; Evans, p. 173, n. 44.
47. Evans, *op. cit.*, p. 171.
48. *Ibid.*, p. 172.
49. *Ibid.*
50. my italics.
51. Evans, *op. cit.*, p. 172.
52. cf. sec. H, pp. 59-61, *see* esp. p. 59.
53. cf. Evans' own discussion, *op. cit.*, p. 162.
54. *Ibid.*, p. 169.
55. *Ibid.*, p. 153.
56. Attfield and Durrant, 'The Irreducibility of "Meaning"', in *Nous*, Vol. VII, No. 3, 1973, p. 292.
57. *See* Evans, *op. cit.*, p. 153.
58. *Ibid.*
59. My italics.
60. By a sortal phrase I mean simply a phrase containing a sortal expression.

61. cf. ch. III, sec. H.
62. It might be suggested that my thesis that 'Demonstrative Identification' (in Evans' sense) presupposes 'Descriptive Identification' is incompatible with my claim that 'Demonstrative Identification' in his sense reduces to sortal identification (which he would call 'Descriptive Identification'). But I claim that there is no such incompatibility. Demonstrative Identification presupposes Descriptive Identification in the sense of 'description', *not* in the sense of 'descriptive picking out' or 'descriptive discrimination'. Demonstrative Identification is reducible to sortal identification (in Evans 'Descriptive Identification') in the sense of sortal (descriptive) 'picking out', 'discrimination'.
63. *See* sec. A.1. above, p. 41; ch. I, pp. 13-14.
64. Geach, *Reference and Generality*, p. 32.
65. *See* his 'Lectures on the Philosophy of Logical Atomism', *op. cit.*, where Russell includes sortals under the heading of *indefinite* descriptions (p. 243).
66. *See,* Evans, *The Varieties of Reference.*, 6.4, p. 171 and for a similar argument, 6.1, p. 143. Essential reference is made to such arguments at 4.1, pp. 89ff.
67. *Ibid.*, 6.4, p. 171.
68. *Ibid.*, 6.1, p. 143.
69. *Ibid.*, 4.1, p. 92.
70. *Ibid.*, 3.2, p. 70.
71. *Ibid.*, p. 71.
72. *Ibid.*, p. 72.
73. *See* pp. 76-78 above.
74. *Ibid.* and *see* Evans, *op. cit.*, p. 172 for 'descriptive thoughts'.
75. David Wiggins, 'Identity, Designation, Essentialism and Physicalism', in *Philosophia*, Vol. V, 1975, pp. 1-30.
76. Evans, *op. cit.*, p. 71.
77. *Ibid.*, p. 67.
78. As Evans rightly points out, one is not, on this view, committed to any 'private language' thesis.
79. Evans, *op. cit.*, p. 67.
80. *Ibid.*, pp. 68-69.
81. *See*, *ibid.*, p. 67.
82. cf. the beginning of this chapter.
83. It might be thought that my thesis that in Demonstrative/Descriptive Identification (*see* e.g. pp. 76-78, 79 above) it is not necessary that the person making the purported *identification* be actually in 'informational contact with', i.e. be 'immediately acquainted with' the object, or that the object be 'present to his senses', is incompatible with my later thesis (*see* ch. III, sec. B.4, pp. 99-100) that a definite sortal expression in an *identifying* role *implies* existence. But being committed to *this* thesis does not entail that the person using such an expression in an identifying role has to be 'immediately acquainted with' or 'in direct perceptual contact with' the object purportedly identified, which is Evans' thesis.

Chapter III

Sortals and the Subject-Predicate Distinction

A. Strawson's Objections to Geach's Explanation of the Subject-Predicate Distinction

Having argued that at one level or aspect of language, the category of the sortal is required for the introduction of names and predicables (or 'predicates') and at another level - the level of speech or discourse - sortal identification is prior to both naming/referring and predicating, I now turn to the question of the relation of sortals to the subject-predicate distinction. This demands that I consider explanations of the distinction offered by other writers since I shall attempt to offer an explanation which has a reference to other explanations and seeks to draw positive value from them whilst offering an account which seeks to establish the role and importance of sortals. Indeed, as will be seen below, I shall distinguish between two senses of 'explanation' and in both senses I shall seek to accent the role and importance of sortals.

I shall first discuss Strawson's objections to Geach's explanation of the distinction. I shall claim that whilst Strawson's objections are answerable there is an important sense in which Geach does not offer us an *explanation* but rather an *explication* of the distinction. I shall seek to offer an explanation in the sense of an account of what *differentiates* an expression occurring in logically subjective role from one occurring in logically predicative role in the context of a subject-predicate proposition; or again, in terms of speech, what *differentiates* a speaker's use of an expression in subjective role from his use of an expression in predicative role, holding that there can be no such explanation *external* to an *explication.* We can only explain in this sense within the context of an explication of the distinction and the distinction is, after all, an *ultimate* one. I do not claim that my explanation, in this sense is particularly original; it is based upon a revision, but an important revision of Strawson's 'Mediating' or 'New' criterion set forth in *Individuals* but which gives due recognition to the particular role and importance of sortals. I differentiate between an *explanation* in the above sense and a *criterion*, and offer a criterion based upon my explanation having first considered Quine's famous explanation and raised the questions: (a) whether Quine's account can be saved from Strawsonian objections; (b) whether it could serve as a *criterion* granted my explanation of the distinction. I shall answer 'Yes' to (a) and, after some agonising, 'No' to (b). I shall then offer an explanation of the distinction in the sense of an answer to the questions:

(1) What makes the subject-predicate distinction *possible*?
(2) What *underlies* the distinction?

which invokes sortals and seeks to make clear the role of sortals in this regard. Following my discussion of sortals and principles of counting and distinction formed from 'matter' terms in Chapter I, I have to acknowledge however that sortals, in the sense I have introduced, are not the *only* category of symbol which may be said to underlie the subject-predicate distinction and render it possible. Even so, I hold that it is the presence and recognition of sortals which underlie and make this distinction possible in a language or mode of discourse in which 'objects' are introduced or spoken of.

A.1. To turn first then to Strawson's objections to Geach's explanation of the subject-predicate distinction. In *Individuals* he presents us with a direct attack on Geach's position as advocated in an early paper; later in *Subject and Predicate in Logic and Grammar* he presents us with what certainly seems to be a veiled attack on Geach's position. I shall discuss and reject both the direct and veiled attack.

At *Individuals*, p. 143, Strawson cites Geach from a paper in *Mind* 1950[1] to the effect that 'the name of an object can ... be used as the logical subject of an assertion about an object .... It cannot, without radical change of sense, be a logical predicate.' Strawson claims that the interest of Geach's statement lies in the fact that it is claimed to be a consequence of certain definitions of 'Subject' (expressions) and 'Predicate' (expressions), so 'we can assess the adequacy of the definitions in the light of their alleged consequence.'[2] These, he holds run as follows:

> A B_2 (i.e. a predicate expression)[3] is an expression which gives us an assertion *about* something if we attach it to another expression that *stands for* what we are making the assertion about;
>
> An A_2 (i.e. a subject expression)[4] is an expression to which there is attached a B_2 (a predicate expression)[5] so that the two together form an assertion *about* what the A_2 (subject expression)[6] *stands for*.[7]

Strawson continues:

> We have to ask whether these definitions have the consequence that an A_2 (a subject expression)[8] can never be a B_2 (a predicate expression)[9] or at least (to restrict ourselves to Geach's general claim) whether they have the consequence that the name of an object can be used as an A_2 (subject expression),[10] but cannot, without radical change of sense be a B_2.[11]

Thus he asks us to consider the assertions:

Raleigh smokes
Socrates is wise

and to consider that whilst we may wish to say that in some contexts the sentences 'Raleigh smokes', 'Socrates is wise', give us an *assertion about* Raleigh, (about Socrates) and to this extent 'Raleigh' seems to qualify as an A_2 (subject expression) and 'smokes' as a B_2 (predicate expression), nevertheless:

> it is also plain that there might be circumstances in which it would be correct to say of someone using the sentence that he was talking *about* smoking and that one of the things he asserted about it was that *Raleigh* smoked or was a smoker. To this extent at least, as far as the word 'about' is concerned the name 'Raleigh' seems to qualify, on the definition, as a B_2 (predicate)[12] expression.[13]

We must first note here that Strawson analyses:

Raleigh smokes

as a composite of two independent, non-heterogeneous elements, 'Raleigh' and 'smokes' but as we have already seen this is distinctly *not* Geach's view. On his view 'Raleigh smokes' is formed from '---- smokes' and 'Raleigh' is inserted in the argument place. 'Raleigh smokes' could only be construed as an assertion about smoking if we could construe '---- smokes' as a name but as Geach rightly points out,[14] this we can never do: '---- *s*', to coin his words on another occasion 'does not even look like a name'. Strawson's case here depends on taking 'smokes' as a complete unit of sense parallel to a name and from which 'smoking' may be derived, a position we have good reason to dispute.

Further, in that 'Raleigh smokes' could be taken as an assertion or proposition about smoking, then: (a) the sense of 'about' is different; (b) the sense of 'smokes' is different. 'Raleigh smokes' ('Raleigh is a smoker') may be said to give us an assertion or a proposition *about* smoking, if 'Raleigh smokes' answers the question 'Who smokes?' but the sense of 'about' here is clearly different since if we are faced with the question 'Who smokes?' and reply 'Raleigh', 'Lord Wilson', 'Sir Alfred Ayer' etc., what we are presented with is a list of persons who smoke; but to be presented with a list of persons who smoke is *not* to say 'something about' smoking in the same sense as to say that 'smokes' or '---- smokes' gives us an assertion or proposition *about* Socrates if we attach it to his name or complete it by his name. In the latter case we are not enlisting things or a thing Socrates does (or in the parallel example, Raleigh does). 'Smokes' in 'Raleigh smokes' is not replaceable by a name whereas in that we have a proposition about smoking the alleged predicate is replaceable by a proper name 'Smoking is done by ----', where '----' takes a proper name. *That* the sense of 'smokes' is different may be seen as follows: on the construction of 'Socrates smokes' as a proposition about Socrates, the truth of the proposition is independent of the question 'Who smokes'. 'Socrates smokes' is not true in so far as any other person and Socrates himself smokes, but on the construction of 'Socrates smokes' as a proposition about *smoking* then the truth of 'Socrates smokes', 'Raleigh smokes', 'Aristotle smokes' *is* relevant to the truth of the proposition. 'Smoking is done by someone' is true in that 'Socrates smokes',

'Raleigh smokes', 'Aristotle smokes' etc., are true. We thus have a different sense of 'smokes' in the two cases. It may be a criticism of Geach not to have pointed this out in the original article, but he would certainly wish to do so. I thus reject Strawson's first criticism of Geach.

I now turn to his second criticism. This turns on the other crucial expression Strawson has italicised, namely, *stands for*. He asks: 'Does this phrase as it occurs there, prevent our saying that "Raleigh" is a B_2 (i.e. a predicate expression)?'[15] and comments:

> No doubt it would do so if we are prohibited from saying that the expression 'smokes' *stands for* smoking or the habit of smoking. But I know of no rule or custom which makes it senseless or incorrect to say this any more than I know of any rule or custom which would make it always senseless or incorrect to say that an assertion made in the words 'Raleigh smokes' was an assertion about smoking ...

and then adds in a footnote (with a view to clinching the case against Geach): 'Geach is indeed *required* to say that "smokes" stands for something. For he commits himself to the view that expressions which are predicable *stand for* properties'.[16]

We must again note that Geach would *not* say that 'smokes' stands for something, but that '---- smokes' stands for something. Having noted this, even if we are somewhat sceptical of the view that '---- smokes' stands for, we must further note that the way in which 'expressions which are predicable' or 'expressions which are predicated' stand for properties is *not* the way in which *names* stand for objects and hence we *are* prevented from saying that 'Raleigh' is a B_2 expression (a predicate expression). The predicate, '---- smokes' stands for the habit of smoking in some individual in that the name of that individual, when put in the argument place, yields a true proposition: in other words '---- smokes' stands for the habit of smoking in that it is satisfied by some individual. Whereas 'Socrates' stands for Socrates in that the man Socrates bears the name, which is clearly a different relation. Thus the sense of 'stands for' is different in both cases and we *are* prevented from saying that 'Raleigh' is a B_2 expression, even though we are not thereby prevented from saying that 'smokes' stands for something. It may be an initial charge against Geach, if I have read him rightly, that he should have made clear the two relevant senses of 'about' and 'stands for' but once these are made clear he is *not* open to the charges Strawson brings.

Strawson's troubles basically arise primarily through failure to appreciate that predicable expressions are incomplete and that this is so can be seen from his comments on the 'Socrates is wise' example on p. 145 of *Individuals*. He there writes:

> There is no absolute prohibition on our saying that an assertion made in the words 'Socrates is wise' is an assertion about wisdom .... The definitions in effect require us to divide the sentence into two parts which together *make up* (my italics) the whole of it; and they *allow* us to make the division in the way we want to make it i.e. between 'Socrates' and 'is wise'. But they do not force

> us thereupon to classify *these parts* (my italics) in the way in which we wish to be forced to classify them.

The point is that they *do* provided we appreciate the nature of the parts.

In *Subject and Predicate in Logic and Grammar*[17] Strawson criticises the doctrine that 'the predicate is "true of" (or "false of") what the subject term stands for' holding that no distinction between subject and predicate is marked by this: indeed, in the end it is his claim that 'the distinction which was to be drawn with the help of the phrase "true of" reduces to the purely grammatical point that in a subject-predicate sentence in English, the subject is a noun or a noun phrase and the predicate a verb or verb phrase'.[18] Strawson does not mention by name writers who hold that 'true of' makes a distinction between subject and predicate 'terms', but this certainly seems to have a veiled reference to Frege and Geach and I think it fair to say that he has Geach primarily in mind. However in that this is so in his critical attack it: (a) commits him to an error concerning Geach's position; (b) attributes to Geach an argument about what follows from a close attention to the use of English, or what is claimed to follow. I can trace no such argument in Geach; indeed he is concerned to propound a logical theory, not a grammatical theory, as was Frege.

Strawson writes:

> It is sometimes implied or suggested, rather than directly stated, that the phrase 'true of' (or 'false of') supplies a clue to the distinction and relation, of subject and predicate. The idea seems to be, roughly, that taking a subject-predicate proposition to be true (or false) is the same thing as taking the predicate to be true (or false) of what the subject stands for; but not conversely. This cannot be more than a rough expression of the idea. Suppose 'Socrates swims' is a subject-predicate sentence, with 'Socrates' as the subject and 'swims' as predicate. No one at all fastidious would allow '"swims" is true of Socrates' as tolerable English. However the following is perfectly allowable:
>
> 'It is true of Socrates that he swims'
>
> as is also:
>
> 'It is true of Socrates that Socrates swims'.[19]

Strawson then goes on to argue that the case is not made out since e.g. 'It is true of *swimming* that Socrates swims'; 'It is true of *swimming* that Socrates does it' is also perfectly admissible 'though unusual' English.[20] This then purportedly shows that there is no case from 'allowable English' to the conclusion that taking a subject-predicate proposition to be true (or false) is the same thing as the predicate to be true (or false) of what the subject stands for, but not conversely: 'allowable English' permits both the original *and* the converse. If we object to this i.e. if we object to permitting 'It is true of *swimming* that Socrates does it' or 'It is true of *swimming* that Socrates swims' then Strawson claims we are invoking a requirement, not so far explicitly stated, on the procedure of the paraphrase, and to invoke this requirement reduces the whole case to a purely grammatical one, which, whilst nevertheless important, was clearly not the intention of the advocates of the position.

That this cannot be regarded as a genuine attack on Geach and if it is to be construed as such may be said to be based on a misunderstanding can already be seen in the invocation of 'fastidious speakers of English'. There is no reference at all in Geach's writings, as far as I can trace, to such fastidious users and it is not clear why anyone concerned with advocating a logical thesis should at all be concerned about the fastidious users of English. Grammatical correctness does not necessarily guarantee or even indicate *logical* correctness; and the phrase 'in case that' is one in current use by logicians but is clearly not good English and doubtless would be rejected by the fastidious users Strawson alludes to. Further, Geach is not claiming that it is 'swims' which is true of Socrates in the example but rather '---- swims' which is true; Strawson has once again failed to acknowledge the fundamentally incomplete nature of predicables which Geach, following Frege, is advocating. Strawson might hold that no one at all fastidious in the use of English would allow '"---- swims" is true of Socrates' but a logical theory: (a) is not concerned with fastidiousness in a given natural language (or its use); (b) the issue is whether a position which yields fruit logically is to be sacrificed on the altar of fastidious language use. Surely not. Further, and crucially, if this is supposed to be an attack on Geach it fails to note that he has written:[21] 'In saying that something is predicated of Peter, I do not mean that this predicate is true of or applies to Peter, but only that in some significant sentence, true or false, it is predicated of Peter.' Geach would not therefore reject the claim that for '"swims" is true of Socrates' as he conceives it, one may write:

'It is true of Socrates that he swims'.

Further, Strawson's case that 'It is true of *swimming* that Socrates swims'; 'It is true of *swimming* that Socrates does it' is also perfectly admissible, 'though unusual' English would also be accepted by Geach but he would rightly reject the claim that *no* 'explanation' has been offered on his account of the subject-predicate distinction on the grounds submitted by Strawson since:

(i) What is acceptable or unacceptable to fastidious users of English is no criterion for rejecting or accepting a logical distinction;

(ii) Strawson's case that no distinction is given by the doctrine that 'the predicate is "true of" (or "false of") what the subject term stands for' since we may equally write:

'It is true of Socrates that he swims'
'It is true of swimming that Socrates does it'

fails since he assumes that in:

'It is true of swimming that Socrates does it'

'swimming' is a predicate in his sense, or, that 'swimming' is a predicable in his sense, which is an incorrect and crucially incorrect assumption. On the contrary 'swimming' is the name of an activity, *not* a predicate (predicable), and 'swimming' is not writable as '---- is swimming' or '---- swims' since no name is reducible to a predicate or predicable.

Even though Strawson's implied challenge to Geach in *Subject and Predicate in Logic and Grammar* may be held to fail on the above grounds it might be said by someone sympathetic to Strawson that there is still a sense in which Geach has not offered an explanation of the subject-predicate distinction, i.e. he does not tell us *why* a name cannot be 'true of' but a predicable can. But an answer to this question is readily at hand from Geach's writing. A name has a *complete* sense and for this reason can never be a predicable; it may however be part of or more correctly, an element in a predicable - as in '---- shaved Peter'. Any proposition concerning which it is claimed that a name, of itself, is a predicable, or, more commonly, 'a predicate' is a misconstruction. Such a proposition is properly construed as a proposition of identity, not as a predication; as in e.g. 'John is James' - 'John is the very same *F* as James'.

B. Strawson's 'Mediating' Criterion for the Subject-Predicate Distinction

Having defended Geach against Strawson's attacks, both explicit and implicit, it is nevertheless fair to say that what Geach offers is an *explication*, an *elucidation* of the subject-predicate distinction. The sense in which he offers an 'explanation' is that he offers us an elucidation. In crude terms he tells us what is meant by the distinction and there is a sense in which he does not offer us an 'explanation'. The sense in which he does not offer us an explanation is that he does not answer the question: 'What differentiates an expression occurring in the role of logical subject from one occurring in predicative role in the context of an subject-predicate proposition?'. Or again: 'What distinguishes a person's use of an expression in logically subjective role from a person's use of an expression in predicative role in the context of a subject-predicate proposition?'. To reply to this by saying that Geach *has* offered us an explanation in that an expression in subjective role stands for something we are forming the proposition about; an expression in predicative role gives us a proposition about something if we attach it to another expression that stands for what we are forming the proposition about is simply to repeat the account, not to answer the question. It simply repeats the distinction; it does not 'explain' it in the sense of answering the above question. And at this point one may profitably turn to Strawson in his 'Mediating' criterion for the subject-predicate distinction as set forth in *Individuals*.

B.1. Strawson writes:[22]

> What we here propose, in effect, is a new or mediating criterion for the subject-predicate distinction. A subject-expression is one which, in a sense, presents a fact in its own right and is to that extent complete. A predicate-expression is one which in no sense presents a fact in its own right and is to that extent incomplete.

It should be noted here that Strawson conflates the issues:

(i) What category of expressions (symbols) may occur in subjective and predicative roles?
(ii) What distinguishes an expression occurring in subjective role from an expression occurring in predicative role?

The 'Mediating' criterion is strictly concerned with (ii); as an answer to (i) it gives rise to at least the following difficulties:

(a) the reference to 'presenting' or 'failing to present' a fact is misconceived. The reference to 'presentation' (or later 'presupposition') of facts relates to expressions in a role, not to categories of expression or symbol. Let me elucidate this with regard to expressions (symbols) which may occur in logically subjective role. Strawson may contend that what distinguishes types of expression which may occur as logical subjects, which may occur in the role of 'making identifying reference' to a particular, is that such expressions implicitly invoke or covertly assert (as he later has it) an *empirical* fact to the effect that there is just one particular which answers to a certain description. Here however we must surely press: 'How can "implicitly invoking" or "covertly asserting" an empirical fact be a feature of an expression which *may play* a certain role?'. The reference to 'implicitly invoking' and 'covertly asserting' an empirical fact - even if these *are* equivalent or even clear notions - has reference to expressions *qua* playing a certain role, *not* to expressions, symbols, as such *qua* their potentiality to play a certain role. To put the matter crudely: to demarcate a type of expression we must look to language, not to what appertains or fails to appertain in the world.

(b) Similarly the reference to failure to present facts in their own right is misplaced as a criterion for *expressions* which may occur in predicative role; failure or alleged failure to present such a fact relates to an expression in a role, not to an expression itself *qua* ability to play such a role.

B.2. Let us then turn to the 'Mediating' criterion as a *role* criterion. How adequate is it? First let us consider the criterion as a criterion for an expression occurring in the role of logical subject.

Strawson offers us an explication of the notion of 'occurring in the role of logical subject' in terms of the role of 'making identifying reference' to a particular. I have already drawn attention to the point that in their descriptions of 'logical subject' logicians have traditionally spoken, not in terms of 'identifying' or 'identifying reference', but in terms of *naming*, *standing for*, *designating* or *referring* to an object. It may therefore be said that Strawson gives an inaccurate and indeed misleading description of the traditional account of 'occurring in the role of logical subject'. To such a charge he would probably reply that he is not concerned *simply* to give a description of the notion of 'logical subject' as traditionally conceived but to explicate it in terms of what he regards as a more *fundamental* notion - that of 'identifying reference' or 'picking out'. I hold that such an attempt to explicate the notion of 'logical subject' is doomed to failure since the notion of 'identifying reference' is itself a confused notion. I argued in Chapter II that 'identifying' is a more fundamental notion than 'referring' (or 'naming', etc.,); that 'referring' presupposes 'identifying'. The notion of 'identifying reference' is confused precisely in that it combines, at one and the same level, two notions the introduction of one of which logically presupposes the other. Proper names primarily serve to refer; they refer or are used to refer to objects not currently present. They *may* occur in an identificatory role but in that they do so they do so by naming. For example, a proper name may serve to reply to the order: 'Identify the man in the dock' - 'He is Ieuan Dafydd Jones'. The witness identifies the defendant by *naming* him. To hold that proper names primarily have an identifying role would entail holding not only: (a) that proper names of their essence have a sense but also; (b) that necessarily they have the sense they do have and; (c) that proper names necessarily are identical with their sense. Whilst I should concur with (a), (b) and (c) are highly problematic: (c) especially so. Indeed one may even go so far as saying that (c) is a thesis which results in nonsense and thus displays a 'type' error. One may go so far as this on the following grounds. First, we may observe that a proper name may be said to *have* a sense and in that it may be said to have a sense it is not easy to see how it may be identical with its sense. Moreover, in that (following a certain interpretation of Frege) sense is required in order for us to understand the reference of a proper name, we may only say that a name is equivalent to its sense in that we are also prepared to say that a sense is required for us to understand the sense of a name - which is nonsense. Again, in that sense is required to establish the identity of a name, the conditions for its continued use as the very same name, it is nonsense to hold that a name is identical with its sense. The conditions for the continued use of a name cannot be said to be identical with - one and the same X as - the name itself. A name itself is not a set of conditions. Yet again, on the same theme, in that sense = meaning and meaning may be analysed out in terms of use, it is nonsense to claim that a name is identical with (one and the same X as) the rules for its use. A name itself is not a rule.

Again, a 'definite description' (to use traditional parlance) primarily serves to identify and in that it fulfils this primary role it does not refer. 'Definite descriptions' may perform a secondary role of referring, but only in that they first serve in an identifying role and in that they so perform a referring role it is arguable that they become quasi names.

Strawson's notion of 'picking out' or 'making identifying reference' thus conflates two distinct and incompatible roles: that of 'identifying' and that of 'referring'. Incompatible in the sense that the one is logically prior and presupposed by the other as I have argued in Chapter II above. This leads Strawson, in Russell's terms, to conflate two radically different classes of symbol viz., 'complete' and 'incomplete' symbols; proper names and definite descriptions: in my terms, to conflate proper names and the definite form of a sortal.

That Strawson's notion of 'identifying reference' in fact generates confusion in his own account of the conditions for an expression to be a 'logical subject expression' can be indicated by a close examination of those conditions. It is to this that I shall now turn with a view to later turning my criticisms of Strawson to positive use in the development of my own account of the subject-predicate distinction.

B.3. *Strawson's account of the conditions for an expression to be a subject expression*

I hold there to be inherent difficulties in Strawson's statement of the conditions which must be met for an expression to be a 'subject expression'. What then are these alleged inherent difficulties? He writes,[23] that subject expressions are able to perform their role 'only because they present or represent facts, only because they presuppose or embody or covertly carry, propositions which they do not explicitly affirm .... They necessarily carry a weight of fact in introducing their terms.' The inherent difficulties are as follow. To *present* a fact or *represent* a fact is clearly not to *presuppose* a fact; to present a fact brings with it affirmation of an existential proposition $(\exists x)Fx$. To *represent* a fact is indeed an unclear notion outside the context of a 'representational' theory of language as, say, encountered in the *Tractatus*, but for all its lack of clarity it surely would at the least entail that some existential proposition had, at some previous time been affirmed and was currently being reaffirmed; whereas to *presuppose* a proposition or fact does not, and necessarily does not, on Strawson's own theory of 'presupposition' entail or in any way involve an *affirmation* or *reaffirmation* of existence. Further, to presuppose a proposition necessarily excludes *embodiment* of a proposition. To follow Strawson's way of speaking, in that expressions of type A (subject expressions) *embody* a proposition they cannot presuppose it, for, on a straightforward understanding, to embody a proposition is for that proposition to be entailed in the use of A or implied by the occurrence of A; whereas if expressions of type A *presuppose* a proposition then, necessarily, their use does not entail it or their occurrence imply it. Strawson himself makes a major distinction between 'entailment' (or 'implication' in the normal accepted sense) and 'presupposition'. Again, if expressions of type A *presuppose* some proposition as a necessary condition of performing their role, they cannot be said to *covertly carry* such a proposition, for to 'covertly carry' some proposition certainly seems to *imply* some proposition.

I thus hold that the all crucial phrase 'necessarily carry a weight of fact in introducing their terms' and the corresponding phrase 'presents a fact in its own right', are not only initially vague but upon investigation conflate 'presupposition' and certain other notions which may be subsumed under the heading of 'implication', thus requiring that 'presupposition' and 'implication' be *alternatives* which they cannot be on Strawson's own account of 'presupposition'. They are incompatible on his own account[24] in the sense that implication can only arise once the requirement of presupposition has been satisfied. To illustrate. On Strawson's view 'The present King of France is bald' can only succeed in making a statement and hence imply some further statement on the condition that it is presupposed that there is a king of France; that the presupposed statement 'There currently exists a king of France' is true. In his later discussion,[25] he admits that 'the variety of terminology ("presenting", "representing", "presupposing" etc.,) may well seem suspicious'; the terminology is, I hold, more than suspicious. It involves the conflation of two notions which on his own account are incompatible in the sense given above.

B.4. *An explanation of the confusion*

I shall now offer an explanation of how it comes about that Strawson comes to the position of holding *both* that an empirical fact is either *presupposed* or *implicitly carried* by a subject term such that these notions are regarded as alternatives. My explanation lies in the confused nature of the all crucial notion of 'Identifying Reference'. Tackling the question of the variety of terminology,[26] Strawson raises the question:

> What precise account can be given of the relations between term-introducing expressions which are, in the relevant sense, complete, and the facts or propositions which confer on them their completeness? How is the content of these facts or propositions determined by or otherwise related to, the actual term-introducing expressions used?

He holds that the variety of cases is too great to allow for a single answer and hence produces examples first of more simple and then of more complex cases. I shall concentrate on the simplest case for my explanation. In the simplest case e.g. in the statement 'That person there can direct you' it is clear, Strawson contends, what the fact is upon which the phrase 'That person there' introduces its term. The term distinguishing fact is that there is just one person there where I am pointing and this term distinguishing fact is *presupposed* by the use of the term introducing expression. He writes:

> If there is no-one at whom I could be taken to be pointing, my putatively term-introducing expression fails of a reference and my statement fails of a truth value. In such cases, then, we have a clear enough sense of presupposition, and

> a clear enough indication of what is presupposed in the use of the term-introducing expression.[27]

Now here we must note: (i) that the expression 'That person' is ambiguous as regards its logical status; (ii) the notion of 'pointing' here is ambiguous as regards its interpretation. 'That person' can be construed as either: (a) a quasi-name which serves to refer and has as its condition of introduction the fact that the speaker can specify what is putatively referred to in the form of such a sentence as e.g. 'The person who is standing there, i.e. between the farmyard gate and the road', even though the phrase 'That person there' is not to be construed as identical in meaning with such a sentence since its use is to refer and not to identify or, in Strawson's and more traditional language, to describe; or; (b) as the definite form of a sortal expression. In this latter case it is writable as (e.g.) 'The person who is standing between the farmyard gate and the road'; in this case the expression 'That person' is writable as and is equivalent in meaning to some such definite sortal expression of the form 'The person who ...'. The notion of 'pointing' can be construed as either simply 'referring' (in that the speaker points he simply refers) or as 'picking out', 'identifying' (in that the speaker points he picks out or identifies). It is clear from the text that Strawson construes 'pointing' as 'referring' (at least basically) since he speaks in term of 'failure of reference' and correspondingly treats 'That person' as a quasi-name. Now in that: (a) 'That person' is treated as a quasi-name; (b) 'pointing' is construed as 'referring', then I claim that he is *right* in holding that 'if there is no-one at whom I am pointing my putatively term-introducing expression fails of reference and my statement fails of a truth value'; that in such a case it is *presupposed* that there is just one person there where the speaker is pointing. Failure of reference implies failure of truth value; failure of reference permits no proposition to be expressed. To hark back to Russell an alleged proposition containing an empty name is no proposition. However in that: (a) 'That person (there)' is treated as the implicit form of the definite form of a sortal expression writable as 'The person who ...' and; (b) 'pointing' is to be construed as 'identifying' in the sense of 'picking out' then it might be said matters are quite different. Such a definite sortal expression plays an identifying (picking out) role and in that it plays such a role or is used to play such a role it does not *presuppose* existence but rather *implies* existence. It might be said that we must resist such a saying on the following grounds. It cannot be said in such a role or use existence is *implied*, that a true proposition of existence is implied, since the definite form of a sortal may occur in such a role, may be used in such a role and yet there be nothing to be actually picked out. It does not fail in such a role in that there is nothing to be so picked out or identified. If one held the thesis that expressions in such a role *implied* existence then one would be in the intuitively absurd position of having to say that e.g. the God of Abraham, Isaac and Jacob existed simply from the fact that someone uses the phrase 'The God of Abraham, Isaac and Jacob' in an identifying, picking out role. Neither can we say that the use of the definite form of a sortal in a 'picking out' or 'identifying' role carries an *implicit* claim to existence: either its use in such a role implies existence or it does not. And it does not. We may draw an analogy here with Russell's 'Definite descriptions'. A definite description does not fail to be a definite description in that there is nothing that satisfies it. A vacuous definite description is still a 'definite description'; it is an empty one. 'The present king of France' in 'The present king of France is bald' does not cease to have sense in that there is nothing answering to it. It has sense in that it is satisfiable, not in that it is actually satisfied. By analogy, the definite form of a sortal in an identifying, i.e. picking out role, still has sense in the case where there is nothing to be so picked out. It can be said to have sense in that it can be used on some occasion to pick out an object. Hence it cannot be claimed that the use of the definite form of a sortal in an identifying, picking out, role *implies* existence - that a true proposition of existence is *implied*; only that *a* proposition of existence is potentially implied.

However, despite such pressure as just exercised, I shall hold that in that a definite sortal expression plays a *genuinely* identifying, 'picking out' role then an expression occurring or being used in such a role may be said to *imply* existence. Where there is nothing picked out or identified by the definite form of the sortal in such a role there is not a *genuine* occurrence or use of such an expression in such a role. We may hardly say that such an expression in such a role succeeds in picking out an object, an individual, and yet hold that the existence of such an individual is not implied. We should note that the 'paradigm', 'standard' for "picking out", "identifying" is where the object is *successfully* picked out. In the cases cited above e.g. 'The God of Abraham, Isaac and Jacob', where there is no implication of existence someone may be said to use such a phrase in a *putatively* identifying role. To refer back to the assertion that 'it does not fail in such a role in that there is nothing to be so picked out or identified' (p. 99), I claim it *does* so fail in that the *norm* for the definite form of a sortal expression performing its role of identifying, 'picking out' is that it is implied that there *is* something so picked out or identified - though of course in the cases where nothing is picked out, no true proposition of existence is implied - a definite sortal expression still retains its status *as* a definite sortal expression having a potential role.

My explanation of how it comes about that Strawson comes to the position of holding *both* that an empirical fact is either *presupposed* or *implicitly carried* by a subject term (or 'subject expression') is that in relation to the notion of 'identifying reference' he confuses the two constructions of 'That person' and takes it that: (a) in that 'That person' is writable as e.g. 'The person who ...', as the definite form of a sortal and performs an identifying, picking out, role, it *implies* existence and this accounts for his speaking of a subject expression as representing a fact, embodying, covertly carrying a (true) proposition of existence; the 'identifying', 'picking out' element in 'identifying reference' demands that a subject expression carry an *implicit* claim to existence; (b) in that 'That person' is not so writable, but construable as a quasi-name, then existence is *presupposed*; the 'referring' element in 'identifying reference' demands that he speaks of a subject expression *presupposing* existence.

B.5. At this point it may be said: Why cannot we hark back to Russell and simply write for 'the definite form of a sortal', a 'definite description' and claim that in one regard Strawson simply confuses a proper name and a definite description and that the notion of 'identifying reference', in one regard, simply arises out of such a confusion? Although a case may be made out for holding that Strawson does indeed confuse proper names and definite descriptions, we cannot simply say outright that in one regard or at one level the notion of 'identifying reference' simply arises out of such a confusion. It could be said to arise out of such a confusion in that it is assumed that a definite description *implies* existence or more correctly - that a proposition containing a Russellian definite description implies the *truth* of some existential proposition. But it does not. And there is good reason for not harking back to Russell and simply writing for 'the definite form of a sortal', a 'definite description'. The difficulty in so harking back to Russell and so re-writing is that for him what can ultimately satisfy a definite description is a particular, an individual, denominated by a 'logically proper name'. I have already given a reason for rejecting Russell's doctrine of logically proper names and shall pursue the matter later in Chapter VI. For me, by contrast, the definite form of a sortal is not satisfiable; it is not, as I have urged, writable as a function. Again, when I write that e.g. 'The person who ...' fails to identify, fails to pick out anything present, this is *not* to be construed as 'fails to pick out anything viz., any (simple) *object*, *x*, so present' but as 'fails to pick out any *F* so present', in this case 'fails to pick out any *person* so present'. I defer from speaking of the definite form of a sortal as a 'definite description' for these reasons.

Russell analyses e.g. 'The author of *Waverley* was a genius' as:

(1) $(\exists x)$ [[x wrote *Waverley* $\cdot$ $(y)(y$ wrote *Waverley* $\rightarrow y{=}x)$] $\cdot$ x was a genius]

It is my thesis that the proposition should be written as:

(2) [$(\exists x)$ person] {[x wrote *Waverely* $\cdot$ ((y) person)(y wrote *Waverley* $\rightarrow$ $y{=}x$)] $\cdot$ x was a genius}

following Quine's writing in *Mathematical Logic*,[28] invoking his 'range indicators'.

The use of Quine's 'range indicators' in (2) accents my point made earlier that we may not embrace Russell's notion of 'an essentially undetermined variable "*x*"' in the sense that it commits one to holding that it is not necessary that every variable is determined. The use of 'range indicators' in the positions in (2) brings out my point that *necessarily* every variable is determined whilst not committing me to the thesis that every variable is *essentially* determined. This latter position would be expressed as:

(3) ($\exists x$ person) {[x wrote *Waverley* $\cdot$ (y person)(y wrote *Waverley* $\rightarrow y{=}x$)] $\cdot$ x was a genius}

and would be rejected anyway on the grounds that it is not a recognisable formulation. Further, (2) is consistent with my claim that I may not, in my case against Strawson, simply invoke what one could reasonably take to be Russell's criticism or at least a Russellian based criticism of the notion of 'identifying reference'.

C. An Attempted Account of what distinguishes an expression occurring in the role of Logical Subject

Having thus criticised and attempted to explain the confusion in Strawson's account of 'subject expression', I shall now attempt to put the implication of these criticisms to some positive use in the form of offering an account of what distinguishes an expression occurring in logically subjective role, utilising his account of 'subject expression' and hence, in one sense of 'explanation', offer an explanation of one element in the subject-predicate distinction. The account I offer is that what distinguishes an expression occurring in subjective role or what distinguishes a person's use of an expression in subjective role in the context of subject-predicate proposition is that *existence is presupposed* by an expression occurring in or being used in such a role. That is, that the occurrence or use of an expression in such a role, presupposes the truth of some existential proposition, as opposed to implies the truth of some existential proposition.

This is in accordance with the tradition that what can primarily occur in the role of 'logical subject' are *proper* names and that the use of a proper name or 'singular term' in a referring or 'designating' role, *presupposes* existence - presupposes that there is just one object so referred to or designated - but with the very important codicil that the very articulation of this requirement itself requires that sortals themselves be not regarded as predicates. That the very articulation of this requirement itself requires that sortals themselves be not regarded as predicates lies in the fact that in that it is said that a proper name in a referring or designating role presupposes *that there is just one object* so referred to or designated, it presupposes that there is just one F, since as I have earlier argued, we only have one object in that we have one F, where 'F' is a sortal. The presupposition of existence for an expression occurring in or being used in logically subjective role within the context of a subject-predicate proposition will thus be a true proposition of the form:

$$[(\exists x)S]\ [\varphi x \cdot ((y)S)(\varphi y \rightarrow y{=}x)]$$

where 'S' is some sortal.

It may now be objected that this account of what distinguishes an expression occurring in subjective role is itself inadequate as a *criterion* for what distinguishes an expression occurring in or being used in subjective role, since whilst it may plausibly be held that referring or naming presupposes existence, referring or naming are not sufficient to mark off an expression (symbol) occurring or being used in subjective role. In reply to this I should first explain that I am not here concerned to offer a *criterion* in the sense of an answer to the question: 'How do we *determine*, within the context of an subject-predicate proposition, *whether* an expression is occurring in or being used in subjective role?' Secondly I am not and cannot be concerned with the question: 'What distinguishes an expression occurring in or being used in subjective role external to the context of a subject-predicate proposition?' for this is surely a nonsensical question. We cannot *hope* to answer this question, i.e. produce an account of what distinguishes an expression in logically subjective role *external* to an understanding of what it is for an expression to occur in, to be used in, subjective role. Yet the objection presented is precisely based on the supposition that we may ask such a question and produce such an account. The question we may ask and which I am here seeking to answer is: 'Granted an expression in subjective role stands for an object about which a proposition is being formed, what distinguishes an expression in such a role?' and I hold we may answer this question by 'presupposition of existence' in the way and sense explained. Of course someone will raise the question: 'Why should we grant that an expression in logically subjective role stands for an object about which a proposition is being formed?' I hold we have good reason to so grant what is, in fact, Geach's account in the light of the failure of Strawson's arguments against Geach considered in Section A of this present chapter. The answer I have given to the question originally raised above must of course be provisional in the sense that I will have to show that such a presupposition of existence can form no part of an answer to the question: 'What distinguishes an expression in predicative role (in the context of a subject-predicate proposition)?' It is to this question that I now turn and once again I shall have reference to Strawson's 'Mediating criterion'.

D. An Attempted Account of what distinguishes an expression occurring in the role of Logical Predicate

D.1. First, as regards Strawson's account of the conditions for an expression to be, in his terms, a 'predicate expression' we must in parallel fashion to that advocated earlier, take his account as relating to role. To take it as relating to a category of expression (symbol) raises difficulties parallel to those already raised in connection with 'subject expressions'.

It is his case,[29] that whilst a subject-expression is one which 'in a sense, presents a fact in its own right and to that extent is complete. A predicate-expression is one which in no sense presents a fact in its own right and is to that extent incomplete'. To re-write this in terms of a 'criterion' appropriate to *role* rather than *expression* we should have the contention that:

> An expression occurring in, being used in, the role of logical predicate in no sense presents a fact in its own right and is to that extent incomplete.

In Strawson's later terms of 'presupposition', an expression occurring or being used in predicative role makes no *presupposition* of fact; in terms of his earlier talk, an expression occurring or being used in such a role does not 'embody' or 'covertly carry', an existential proposition which it does not explicitly affirm;[30] it does not covertly present any fact;[31] it does not carry any weight of fact.[32] Difficulties for assessing Strawson's thesis admittedly arise, as I have already argued, from a conflation of 'presupposition' and certain other notions which may be said to come under the heading of 'implication'. Having already discussed these difficulties and offered an explanation as to how they arise, I shall not pursue the matter again here and shall take it that his principal thesis is to be expressed in a theory of *presupposition* viz., that he is claiming that an expression in predicative role makes no *presupposition* of existence.

I shall contend that granted his *own* account of a 'predicate expression' Strawson cannot *show* that an expression occurring in predicative role makes no presupposition of existence; that an expression occurring in the role of logical predicate of its nature *or essence* makes no presupposition of existence. On his account of the nature of logically predicative expressions the most he can claim is that it is *contingently* the case that expressions in such a role make no such presupposition and thus, within the limits of his own theory, he cannot be said to have offered an account of what the *essential* difference is between expressions occurring in a logically predicative and logically subjective role. To put the matter the other way round. On Strawson's account of 'logically predicative expressions' or 'predicate' expressions there is, as we shall see, no *essential* difference between such expressions and what he designates 'subject' expressions and correspondingly no *essential* difference between an expression in predicative role and one occurring in subjective role. That this is so becomes clear in his later writings and this is an important deficiency in his various accounts of the subject-predicate distinction as he acknowledges this distinction to be fundamental and the difference at both the level of expression and role has been taken to be fundamental. As regards the 'Mediating criterion' itself, it is my case that it expresses what is a necessary truth as a purely contingent one and that therefore it cannot present us with an account of the difference in nature between an expression occurring in subjective role and one occurring in predicative role and indeed implicitly denies that there is an *essential* difference. I shall attempt to explain this criticism within the context of what I offer as an account of what distinguishes an expression occurring in predicative role within the context of an subject-predicate proposition.

D.2. My thesis is that an expression occurring in predicative role in such a context *cannot be said* to make any presupposition of existence as opposed to it being merely false that it makes no such presupposition. As I have previously contended a predicative expression is, of its nature, incomplete. Geach has argued:

> A *predicable* is an expression that gives us a proposition about something if we attach it to another expression that stands for what we are forming the

> proposition about: the predicable then becomes a *predicate* and the other expression its *subject* ...[33]

Rather more precisely we should write: 'a predicable is an expression that gives us a proposition about something if we *complete* it by another expression that stands for what we are forming the proposition about ...'. This re-writing makes it absolutely clear that an expression occurring in predicative role is of its essence *part* of a proposition; in traditional parlance a 'propositional fragment'. But a 'propositional fragment' cannot intelligibly be said either to presuppose *or* fail to presuppose existence precisely because it is *fundamentally* incomplete. Of itself it can say nothing. It is just in that Strawson fails to recognise the radically incomplete nature of logically predicative expressions and their corresponding role as propositional fragments or, alternatively, denies that they have such a nature, that he fails to offer us a criterion for an expression occurring in the role of logical predicate which will enable us to draw a radical or fundamental distinction between predicative and subjective roles.

It might be thought that even though an expression in predicative role cannot be said to presuppose existence for the reason given it may even so be said to *imply* existence. Such a thought must be rejected on two grounds. (i) It is difficult, indeed impossible, to see how an incomplete expression, which in that it performs a certain role is a fragment of a proposition, can equally be said to *imply* anything. The relation of 'implication' holds between propositions and to regard a predicate as implying would precisely be to regard it as an *element* in a proposition, as a complete whole which may 'say something'. (ii) To regard an expression as occurring in predicative role as *implying* existence would have the unacceptable consequence that all predicates were instantiated. And Strawson is surely right in wishing to reject this thesis.[34] From the consideration that all predicables are *instantiable*, it does not follow that all 'predicates' are *instantiated*. The thesis that all predicables are instantiable is a purely formal remark on the nature of predicables; it draws our attention to the point that a predicable of its essence contains an empty place which may be filled by the name of an object and hence create a proposition, true or false. It neither implies nor presupposes the existence of any object nor can be said to imply or presuppose. The following worry might arise at this juncture. If one holds, as I am holding, that the question of presupposition or implication of existence cannot arise in the case of an expression in predicative role, is not one then committed, in effect, to saying that there is no such thing as the instantiation of a predicable (or 'predicate')? - which is absurd. One is not thus committed. For there to be instantiation of a predicable, 'predicate', is for there to be some true proposition formed by replacing the blank in '---- is *P*' by the proper name of an object (an *F*). Whilst of course 'is *P*' may not be said to be instantiable since it contains no empty place to be filled by the name of an object, '---- is *P*' may be said to be instantiable for this very reason.

Worries may still be aired concerning the position I am advocating and even that of Strawson. It might be said that in that for an expression to occur in predicative role is for an expression to occur in the role of 'saying something about' an object,[35] then surely this commits one to saying that the existence of an object is presupposed. How can one say (by reference to Strawson's earlier examples) that 'smokes' or 'is smoking' gives us a true or false predication of Socrates in the proposition 'Socrates smokes (is smoking)' if there is no Socrates? Such a consideration however does not show that an expression in predicative role presupposes existence or can be said to presuppose existence for an expression in predicative role is of its essence part of a *proposition* and hence what can be said to presuppose (or fail to presuppose) existence is the whole proposition. Whilst in one sense one must concur with Strawson that it is wrong to hold that an expression in the role of logical predicate presupposes existence, it is difficult to see how *he* can subscribe to an account of predicative role in terms of 'saying something about' etc.,[36] and yet consistently deny that 'predicate terms' presuppose or imply existence. That he is in some difficulty on the point can be seen as follows. Whilst holding that a 'predicate expression' does not presuppose existence, he also wants to hold that it is a *proposition* which is either true or false of something, not a 'predicate'. The possibility of allocating a truth value to a proposition of a subject-predicate kind depends upon there being something which the *proposition* is about, not upon there being something which the *predicate* is about or is true or false of. What therefore presupposes is the whole proposition and what the whole proposition presupposes can be taken as what is presupposed by the subject 'term'. Now the thesis that only a *proposition* can be true or false and not a predicate relies on the principle that a predicate is of its essence or nature incomplete that it is of its essence or nature, *part* of a proposition. This entails accepting that a 'predicate' expression is of the form: '---- is *P*' and *not* either of the form '*P*' or 'is *P*'. But just this Strawson: (i) does not; (ii) cannot hold. Indeed it is clear from his writing in *Individuals* that he holds that a predicate expression is of the form: '*P*' (verb) or 'is *P*'. Thus I claim he cannot consistently show that a predicate or predicate expression does not of its nature or essence presuppose (or imply) existence: the most he can show is that a predicate or 'predicate expression' such as 'is wise' or even 'wise'; 'smokes' or 'is smoking'; is *accidentally* incomplete and to that extent it is an *accident* that they do not presuppose (or imply) existence. In that he is concerned to present us with a 'criterion' for a logical distinction of a fundamental kind for him to have to hold that it is an accident of predicates or 'predicate expressions' that they do not presuppose (or imply) existence is clearly unsatisfactory.

Whilst I have maintained that Strawson's account of what distinguishes an expression in predicative role is unsatisfactory, consideration of his position yields positive results in that it enables me to proffer an account of what distinguishes an expression occurring in predicative role in the context of an subject-predicate proposition - namely that in regard to an expression in such a role the question of existence cannot intelligibly be raised. Expressions, symbols, in predicative role can be said neither to presuppose nor imply existence: what can be said to presuppose or carry an implicit claim to existence is the whole proposition and the whole proposition presupposes existence in that it is a proposition about, or purports to be a proposition about, what is referred to or designated by an expression in a referring or naming role.

To put this distinguishing feature of predicative expressions in terms of a person's use of an expression in predicative role: a person employs an expression *P* in predicative role in the context of a subject-predicate proposition in that he acknowledges that the question of presupposition of existence or implication of existence cannot intelligibly be raised.

Parallel to the objection raised earlier in connection with 'subjective role' it will be said that this account, 'explanation', of an expression occurring in predicative role is unsatisfactory. In like fashion, whilst it may be held that the question of presupposition (or implication) of existence may not be raised in connection with a predicable expression playing a role, a predicable expression playing a role is *not* sufficient to mark it off as a *predicate.* A predicable after all may occur in a non-predicative role. Here again it must be reiterated that I am not concerned with, and cannot be concerned with, the question: 'What distinguishes an expression in predicative role external to the context of a subject-predicate proposition?' We cannot hope to produce an account of what distinguishes an expression in predicative role *external* to an understanding of what it is for an expression to occur in, be used in, predicative role. To seek to do this, one might perhaps add, is tantamount to denying that subject-predicate distinction is an *ultimate* distinction. The question one may answer and which I am here seeking to answer is: 'Granted an expression in predicative role gives us a proposition about something if we attach it to another expression that stands for what we are forming the proposition about (or that stands for what the proposition is being formed about), what distinguishes an expression in such a role from an expression in subjective role?' And I hold that we may answer *this* question by: 'The question of presupposition (or implication) of existence cannot be raised'.

D.3. Having offered, or at least attempted to offer, an explanation of the subject-predicate distinction in the sense of offering or attempting to offer an answer to the question: 'What differentiates an expression occurring in the role of logical subject from one occurring in predicative role in the context of an subject-predicate proposition?' (cf. sec C. above: pp. 102-103), the nagging objection will be pressed that, in one clear sense, I still have not offered an *explanation* of the difference between an expression occurring in (being used in) subjective role and an expression occurring in (being used in) predicative role. The clear sense, it will be said, in which I have not offered an explanation is that of offering an account of the distinction which does not itself invoke an understanding of the distinction. The most I have done is to offer an explanation of the difference (the distinction) in the sense of a description of the difference (distinction). But to attempt to explain the difference (the distinction) in the sense demanded is theoretically impossible. It would be tantamount to denying that the subject-predicate distinction was an ultimate distinction, as I have just commented above. Whilst my 'explanation' of the distinction has to invoke an understanding of the distinction as set out in Geach's explication, it emphatically does not *repeat* such an understanding or explication. It offers an explanation of the distinction, and not merely a description of the distinction, in that it answers, or attempts to answer, the question posed. And this is the most that can be asked for. Finally, it is to be noted, that in that Strawson thinks *he* has offered an explanation of the distinction in the sense of what differentiates a 'subject expression' from a 'predicate expression' (to use his terms) *independently* of invoking an understanding of the distinction, he too is mistaken, for it can be held against him also that 'presenting a fact in its own right'; 'failing to present a fact in its own right'; are not sufficient *of themselves* to demarcate a 'subject expression' from a 'predicate' expression. It is necessary to emphasise the point that we may only offer an explanation of the subject-predicate distinction in the sense that either I myself or Strawson seek to offer an explanation against the background of an *explication* of the distinction. And such an explication is furnished by Geach and, if I am right, has withstood the attacks of Strawson considered earlier in this chapter.

E. A more Precise Account of the Relation between Sortals and the Subject-Predicate Distinction

Having offered an 'explanation' of the subject-predicate distinction in the sense of offering an answer to the question: 'What differentiates an expression occurring in the role of logical subject from one occurring in predicative role?', it is now time to present a more precise account of the relation between sortals and the subject-predicate distinction.

I have argued earlier in Chapter II that referring, as a general mode of speech activity, presupposes, rests upon, identifying - sortal identifying in the sense of sortal 'picking out'. It will now be said that in the light of: (a) the consideration that we are unable to offer an explanation of 'occurring in the role of logical subject' or a person's use of an expression in such a role simply in terms of 'referring': 'naming' or 'designating' - such accounts simply give us an explication of the notion; (b) the consideration that an expression occurs in the role of logical subject or a person uses an expression in such a role on an *occasion*; and similarly in regard to 'occurring in the role of logical predicate', I need to present a tighter, more precise account of the relation between 'sortal identifying' and 'occurring in', 'being used in' subjective and predicative role. This I shall now attempt to offer.

I (A) *Subjective role*

(i) It is a presupposition of an expression occurring in the role of logical subject that what is putatively referred to, designated, as that about which the proposition is being formed is: (a) specifiable by reference to a sortal; (b) pickable out, discriminable by reference to the definite form of a sortal;

(ii) but that it cannot be held that the occurrence of an expression in such a role on an occasion demands that the object of reference, that about which the proposition is being formed, (a) is *correctly* specified or (b) is successfully picked out under its correct specification. All that is required is that it is specifiable by reference to some sortal; not that it is specified by reference to some *one* sortal i.e. the sortal under which the object of reference correctly 'falls'. To refer back to Chapter II, we cannot say that e.g. 'Socrates' fails of reference and that no proposition about Socrates is being formed in that the object of the speaker's reference and the subject of his intended proposition is incorrectly specified as e.g., a cat. As regards (b), failing to pick out an object under its correct specification does not entail failure of reference or failure to form a proposition concerning that object, as witness examples from Kripke earlier.

I (B) *A person's use of an expression in Subjective role*

(i) It is a presupposition of a person's use of an expression in subjective role that:

(a) he can specify in some way i.e. by reference to a sortal, that which he is putatively referring to or designating as that about which he is forming a proposition;
(b) that he can pick it out, discriminate it by reference to the definite form of a sortal;

(ii) but that it cannot be held that the use of an expression in such a role on an occasion demands that the object of his reference, that about which the proposition is being formed:

(a) is correctly specified by him;

or

(b) is successfully picked out by him under its correct specification (for reasons offered above and earlier in Chapter II).

II (A) *Predicative role*

(i) It is a presupposition of an expression occurring in predicative role i.e. in the context of giving us a proposition about something if it is attached to another expression that stands for what the proposition is being formed about, that what the proposition is being formed about is:

(a) specifiable by reference to a sortal;
(b) pickable out by reference to the definite form of a sortal;

(ii) but that it cannot be held that the occurrence of an expression in such a role on an occasion demands that what the whole proposition is being formed about:

(a) is *correctly* specified;
(b) is successfully picked out under its correct specification.

Once again all that is required is that that about which the proposition is being formed is specified by reference to some sortal; not that it is specified by reference to some one sortal i.e. the sortal under which the subject of the predication correctly 'falls'. We cannot say that we have no predication concerning Socrates in that Socrates is incorrectly specified as a cat, for it is contingent that 'Socrates' has the use of being the proper name of a man and we may still have a predication concerning Socrates even though Socrates is not successfully picked out under a correct specification. We may still hold that 'Socrates is wise' gives us a true predication concerning Socrates even though Socrates is successfully picked out on an occasion by e.g. 'The cat which ...'.

II (B) *A person's use of an expression in predicative role*

(i) It is a presupposition of a person's use of an expression in predicative role i.e. in the context of his using the expression to give us a proposition about something in that he attaches it to another expression that he uses to stand for, refer to, what he is forming his proposition about, that:

(a) he can specify in some way i.e. by reference to a sortal what he is forming his proposition about;
(b) he is able to pick out, discriminate, what he is forming his proposition about by reference to the definite form of a sortal;

(ii) but that it cannot be held that his using an expression in such a role on an occasion demands that what he is forming his whole proposition about is:

(a) correctly specified by him;
(b) is successfully picked out by him under its correct or even a correct specification (for reasons offered in II (A) above) and earlier in Chapter II.

The essence of this attempted more precise account of the relation between 'sortal identifying' and 'occurring in', 'being used in' the roles of logical subject and logical predicate is that *both* roles presuppose sortal identifying in the sense of both: (i) specification by reference to some sortal; (ii) 'picking out', discrimination by reference to the definite form of some sortal, but they do not presuppose specification by reference to some one sortal or again discrimination by reference to the definite form of some one sortal, and this account coheres with my thesis that it is essential that every variable is determined as opposed to the thesis that I earlier rejected - namely that every variable is essentially determined. To hold that every variable is essentially determined would be to hold that e.g. the occurrence of an expression in the role of logical subject presupposed sortal identifying in the sense of: (i) specification by reference to *some* one sortal; (ii) discrimination, 'picking out' by reference to the definite form of *some* one sortal since on this thesis as every variable is essentially determined so every proper name is essentially determined.

F. A Contrast with Quine

Having thus offered, by reference to Geach, an explication of the subject-predicate distinction and by reference to an importantly revised form of Strawson's 'Mediating criterion', an explanation of the distinction in the sense of an answer to the questions: 'What differentiates an expression occurring in subjective role or being used in a subjective role from an expression occurring in, being used in, predicative role within the context of an subject-predicate proposition?', it is now time to compare and contrast my explanation with the famous explanation offered by Quine. No discussion of the subject-predicate distinction is complete without reference to Quine. Whilst I cannot accept Quine's account for at least the reasons to be enlisted below, there is one aspect of that account with which I crucially concur and which is compatible with one element in my own account - namely that predicative position does not admit of quantifiers. Just this aspect of Quine's explanation has been severely attacked by Strawson. I shall defend Quine against the onslaughts of Strawson and hence defend one element in my own account, namely that an expression in predicative role may not presuppose and may not entail existence.

F.1. Quine holds that:

> In terms of logical structure, what it means to say that the singular term "purports to name one and only one object", is just this: *The singular term belongs in positions of the kind in which it would also be coherent to use variables* '*x*', '*y*', etc., (or in ordinary language, pronouns).[37]

> ... the mark of a name (subject term) is its admissibility in position of variables.[38]

> What distinguishes a name is that it can stand coherently in the place of a variable in predication .... Predicates are not names: predicates are the other parties to predication.[39]

> When we schematise a sentence in a predicative way "*Fa*" or "a is an *F*", our recognition of an "*a*" part and a "*F*" part turns strictly on our use of variables of quantification; the "*a*" part represents a part of the sentence that stands where a quantified variable could stand, and the "*F*" represents the rest.[40]

and the same doctrine is present in *Word and Object.*

Now I cannot accept this as an explanation of the subject-predicate distinction for at least the following reasons:

(1) it would commit me to the thesis that names, ordinary proper names, may not be regarded as a prime vehicle of reference which is inconsistent with a position already advocated. To hold that names are reducible to pronouns is unsatisfactory:

(a) in that pronouns are substitutable for names in a way in which names are not substitutable for pronouns;

(b) a proper name of its essence may be used and is used to refer to the same *F* on a number of different occasions; a pronoun of its essence is not so used. Pronouns of their nature have no 'nominal essence'.

(2) it would likewise commit me to the thesis that we may understand the use of pronouns *independently* of some sortal which serves to specify the reference of the pronoun; that the use of pronouns and our understanding of pronouns does not depend upon, rest upon, the use of expressions (symbols) which serve to specify the reference of pronouns. But if this were so we could never understand pronominal reference. I have argued for this in Chapter II and I shall be returning to the matter again and its importance in my final discussion of Quine's theory of 'Canonical Notation' in my penultimate chapter.

(3) It would commit me to denying the thesis I have advocated all along; namely that necessarily every variable is determined.

(4) It would commit me to what I, and indeed others, would claim are absurdities i.e. to Quine's doctrine 'that to be is to be the value of a variable' (a proper name variable) yielding such results as 'There is Socrates'; 'There is 2'.

(5) Finally, it would commit me to the 'full version' of Quine's doctrine and theory of canonical notation i.e. to the elimination of singular terms altogether in favour of the variables of quantification and the position beautifully presented by Strawson in *Subject and Predicate in Logic and Grammar*.[41] It is valuable to quote Strawson at some length here since he encapsulates Quine's position most succinctly. He asks: 'What reason could there be for presenting it (viz., the doctrine that names can, and predicates cannot, stand in positions accessible to variables of quantification),[42] as the *sole* ground of the distinction?' and proceeds to write:

> This question again has interest only if such presentation is not merely an internal feature of the presentation of the grammar of a limited logical language. But it seems likely that Quine regards it as more than this. We may get light on his reasons by considering the character of that ideally austere language favoured by Quine in which subject terms have no place at all. What

> in that language takes the place that would be theirs if they had a place, is a certain combination of existential quantifier, predicates, and variables. Where, in a language with names, we have '*Fa*' in Quine's nameless language we have something like '($\exists x$)(uniquely - *Ax* and *Fx*)'. Instead of 'Socrates swims' we have 'Something is uniquely Socratic and swims' or 'Something which is uniquely Socratic swims'. Instead of 'Socrates' *tout court* we have 'Something which is uniquely Socratic'. As far as this language is concerned, names (and subject terms in general) are to be seen simply as *abbreviations* for such constructions as these .... A name (or subject term) is distinctly characterised as an expression which can stand in the place of a bound variable because it simply abbreviates such a construction as has just been described, with an extra bound variable awaiting its predicate. A name can go in the place of a variable because a bound variable can go in the place of a variable.[43]

I may not accept such a position since: (i) I could not allow, simply '($\exists x$)' in the formulation '($\exists x$)(uniquely - *Ax* and *Fx*)'; (ii) it is in any case objectionable to write 'uniquely - *Ax*' e.g. 'uniquely Socratic' since this requires that a name is turnable into a predicate. Here one is faced with the embarrassment that strictly 'Socratic' should be written as 'socratic' (small case 's'), but once this is done then the impetus is lost for there is no necessary connection between 'Socratic' and 'socratic'. 'Socratic' in the formulation is neither a name nor a predicate. We may justly hold, on the general grounds of the absolute distinction between names and predicables (predicates) that no name is writable as a predicate (predicable) and remains the *same F*, the *same* term. Further, such general logical consideration apart, it is implausible to hold that a name is reducible to a set of predicates since clearly we may understand the name in question and yet be unable produce the alleged list of predicates which uniquely distinguish the name. Again, the thesis that a name is writable as a set of predicates fails on the count that such a thesis would render a sortal itself as a predicate since I have argued that a sortal is required for the introduction and continued use of a name and that sortals are not reducible to predicates has been a principal tenet of this book so far.

F.2. To turn now to Strawson's attacks on Quine; first in 'Singular Terms and Predication'. Here Strawson contends outright that predicative position *is* accessible to quantifiers. He writes:[44]

> Predicative position is supposed to be inaccessible to quantifiers. But is it? Betty is a better date than Sally. Betty is willing and pretty and Sally too is willing and pretty. But Betty is also witty and Sally is not witty. Surely, it seems, 'willing', 'pretty' and 'witty' are here in predicative position. But is their position inaccessible to quantifiers? As a date, Betty is everything that Sally is (i.e. willing and pretty) and something that Sally isn't (i.e. witty). Or, if you like, there is nothing that Sally is that Betty isn't and something that Sally isn't that Betty is.

But, we may ask: 'What exactly does Strawson's claim commit him to and are there serious objections to such a commitment or commitments?' I hold that there indeed are.

In relation to 'Betty is *everything* that Sally is', we may raise the question 'Every *what* that Sally is?', for to speak of 'Everything' is to speak of 'Every *F*', where *F* is a symbol which furnishes us with a principle of counting and distinction. The natural reply to this question is that Betty is or has every quality that is a quality desirable in a date that Sally is or has. This reveals that in that predicative position is open to quantifiers one is committed to the existence of singular qualities desirable in a date and, more generally, to the existence of singular qualities. That one is so committed becomes clear in Strawson's 'redescription' of his position which presents us with what he acknowledges to be a 'logically more candid form'.[45] To be so committed to the independent existence of such qualities has, I hold, involved Strawson in the move from:

(a) 'Betty is willing, pretty and witty'

to

(b) 'Betty *has* willingness, prettiness and wit(wittiness), and concomitantly to holding that such terms as 'willing' etc., in predicative position, are construable as *names*, writable as 'Willingness' etc.,: that they are the names of individual qualities falling under the species 'qualities desirable in a date'.

Having made these observations on what Strawson is committed to, I shall defer examination of them until I have presented the context of his 'redescription' since he is the first to acknowledge that the redescription is logically more candid and bases his new criterion for the 'relative concepts of referential and predicative position' on it. The 'redescription' is as follows:[46]

> Prettiness is a quality desirable in a date and Betty has prettiness and Sally has prettiness. Similarly with willingness. Wit is a quality desirable in a date and Betty has wit and Sally has not. Everything (i.e. Every quality)[47] which Sally has and which is a quality desirable in a date is something (i.e. some quality)[48] which Betty has; but there is something which is a quality desirable in a date and which Betty has, which Sally does not have.

This 'redescription' is more candid since: (i) the terms which yield their position to quantifiers have the grammatical form of substantives and not of adjectives; (ii) there is something *explicit* in this account which was only implicit in the former and this is made explicit by the use of the expression 'quality desirable in a date'. Strawson follows this 'clue' which he claims yields the following result:

> 'Prettiness' occupies singular term or referential position because it is joined with such an expression as 'desirable in a date', which, relative to it occupies predicative position. Generally, whenever, explicitly or implicitly, two terms are joined of which the first stands to the second in that characteristic relation in which 'prettiness' (or 'pretty') stands to 'desirable in a date', then, relative to each other, the first is a candidate for referential and the second for predicative position.[49]

From this result Strawson is able to draw up a new criterion for the relative concepts of referential and predicative positions which he designates the *Type* criterion viz.:

> 'Pretty' has predicative significance relative to 'Betty', referential significance relative to 'desirable in a date'. We have a series consisting of (1) 'Betty' (2) 'pretty' or 'prettiness' (3) 'desirable in a date' such that any earlier term in the series is a candidate for referential position relative to the immediate succeeding term and any later term is a candidate relative to the immediately preceding term.[50]

This, after reflection, is generalised as follows:

> Two terms coupled in a true sentence stand in referential and predicate position respectively, if what the first term designates or signifies is a case or instance of what the second term signifies. Items thus said related (or the terms that designate or signify them) may be said respectively to be of lower and higher type; and this is why I called the new criterion one of type.[51]

Strawson further claims that this criterion implies that whilst a term designating a particular can never occupy predicative position, a term signifying a kind or property may occupy referential or predicative position, depending on whether it is, or is not, coupled with a term of a still higher type, and this is of course required for any satisfactory explanation of the subject-predicate distinction.

Let me now tackle the issue of the existence and independent existence of qualities which I mentioned earlier. It seems clear that Strawson is committed to the existence of singular qualities; but is such commitment warranted? I shall hold that it is not. I shall hold that the move which 'ushers in abstract singular terms'[52] is erroneous. To posit the independent existence of such qualities or their singular abstractions assumes the validity of the following move:

Taking 'pretty' as our example, from the fact that (1) Sally is pretty; (2) Betty is pretty; (3) Angela is pretty have *something* in common in virtue of 'pretty' being true of all three girls mentioned, it follows that they have some *one* thing i.e. just one thing in common, namely 'prettiness'. This move however is plainly invalid. From the fact that A, B and C have *something* in common it by no means follows that they have some one thing in common; they may have a conjunctive or disjunctive set of features in common or exhibit a 'family resemblance' to one another, as Wittgenstein has suggested. Further, to contend that the one thing they have in common in virtue of which 'pretty' is true of all three girls mentioned is *prettiness* clearly offers us no explanation of how it is that 'pretty' holds true of all three. To say that all three are pretty in that all three share the single quality *prettiness* merely invokes that which it seeks to explain, as Plato's 'Forms' in his classical theory did. Yet again, on the explicit assertion that e.g. 'Prettiness' is an abstract singular term, how may it, since it *is* abstract, be said to refer or designate? We thus have, I hold, no argument for positing the independent existence of *prettiness* or for that matter any other singular abstract attribute or quality. It is not the case that there is some one thing which all three girls have in common and even if it were, this some one thing could not be *prettiness*. Similar considerations apply in the case of willingness and wit(wittiness).

To pursue the argument from another standpoint. From the fact, if it is a fact, that 'pretty' has a single sense in the following: 'Sally is pretty', 'Betty is pretty', 'Angela is pretty', (where 'Sally' etc., are the proper names of girls), it doesn't follow: (i) that there is some one thing which is its sense; (ii) that it has a single reference. Singleness of sense for 'pretty' does not imply that there is something which is its sense and which is single; neither does such singleness of sense imply singularity of reference. We may plausibly hold that 'pretty' has the same sense in all three propositions in that it is intersubstitutable and that this in turn is determined by all three girls exhibiting some one or more of a disjunctive set of characteristics; and further that the reference of 'pretty' in all three cases is the girls concerned in that the propositions are true. Strawson however requires that there is some *one* thing which is the sense of 'pretty' and that this single X, viz., the quality *prettiness* is the reference of 'prettiness', thus, it seems, conflating sense and reference. It is necessary for Strawson that 'prettiness' (for example) has a single reference and indeed that 'pretty' likewise. He never considers the view that whilst 'pretty' might be said to have a single sense, it has multiplicity of reference. He holds that in that 'pretty' (for example) has a single sense in such cases as instanced, it must have a single reference and hence he has no *hesitation* in moving from *pretty* to *prettiness*. This move of positing the independent existence of *prettiness* is, I contend, quite illegitimate for it relies on the move from: 'A, B, C are pretty' to 'There is some one thing which A, B C share, namely, *prettiness*' - a move which I have shown earlier to be illegitimate. To hold such a thesis of the independent existence of *prettiness* (or *willingness* or *wittiness*) commits one to holding that for prettiness etc., to exist is not for there to be pretty A's or B's e.g. pretty girls, shrubs, flowers, cats, such that for prettiness etc., to exist is not for there to be propositional functions of the form: '*x* is pretty' which are true for some values of *x*, where '*x*' ranges over the proper names of girls, shrubs, flowers, etc., but for there to be an abstract quality. But since, as St. Augustine long ago argued, one only arrives at abstract qualities via concrete predications, one has no good reason to accept this view.

It is a concomitant of Strawson's view, and indeed essential to his positing the independent existence of singular abstract qualities, that he regards e.g. 'prettiness' as a singular name which may occupy singular or referential position and that we may replace the general term (predicate) 'pretty' by the singular name 'prettiness'. The argument against replacing the general term (predicate) 'pretty' by the abstract singular term 'prettiness' I have already advanced earlier, so I now turn to what I contend is a disastrous consequence of holding that 'prettiness' is a name which may occupy predicative position and hence be accessible to quantifiers. Recall that he has written:

> Generally, whenever, explicitly or implicitly two terms are joined of which the first stands to the second in that characteristic relation in which 'prettiness' (or 'pretty') stands to 'desirable in a date' then relative to each other, the first is a candidate for referential and the second for predicative position.[53]

If 'prettiness' is a singular name in 'Sally has prettiness' how may 'prettiness' be said to *characterize* Sally? If it is a singular name (and a singular abstract name) standing for a single abstract object, how may the relation of 'being a characteristic of' enter the matter? A singular name, even of a supposedly abstract object, may *not characterize* anything or stand in a characterizing relation to anything; to perform such a function or stand in such a relation, 'prettiness' must at the very least be construed as a general name. But to so construe 'prettiness' would be to undercut Strawson's case that we may coherently admit quantifiers into predicative position such that we may write e.g. 'Every (quality), namely, prettiness and willingness, which Sally has and which is a quality desirable in a date is some (quality) namely prettiness, willingness, which Betty has; but there is something i.e. some quality namely wit (wittiness) which is a quality desirable in a date and which is a quality which Betty has, which Sally does not have', for here 'prettiness' etc., are singular names of abstract objects. A general name names or stands for many (several) objects, but 'prettiness' may only name one object.

Since the relation of 'being a characteristic of' may not enter the matter one is faced with the problem of what the relation *is* between 'prettiness' on the one hand, and Betty, Sally, etc., on the other. It must be acknowledged that Strawson later on attempts to say what this relation is in terms of his doctrine of 'grouping' but this in turn is open to serious difficulties. He writes:[54]

> And it will scarcely be denied that the meaning of 'pretty' is such that it may be said to *group* such particulars in accordance with a certain kind of principle. The term may be said to group all those particulars whose designations may be coupled with it to yield true statements.

We must note here that whilst he starts off by claiming that it is the *meaning* of 'pretty' which groups particulars in accordance with a certain kind of principle he moves on, without apologia, to say that it is the term pretty i.e. *prettiness* which so groups particulars and so it certainly seems that we have no option but to say that he must hold that the meaning of 'pretty' is a single abstract entity which is also its designation. An unplausible thesis which has been detected earlier. Whilst there may be some sense in which the meaning of 'pretty' may be said to group particulars 'according to a certain principle', there is *no* sense in which 'prettiness' as an abstract singular term may be said to so group. We may raise the question: 'How many pretty *F*'s are there in this room?' or, again even: 'How many *F*'s in this room satisfy the description "pretty"?'; but distinctly *not* 'How many prettiness in this room?' We could, of course, with certain reservations, ask 'How many cases of prettiness are there in this room?'; but then 'prettiness' would cease to be the name of a single abstract object.

In parallel fashion it may seem that we are faced with the problem of what is to *prevent* us from construing 'desirable in a date', in 'Prettiness is desirable in a date', as a singular term 'Desirableness in a date', since Strawson has allowed that for 'pretty' we may write 'prettiness' and indeed it will be pressed that we must construe 'desirable in a date' in such a way in that he admits quantifiers into predicative position. If this is so, how then may we speak of 'prettiness' standing to 'desirable in a date' in a characterizing relation? Strawson will protest that there *is* something which prevents us from so writing the latter, namely, that 'desirable in a date' is a general abstract term, unlike 'prettiness'. This prevention is bought at some cost for: (a) we may then ask: 'How is it possible to quantify into the predicate place?'. At this point Strawson, it seems, has to admit that we may only so quantify in that we admit variables other than proper name variables which makes his criticism of Quine rest on an assumption which Quine himself would reject. There may be justification for such a thesis but Strawson offers no such justification here. (b) Even if some such justification is forthcoming, the prevention is bought not only at the cost of initially begging the question against Quine, but more importantly, at the cost of rendering his *own* Type Criterion for the subject-predicate distinction itself impossible. It does so in the following way:

In that 'prettiness' stands to 'desirable in a date' in a characterizing relation and this is consistent with 'desirable in a *date*' being writable as 'desirableness in a date', as an abstract general term, then 'pretty' cannot stand to Betty in such a characterizing relation since 'pretty' is writable as 'prettiness', as an abstract singular term. Yet the idea of a 'characterizing relation' as exemplified by 'prettiness is a quality desirable in a date' is absolutely central to Strawson's exposition of the Type criterion. This central feature of the exposition can only be maintained at the cost of *denying* that 'pretty' is writable as 'prettiness' i.e. denying that predicative position is accessible to quantifiers in such propositions as 'Betty is pretty'. I thus conclude that Strawson's account of the Type criterion is inconsistent in that it demands both of two mutually incompatible requirements: it requires *both* (a) that (for example) *prettiness* should be able to stand in a characterizing relation to Betty *and* (b) that in 'Betty is pretty' and 'Prettiness is a quality desirable in a date' we have the *very same term* occurring. (a) can only be met at the cost of denying (b), since for (b) to hold it must be the case that 'prettiness' is a singular abstract term, yet no singular abstract term may characterize; (b) can only be met at the cost of denying (a) since for (a) to hold it must be the case that 'prettiness' is not a singular abstract term. In that Strawson allows that predicative position is accessible to quantifiers in a way which is appropriate to Quine he cannot speak of 'that characteristic relation'[55] at all; in that he allows that predicative position is accessible to quantifiers in a manner which is impertinent to Quine and could produce a justification of such a position, he cannot speak of 'that characteristic relation' which holds between 'prettiness' and 'desirable in a date' on the one hand, *and* also between 'Betty' and 'prettiness' on the other.

I thus hold that the Type criterion may not be rationally advocated and that Strawson may only admit quantification into the predicate place by construing predicates as a type of name yet no predicate *is* so reducible both: (i) for the general reason that names and predicables (predicates) are absolutely distinct; (ii) for the specific reasons offered in the discussion above.

F.3. To turn now to the discussion in *Subject and Predicate in Logic and Grammar*.[56] Here Strawson considers the two simple sentences 'Socrates swims'; 'Socrates is brave'. These yield by generalisation the sentences: 'Someone swims' and 'Someone is brave' or 'There is someone who swims' and 'There is someone who is brave', but 'according to the doctrine in question' i.e. Quine's, such sentences cannot yield by generalisation 'Socrates does something' (or 'There is something Socrates does') and 'Socrates is something' (or 'There is something Socrates is') for 'to allow that they did would be to allow quantification into predicate place'. To advance his thesis here that one may allow quantification into the predicate place Strawson holds that: 'if we are prepared to abandon the ground level form of the sentences to specify our concepts by nouns instead of verb phrases' and:

> to introduce some suitable two-place predicate, *then* we can proceed to generalise, to eliminate specificity in favour of non-specificity, in a way perfectly consistent with the doctrine. Sentences on the lines of:
>
> (A) Socrates { performs / does / exemplifies } swimming
>
> or of
>
> (B) Socrates { possesses / has / exemplifies } bravery

> have attribute-naming nouns, 'swimming' or 'bravery', in referential or subject-position, coupled by a *new two-place predicate* with the name 'Socrates', also in subject-position; and the position of these names can safely be yielded to the non-specific 'something' or 'there is something which'. For when the names go, they leave a predicate behind; and (the phrase is Quine's) '*predicates are wanted in all sentences*'.

Deferring my comments on this emphasis on Quine's remark till later, I shall first argue that we may not re-write such sentences as 'Socrates swims' or 'Socrates is brave' in terms of 'Socrates performs etc., swimming'; 'Socrates possesses etc., bravery' and then turn to the question of why we should or should not be prepared to abandon the 'ground level' form of the sentences.

To hold that we may so re-write or advocate (A) or (B) above as more fundamental forms is to hold that a predicate is reducible to or eliminable out in favour of a name. Strawson in effect admits this when he speaks of 'attribute-naming nouns'. But this, as I have urged, is an error. Let me turn to the original examples. The case for saying that by generalisation *from* 'Socrates swims' or 'Socrates is brave' *we may obtain* 'Socrates does something' (or; 'There is something Socrates does'); 'Socrates is something', (or 'there is something Socrates is') may be spelt out and to be consistent with my general thesis concerning sortals, should be spelt out as follows:

(1) 'Socrates does something' - 'Socrates performs some *A*, some activity namely *Swimming*'; 'There is some activity, some *A*, namely *Swimming* which Socrates does';

(2) 'Socrates is something' - 'Socrates is some *V*, some virtue, namely *Bravery*'; 'There is some virtue, some *V*, namely *Bravery* which Socrates is'.

Now (1) clearly relies on the thesis that we may substitute the name 'Swimming', the proper name of an activity, for the predicate 'swims' (technically '---- swims') and I have already argued against such a position both in general and in particular in my discussion of Strawson above. As it stands (2) is strictly nonsense; the 'is' here, to yield sense, must be construed to read e.g. 'exemplifies' but on such a construction we: (a) once again have the thesis that a predicate is writable as or reducible to a name, and indeed to a proper name, for 'Bravery' is the proper name of a virtue; (b) are faced with the problem of how what a proper name stands for may be exemplified. *No* name is exemplifiable, let alone a proper name.

I now turn to the question of why we should or should not be prepared to abandon the 'ground level' form of such sentences. What argument is there to support the view that the 'ground level' form of such sentences is to be abandoned in favour of sentences such as those depicted in Strawson's diagram given above (*see* p. 119), and in dealing with this question we may also deal with Strawson's emphasis on the passage from Quine, which I said I would defer.

As I read it, Strawson's argument for the case that we should abandon the ground level form of such sentences in favour of an account in terms of sentences invoking a two place predicate '---- is ----', '---- performs ----', '---- exemplifies ----', together with proper names and 'attribute-naming nouns', is that what is wanted in all sentences of a propositional kind is:

> an indication of propositionality, the verb form or whatever does its duty. It is no more necessary that sentences should contain predicates in the full sense which entails concept-specification than it is that they should contain subjects in the full sense which entails individual specification.[57]

Quine, in his comments that 'predicates are wanted in all sentences' is allegedly in confusion since he takes it that a requirement of being a sentence (of a propositional kind) is that it should contain a predicate 'in the full sense' which entails concept specification, but this is not so. Hence all we *need* for the formation of sentences of a propositional kind, which by contextual implication, Strawson means sentences of a subject-predicate propositional kind, are sentences which contain an indication of propositionality, the verb form, and *this* we have in the proposed new basic level form as instanced in: 'Socrates performs etc., swimming'; 'Socrates possesses bravery'. Since from these sentences we may by generalisation write: 'Socrates does something' or ('There is something Socrates does'); 'Socrates is something' (or 'There is something Socrates is'), so 'Socrates does something' (or 'Socrates is something') is just as good and *just as direct*, a generalisation from 'Socrates swims' (or 'Socrates is brave') as 'Someone swims' (or 'Someone is brave') is.[58]

This however will not do. Leaving aside difficulties which may arise specifically for Strawson's thesis advocated earlier that 'predicates, in our fundamental class of sentences have the double function just remarked on: the function of specifying the assigned concept *and* the function of indicating propositional assignment',[59] it will not do to say that *all* that is wanted in sentences which express propositions of a subject-predicate kind, is an indication of propositionality - 'the verb form or whatever does its duty'. Whilst the verb form is a necessary, though not sufficient condition of propositionality - we may have the verb form in clauses within a sentence yet a clause itself may not be taken to 'indicate propositionality' - the verb form, where it is the main verb within a sentence, is a necessary but not sufficient condition for determining that the sentence is, expresses, a proposition of a subject-predicate kind. The verb form, in crude, is required in all sentences of a propositional kind as 'an indication of propositionality'; the mere presence of a verb form then does not and cannot distinguish sentences, propositional sentences of a subject-predicate kind. This would seem so obvious that it is almost incredible that Strawson should have missed it. An explanation is called for. I give this as follows. Whilst it is not true to say with Quine that 'predicates are wanted in all sentences' in the sense of predicates 'in the full sense which entails concept specification' in order for a sentence to be of a propositional kind and Strawson is correct in this, he unfortunately moves from: (a) talking about what is wanted in *all* sentences of a propositional kind (viz., an 'indication of propositionality') to (b) a thesis about *all that is wanted* in sentences of a subject-predicate kind to distinguish them as sentences of such a kind. We may explain his failure to see or acknowledge that the verb form is *insufficient* to distinguish sentences of a subject-predicate kind because he engages in an 'operator shift' from:

(i) All sentences of a propositional kind require the verb form as an indication of propositionality;

to:

(ii) All that is required in sentences of a subject-predicate propositional kind to distinguish them as sentences of that kind, is the verb form, an indication of propositionality.

If Strawson were to reply here that it is not *any* verb form which is required in sentences of a subject-predicate kind, but a verb form which is a one-place predicate, as opposed to a two-place predicate, then he is faced with the objection that it is precisely a demand of his thesis that: (a) we may quantify into the predicate place; (b) all we require for sentences of a subject-predicate propositional kind, the verb form required, *is* two-place.

On these grounds then I see no good reason for and good reason against abandoning what Strawson calls the 'ground level' form of such sentences. To hold that the 'ground level' or 'basic level' form of such sentences is really a sentence invoking a two-place predicate is: (i) based on a slip in reasoning which yields the result of one not being able to distinguish subject-predicate propositional sentences from other propositional sentences; (ii) commits one to the implausible thesis that predicates are at least writable as names, even 'proper' names.

Rejecting Strawson's arguments to the contrary then, I hold with Quine that we may *not* quantify into predicative position; predicative position is inaccessible to quantifiers. Invoking my earlier position concerning sortals, I hold the following *vis-à-vis* the Strawson/Quine dispute:

(A) By generalisation such sentences as 'Socrates swims'; 'Socrates φ's'; 'Socrates is brave'; 'Socrates is φ'; yield:

'There is something, some *F*, who (which) φ's' (where '*F*' is a sortal);
'There is something, some *F*, that is φ' (again, where '*F*' is a sortal).

(B) but that they do *not* yield:

'There is some *A* that *x* does';
'There is some *V* that *x* is'.

However from the fact that I have supported Quine in the Strawson/Quine dispute it does not follow, as is clear from what I have said earlier concerning my own explanation of the subject-predicate distinction and my comments in F.1. above, that I accept Quine's explanation of the subject-predicate distinction. It has been necessary for me to attempt to refute Strawson since I could not consistently hold both: (i) that the question of presupposition or implication of existence cannot arise in relation to an expression in predicative position and; (ii) that one may quantify into predicative position.

G. The Issue of a *Criterion* for the Subject-Predicate Distinction

I now turn to the issue of a *criterion* for the subject-predicate distinction. I hold that we must distinguish between an *explanation of*, in the sense earlier expounded, and a *criterion for* since to produce a criterion for is to answer the question: 'How do we *determine* whether an expression is occurring in subjective or predicative role in the context of an subject-predicate proposition?' Or, to speak in terms of 'positions', 'How do we *determine* whether an expression is occurring in subjective or predicative position in the context of an subject-predicate proposition?'.

G.1. Having defended Quine against Strawson to the effect that predicative position does not admit of quantifiers, it might be thought that Quine's explanation of the subject-predicate distinction in a form suitably amended to cater for my position on sortals, might be adopted to yield a *criterion* which I could advocate. This would be on the principle that positing X as a *criterion* for something does not entail offering X as an *explanation* for that thing. Such a thought would be expressed as follows:

(A) '*x*' occurs in subjective position in the context of a subject-predicate proposition *iff* from:

(i) '*x* is φ'
(ii) '*x* φ's'

we may derive an existential proposition:

(i)(a) 'There is something, some *F*, that is φ'
(i)(b) 'There is something, some *F*, who, which φ's'

(B) whereas φ occurs in predicative position in such a context *iff* from '*x* is φ' or '*x* φ's' we may not derive any such corresponding existential proposition 'There is some φ that *x* does' or 'There is some φ that *x* is'.

Now whereas at first blush (B) could arguably serve as my criterion for whether φ is occurring in predicative position or role, it is not a fully satisfactory account. My account (explanation) of an expression in predicative role, namely that the question of presupposition or implication of existence cannot arise in relation to such a role entails that from '*x* is φ' or '*x* φ's' we may not derive any corresponding existential proposition 'There is some φ that *x* does', 'There is some φ that *x* is', entails more than this, namely, that it is absolutely impossible that we should derive any such corresponding existential propositions. To claim that we may derive any such corresponding existential propositions fails to acknowledge the fragmentary nature of a predicable and hence a predicate. To cater for this aspect of my explanation I initially put forward as a criterion for occurring in predicative role the following:

I. An expression occurring within the context of an subject-predicate proposition is occurring in predicative role *iff* it is not the case that we may intelligibly derive an existential proposition: 'There is some φ that *x* does' or 'There is some φ that *x* is' from '*x* is φ' or '*x* φ's'.

This may be expanded as:

II. An expression occurring within the context of an subject-predicate proposition is occurring in predicative role *iff necessarily it is false that*, as opposed to *it is necessarily false that*, we may derive an existential proposition 'There is some φ that *x* does' or 'There is some φ that *x* is' from '*x* is φ' or '*x* φ's'.

The insertion of 'necessarily it is false that' marks the point above that it *is absolutely* impossible that we should derive any such corresponding existential propositions.

It is possible to express the criterion just given in a less strong fashion, namely in the form: 'An expression ... in predicative role *iff* it is *impossible* that we may derive an existential proposition ...'. It depends on exactly how we are to take Quine's position that predicative position is inadmissible to quantifiers. If we read this as it is false that quantifiers may occupy predicative position, then my criterion would be expressible in the way just given. If we read it as it is impossible that quantifiers should occupy predicative position, then my criterion would have to be expressed in the form given in (II) above. I consider it correct to express Quine's position in terms of it being impossible that quantifiers should occupy predicative position since: (i) he speaks in terms of predicative position being *inadmissible* to quantifiers, not simply 'not admitting' quantifiers; (ii) in connection with his explanation of 'logical subject' or 'subject term' he speaks of a name *coherently* standing in the place of a variable in predication. That is, Quine is concerned with *coherence*, not truth or falsity. On these grounds I give (II) above as the criterion for occurring in predicative role or position.

G.2. Whereas (B), at first blush, could arguably serve as my criterion for whether an expression is occurring in predicative role or position, (A) could not, even at first blush, serve as my criterion for an expression occurring in subjective role. The reason for this is clear. Quine actually analyses out a subject-predicate proposition such as 'Socrates swims' in terms of an existential proposition. Ordinary proper names are eliminable in favour of the variables of quantification. This position is clearly incompatible with my thesis that an expression in subjective role *presupposes* a true proposition of existence to the effect that there is at the least and the most one F such that ... I thus offer as a criterion for 'occurring in subjective role' an adaptation of my earlier explanation of an expression occurring in subjective role:

An expression occurring within the context of an subject-predicate proposition is occurring in subjective role *iff* it is a *condition* of any proposition being expressed that there should be a true existential proposition to the effect that there is at the least and the most one F such that ... viz., a true existential proposition:

$$[(\exists x)S]\ [\varphi x \cdot ((y)S)(\varphi y \rightarrow y{=}x)]$$

G.3. In G.l. above, I considered the thought that I might use Quine's explanation in a suitably amended form (as later given), as my criterion. This I said would be on the principle that positing X as a criterion does not entail offering X as an explanation. I did not there challenge this principle but I do so now. We may offer as a criterion for someone being a member of the Church of England that he has been baptised according to the formularies of the Book of Common Prayer but if we are then asked 'What differentiates a member of the C of E from e.g. a member of the Church in Wales or a Roman Catholic?', *part* of our answer will be 'having been baptised according to the formularies of the Book of Common Prayer'. In other words what appears in one context as a criterion appears in another as at least part of an explanation.

H. An Account of the Relations between Referring, Identifying and Existence

In Section E I offered an account of the relations between sortals and the subject-predicate distinction; it is now necessary to offer some account of the relations between referring, identifying and existence for these are key concepts invoked in that distinction. I have argued that: (1) referring, and hence occurring in the role of logical subject, rests on sortal identification in the sense: (a) being picked out by reference to the definite form of a (some) sortal; (b) specified by reference to a (some) sortal; (2) that referring and hence occurring in the role of logical subject, presupposes existence. How then are we to spell out the relations between referring, identifying and existence?

H.1. We might offer the following as a first shot.

(A)(1) Referring rests on, presupposes, sortal identifying in the sense of:

(a) sortal 'picking out';
(b) sortal specification;

(A)(2) Sortal identification in the sense of sortal 'picking out' implies some existential proposition of the form:

$$[(\exists x)S]\ [\varphi x \cdot ((y)S)(\varphi y \rightarrow y{=}x)]$$

(A)(3) *ergo* referring implies existence.

This attempt to establish a relation between, referring and existence via identifying presents me with a paradox; for the conclusion that referring implies existence is downright incompatible with the thesis that referring presupposes existence - which I have already advocated.

(B) I am forced to acknowledge, it seems, that referring *both* implies and *presupposes* existence. This paradoxical situation may be fully spelt out as follows:

(B)(1) Referring presupposes existence; occurring in the role of logical subject presupposes existence;

(B)(2) Referring in general presupposes, rests on sortal identification in both the senses (a) and (b) introduced in (H) above, *ergo* so does occurring in the role of logical subject;

(B)(3) Sortal identification implies some proposition of existence of the form given in (A)(2) above;

(B)(4) *ergo* referring both presupposes existence from (1) and implies existence from (2) and (3) since sortal identification implies existence and referring presupposes sortal identification.

(Schematically: A presupposes B
B presupposes C
C implies B
A implies B [and yet also presupposes B])

This argument rests on the general principle that in that a proposition presupposes another proposition it presupposes all that that other proposition itself implies. I do not deny the general value of this principle but I do not think that it can be used here to generate a paradox for my position for the following reasons:

(i) I am not talking about *propositions* but about activities, operations, on the one hand, and propositions of existence on the other;

(ii) More importantly, (A) above seeks to establish a relation between referring and existence via identifying. I claim that it is a mistake to seek to establish any relation between referring and existence via identifying since it is within the context of 'modes of speaking' that referring is posterior to, presupposes identifying. Referring is posterior to, presupposes identifying *as* a mode of speaking or as a mode of linguistic activity; but it is *not* within the context of 'modes of speaking' or 'modes of linguistic activity' that referring presupposes existence. Existence itself is not a mode of speaking and neither is a proposition of existence. It would be nonsense to say e.g. that existence is prior to *referring or identifying* or posterior to *referring or identifying* i.e. to such modes of speaking or linguistic activity. This is not to say that we may not ever speak of relations between modes of speaking or modes of linguistic activity and existence; we clearly may, but what we may *not* do is present, as in (A) and (B) above, a form of argument which is a combination of both: (a) relations between modes of speech (or modes of linguistic activity); (b) relations between modes of speech (or modes of linguistic activity) and existence, for *there is no context which covers both.*

In the light of this I hold that the most that can be offered is an account of the relation as follows:

At one level - the level of modes of speech or linguistic activity:

(1) Referring presupposes, rests on sortal identifying; as a mode of speech activity it presupposes another mode of speech activity; and likewise 'occurring in the role of logical subject' presupposes, rests on, another mode of logical operation;

At another level:

(2)(a) Sortal identifying, sortal 'picking out', sortal discrimination, (mode of speech activity), implies a proposition of existence;

(2)(b) Referring presupposes existence and likewise occurring in the role of logical subject presupposes existence;

but from these accounts we may not go on to draw any conclusion such as 'Referring implies existence' or seek to establish a relation between referring and existence via identifying.

At this point it might be interjected that there is another paradox waiting in the wings. The argument would run as follows:

(i) Referring presupposes sortal identifying;
(ii) Referring also presupposes existence;
(iii) *ergo* sortal identifying presupposes existence.

Yet I have said that sortal identification implies existence; hence I am, it seems, committed to holding both that sortal identification presupposes and implies existence. This argument also fails for the same reason that the earlier one did; it assumes that there is a context which covers *both*: (a) relations between modes of speech activity or modes of linguistic activity; (b) relations between modes of speech or linguistic activity and existence. The above argument seeks to establish a relation between referring and sortal identifying via existence and the failure of this argument, when taken in conjunction with the failure of the earlier argument, indicates that we may not seek to establish relations between the two levels (1) and (2) above. Indeed, one may go even further than this and add that the fact that paradoxes arise *shows that* it is a fundamental mistake to seek to establish relations between referring and existence via identifying or relations between referring and sortal identification via existence; to engage in such activity reveals a type error. We must stick to spelling out relations *internal* to each of the levels (1) and (2) above and not attempt to cross type barriers. Hence *necessarily* the *most* that can be offered is the accounts set out in (1) and (2) above. It is also to be noted that there is another argument against holding that sortal identification in the sense of sortal 'picking out' or sortal discrimination *presupposes* existence. This is the consideration that the definite form of a sortal in such a role is required, as I have earlier argued, for the very introduction of an expression in a referring role which *itself* presupposes existence and hence sortal identification in this sense cannot be said to presuppose existence.

H.2. Having thus set out all that it is possible to set out - namely, at one level the relation between (1) referring and sortal identifying and at another level (2) the relations between: (a) sortal identification in the sense of sortal 'picking out', sortal discrimination, and existence; (b) referring and existence; it will now rightly be observed that I have yet to make it clear what the relation is between sortal identification in the sense of *sortal specification* and existence. This I shall now attempt to do.

It cannot be said that ability to specify what something is *entails* existence. One may be able to produce a quite detailed account of what something is e.g. what God is, what we understand by 'God' (the 'nominal essence' of God) - or again to take the case of an individual - Oberon, the Red Queen, Red Ridinghood, Beelzebub - we may be able to produce quite a detailed account of the 'nature', offer an account of the 'nominal essence' of such individuals in the form:

'*a* is the *F* which ...'

but from our ability to produce such an account it clearly does not follow either that the sort of thing (God) or the individual thing (e.g. Beelzebub) exists. There is however an ambiguity in 'ability to specify what something is'. This may be read as either: (i) a given person's ability to specify on an occasion or series of occasions what something is; (ii) *our* ability to specify what something is. And (ii) can be written as *the* ability to be so specified, or, more simply, to be so specifiable. As regards the first reading, from the fact that a person is able to *correctly* specify, say, what something is it clearly does not follow that that thing exists. A person may correctly specify what God or the Devil is, that is he may give an account of God's nature or the Devil's nature which is fully in accordance with Christian teaching but from this it does not follow that God or the Devil exists; and similarly in the other cases already mentioned above. Thus not even *correct* specifying entails existence. Neither does a person's ability to correctly specify what something is, be said to *presuppose* existence; in analogous fashion we may invoke the 'God' and the 'devil' cases. Thus in that either (i) the definite form of a sortal or (ii) the indefinite form of a sortal occur or are used in a specificatory role by a person on an occasion, existence can be said neither to be presupposed nor entailed. Can existence however be said to be implied in the sense that there is an *implicit* claim to existence? I earlier said that there was such an implicit claim to existence in the case of a sortal occurring or being used in a 'picking out', discriminating, role. I do not hold the same thesis here for the following reason. In that a sortal occurs or is used on an occasion in a specificatory role, to serve to specify what something is, it cannot be said that there is an implicit claim to existence, rather in that it so serves to specify, is used by someone to specify what something is, it serves to specify *what it is for that thing to exist*. In that it correctly specifies what the X in question is, it offers a correct account of *what it is for that thing to exist*, though whether there *is* such a thing remains an open question. Thus the question of whether there is an implicit claim to existence is impertinent.

As regards the second reading of 'ability to specify what something is', it is also true to say that from the fact that X is specifiable, it does not follow that X exists. This is clear from examples already given; but in that a sortal in general serves to specify what something is it specifies *what it is for something to exist*. Our ability to specify what something is, the ability to be specified or more simply, to be specifiable by reference to a sortal, is what it is for that thing to exist, though whether there *is* such a thing remains an open question.

H.3. Having put my cards on the table as regards sortal identification in the sense of sortal specification and existence; in the crudest possible terms to exist is to be specifiable by reference to a sortal, it will now be commented that I seem to be advocating Aristotle's thesis that 'exists', in predicative position, as in e.g. 'Aristotle exists' is a redundant predicate. On this thesis for x, an individual, to exist is for x to be F, where 'F' is a predicate; and in that I am advocating Aristotle's position I am open to the charges that are levelled against him - most notably that he failed to recognise the quantifier sense of 'is' (sc. 'exists') and attempted to explain the quantifier sense in terms of the predicate sense. To elucidate my position *vis-à-vis* Aristotle. First, I do not wish to claim that for an individual to exist is for an individual to be an *F*, where '*F*' is construed as a *predicate* in the sense which I have propounded and discussed, for I have explicitly argued that sortals are *not* predicates or reducible to such. Neither do I advocate the thesis that for an individual to exist is for an individual to be an *F*, where '*F*' is a 'predicate' in any of the senses of 'predicate' set out by Strawson in his lists in *Individuals*. However in that Aristotle is advocating the thesis that '*x* exists' is writable, or primarily writable, as '*x* is an *F*', where '*F*' is a sortal, then my thesis is to that extent Aristotelian. Secondly, whilst Aristotle did indeed fail to recognise the quantifier sense to press this charge against a pioneer is to fail to do full justice to him, as Owen has argued.[60] I do recognise the quantifier sense and whilst it can be argued that there is a sense in which the 'redundant predicate' sense of 'is' (sc. 'exists') rests on the quantifier sense in a way to be expounded below, there is also a sense in which the quantifier sense rests upon the redundant predicative sense in a way to be developed below. Thirdly, I shall argue that sortals are required for an account of the quantifier sense of 'exists'.

In what sense can the 'predicative' sense of 'is' be said to rest on the quantifier sense? Consider the following examples. For Aristotle, for Health to exist is for health to be a state; for Rheumatism to exist is for rheumatism to be a disease; for length to exist is for it to be a magnitude. It might be thought that the quantifier sense is indirectly invoked as follows: whilst (e.g.) for Health to exist is for health to be a state; for an *individual* health to exist is for *there to be* ($\exists x$) someone or something who (that) is healthy; whilst for Rheumatism to exist is for it to be a disease, for a particular case of rheumatism to exist is for *there to be* ($\exists x$) something which is rheumatic; for a particular length to exist is for it *to be* ($\exists x$) the length of something. Aristotle would not allow this. For him, for the individual case of health to exist, for there to be an individual state of health, is precisely for 'is healthy' to be truly predicated of someone or something. For there *to be* ($\exists x$) the individual length is for that length to be truly predicated of some individual. He in effect analyses out the quantifier sense of 'is' in terms of the redundant predicative sense. In what sense then, if any, may the predicative sense of 'exists' be said to rest upon the quantifier sense? The following will be advanced. In order for e.g. 'is healthy' to be truly predicated of someone or some length to be truly predicated of something, there must *be* ($\exists x$) some individual of which they *are* truly predicated. This I think has to be conceded, but it is of central importance to my thesis that in that the 'predicate' sense of 'exists' can be said to rest on or presuppose the quantifier sense, it can only be said to so rest or presuppose in that we can write for 'Some individual', 'Some *F*', even though it is contingent which *F*.

In what sense can the quantifier sense be said to rest on the 'predicative' sense? The sense in which this may be said to be so is that our ability to formulate propositions of existence invoking the quantifier requires that we are able to say what something is - in general what things are - and this involves the occurrence or use of a sortal in specificatory role. The quantifier sense rests on the 'predicative' sense in that it is only in that we can say what something is that we can formulate the appropriate quantificational form. For example, it is only in that we are able to say that Jehovah is a *God*, that Beelzebub is a *devil*, Oberon is a *fairy*, Jemima is a *cat* that we are able to express the form: $(\exists x)Fx$. In crude, if we are unable to say in any way what '*x*' is, then we are unable to say what it would be to propound its existence; to propound that there is such an individual. Again, it is only in that we are able to say e.g. that a cat is an animal, that a snake is a reptile, that a robin is a bird etc., that we are able to express the form: $(\exists x)Fx$. This will immediately be denied. It will be said that e.g. 'Cats exist' is simply writable as 'There are cats' $(\exists x)Fx$ without any reference to what a cat is and I can assert that there are cats without knowing what a cat is. To claim however that 'There are cats' is simply writable as $(\exists x)Fx$ - 'There exists an object *x* such that it is a cat' - is open to the objections I raised in Chapter I. There are *no* such X's as 'absolutely simple' objects. And to claim that one may assert that there are cats without knowing what, at least in general, a cat is meets with the objection that to assert that there are X's entails that one could pick these X's out if called upon to do so or one could at least be called upon to describe them. But if there were no sense in which one knew what these X's were one could not carry out such an activity. Hence I hold that for (e.g.) 'There are cats' we must write: 'There is an *F* (sortal) such that it is a cat'; in this case, 'There is an animal such that it is a cat': in my notation $[(\exists x)F]F_1x$.

My reformulation of the quantifier sense will call forth the following objections immediately: (i) in my reformulation, I myself allow sortals in predicative position; 'cat' in the formulation 'There is an animal such that it is a cat', $[(\exists x)F]F_1x$, occurs in predicative position. Yet I have all along contended that sortals are not predicates. (ii) why cannot we simply say that 'object' in 'There is an object such that ...' $(\exists x)Fx$, is *simply* the formal expression of a sortal?; that it is a 'formal concept' representing a sortal or a range of sortals? - or which is manifested by a sortal or range of sortals?

My reply to (i) is that my *general* thesis does not commit me to saying that sortals may never occur in predicative position or as predicates; in the example just given I have to concede that 'cat' does. But this does not controvert my thesis that sortals are not predicates since: (a) in that 'cat' occurs in predicative position above, it is writable as 'feline' and in that it is so writable it ceases to be a sortal; (b) a condition of 'cat' occurring in predicative position is that it rests on, presupposes, the occurrence of *some sortal*, in this case 'animal'. That is, a condition of a sortal expression appearing in predicative position, is that there already *is* a sortal *not* in predicative position. Thus it cannot be said that sortals *in general* as a class, in principle are reducible to predicates, may *always* occupy predicative position. My reply to (ii) is that: (a) I have already commented on the difficulties which arise in connection with the 'formal concept' "object" as a formal concept; (b) if we say that 'object' simply does duty for a sortal or range of sortal then we may intelligibly replace 'object' by '*F*' or 'There is an object' by 'There is an *F*' but *just this* we may not do since the use of 'object' in 'There is an object such that it is *F*' demands that always and necessarily we write a sortal as a predicate, in predicative position.

It may further be commented that my particular account of e.g. 'Cats exist' in terms of 'Cats are animals (of a certain sort)' and my revision of $(\exists x)Fx$ in terms of $[(\exists x)F]F_1x$, lands me in a well known difficulty with which Aristotle was faced in his analysis of general propositions of existence. I am not unaware of this and shall be returning to the matter later (section I.3).

I now turn to my third thesis set out above - namely that sortals are required for an account of the quantifier sense of 'exists'. My case for this is that one may only introduce the notion of 'an object' in that one may introduce the notion of an F (sortal). Yet the quantifier sense of 'exists' as traditionally given and expressed fails and necessarily fails to recognise this. It introduces us to the notion of 'an object' which necessarily is of no sort since all sortals occupy predicative position. Let me expand on this. I am holding that: *Necessarily* to be an individual, an object, is to be specifiable by reference to some sortal *as opposed* to: to be an individual, in the sense of an object, is *necessarily* to be so specifiable. That is, *Necessarily it is true that* 'There is', $(\exists x)$ has a reference to some sortal, *as opposed to*, concerning 'There is, $(\exists x)$, it is *necessarily true* that *it* has a reference to some sortal. The *first* account yields that 'There is'; $(\exists x)$, is writable as 'There is an F (sortal) such that ...' $[(\exists x)F]$; the *second* yields that 'There is'; $(\exists x)$, is writable as: 'There is an object x such that *it* is F'; '$(\exists x)Fx$'. Invoking the language of Quine's 'range indicators' I am holding: *Necessarily* to be a value for a variable is to fall within the scope of a range indicator, *as opposed to* (say) to be an individual *is necessarily* to fall within the scope of a range indicator. On the first account we have the position that it is *impossible* to eradicate range indicators; on the second it is theoretically possible to eradicate range indicators since to fall within the scope of a range indicator is merely a necessary predicate of a value for a variable. On this account we can have the notion of 'a value for a variable' *independently* of *any* range indicator, for we have already spoken of a necessary predicate of *it*, and this leads to the treatment of range indicators, sortals, as predicates.

Not only do I hold that: (a) *Necessarily* it is true that 'There is'; $(\exists x)$, has a reference to some sortal but also that; (b) *Necessarily* for an individual, a particular object, to exist is for it to be specifiable by reference to some sortal *as opposed* to: for an individual, an object, to exist is *necessarily* for *it* to be specifiable by reference *to* some sortal. In other words the contrast I am drawing is between: (i) *Necessarily* to be a particular, an object, is to be of a certain sort; and; (ii) to be a particular, an object, is *necessarily* to be of a certain sort. The first gives us a necessary truth, a necessary proposition, in the sense that 'necessarily' operates on the whole proposition: 'To be a particular ...'; the second gives us a *predicate* concerning any particular. To revert again to Aristotle. He holds e.g. that 'Socrates exists' comes out as 'Socrates is an F'. In that I subscribe to (i) above I am holding that 'Socrates exists' comes out as: *Necessarily* 'Socrates is an F'. This does not entail that Socrates is *necessarily* F since '*Necessarily* Socrates is F' does not entail 'Socrates is *necessarily* F'.

If it is now pressed: On what grounds do you hold: (a) *Necessarily* it is true that 'There is'; $(\exists x)$, has a reference to some sortal?; (b) *Necessarily* it is true that for an individual (an object) to exist is for it to be specifiable by reference to some sortal? (as opposed to holding the relevant contrasting theses), I would appeal to what I have already said in Chapter I. To reiterate. It is that sortals, in their own right, furnish principles of counting and distinction for objects: it is in that we have the notion of a sortal, that we have the notion of 'an object' and in the light of my earlier argument it is *only* in that we have the notion of a sortal that we have the notion of 'an object'. To hold the contrasting thesis commits one to the view that we may have the notion of 'a particular', 'an object' *independently* of any sortal. What *this* commits one to, or might be held to commit one to, will be the subject of much discussion throughout the rest of this book.

I have commented earlier (H.2) that sortal specification cannot be said to entail existence and by a parallel consideration intimated that sortal specification cannot be said to *presuppose* existence. This latter claim may still produce a sceptical worry. 'Surely in that e.g. "man" serves to specify what Socrates or Athanasius is, or "cat" what Jemima is, then it is presupposed that Socrates, Athanasius and Jemima exist?' However the sceptical worry expressed here supposes that there could be such a thing as Socrates, Athanasius and Jemima existing *independently* of there being some *F* (whatever it is) that Socrates or Athanasius and Jemima is. That is, it supposes that Jemima, Athanasius, Socrates, exist *independently of any specification*, *independently* of *any* use of a sortal in specificatory role; but just this, I have argued is absurd. One who is tempted to such a thesis would argue that '*We* classify things according to our interest in them' i.e. would be tempted to a 'relativist' position as Copi is in his paper 'Essence and Accident'.[61] On such a position we are faced with the question: 'What (Which) things?' and in seeking to answer *this* the relativist is engrossed in a contradiction or has to posit the existence of 'pure' objects as I have argued in my paper on 'Essence and Accident'.[62]

I have already had considerable recourse to the distinction between: (A) *Necessarily A* is *P*; and (B) *A* is *necessarily P* and I shall be having considerable recourse to it anon, but for the present it is of value to note that: (I) it is in that there has been confusion of: (a) *Necessarily* it is true that 'There is'; $(\exists x)$, has reference to some sortal; and; (b) concerning 'There is'; $(\exists x)$, it is necessarily true that *it* has a reference to some sortal, that we get the position that the quantifier sense of 'exists' is writable as $(\exists x)Fx$ as opposed to $[(\exists x)F]F_1x$: (II) it is in that there has been a confusion of (c) *Necessarily* it is true that for an individual object to exist is for it to be specifiable by reference to some sortal and (d) for an individual, an object, to exist is *necessarily* for *it* to be specifiable by reference to some sortal, that we get the position that a sortal is construed as a *predicate*; it being true of an individual, an object, that *it* is necessarily specifiable by reference to some sortal. More generally, in relation to my later discussion of Strawson in 'Particular and General',[63] it is true of every particular that *it* necessarily falls under some general *term*; that *it* is necessarily an instance of something general, *as opposed to*: *Necessarily* every particular is an instance of something general.

To conclude this section: in the light of the arguments presented I hold that sortals *qua* performing a specificatory role or function may not be said either to imply or presuppose existence; they are required for the articulation of propositions that are implied by a sortal in a 'picking out' role or presupposed by a name in a referring role.

H.4. In the above section I have attempted to spell out the relation between sortal specification and existence. It will have been noted that this has been limited to sortal specification in the sense of the indefinite form of a sortal - a god, a cat etc. What then is the relation between sortal specification in the sense of the use of the *definite* form of a sortal and existence? I would first draw the reader's attention to what I said in connection with the use of the definite form of a sortal in a 'picking out', 'discriminating' role. At the end of H.1. I argued that the definite form of a sortal in such a role could not be said to *presuppose* existence since the definite form of a sortal in such a role is required for the very introduction of an expression in a referring role which itself presupposes existence. In parallel fashion here I contend that definite form of a sortal in specificatory role cannot be said to presuppose existence since the use of a sortal in such a role is required for the very introduction of 'referring', which itself presupposes existence. In order to refer to something, an object, we must be able to identify it by reference to the definite form of some sortal, as I argued in Chapter II. But there is yet a further argument against saying that the definite form of a sortal in specificatory role presupposes existence. Following what I have said in H.3. above, what it is for an individual object *to exist* is for it to be specifiable by reference to the definite form of a sortal. I have, in effect, argued above that e.g. for Jemima to exist is for Jemima to be an *F*, but to be an *F* is to be a particular *F* - an *F* specifiable by reference to the definite form of some sortal.

I have further argued earlier that the definite form of a sortal in a 'picking out', 'discriminatory', role implies some existential proposition: what about the definite form of a sortal in a specificatory role? Can this use also be said to imply some existential proposition? My answer to this is clearly in the negative; such a use cannot be said to imply any existential proposition since in that we specify what something is by the use of the definite form of a sortal we are saying what it is for that thing to exist. 'Jemima is the tabby cat which ... (plus a list of the features which serve to distinguish Jemima from other cats)'; for Jemima to exist is for Jemima to be the cat with some distinguishing features A, B, though it might well be argued that it is *contingent* which ones. It is conceivable that Jemima could still exist, remain the same cat, without just those features but she could not exist without *any* distinguishing features nor could she remain the same cat without *any* such distinguishing features. All I need for my thesis is that for Jemima to exist is for Jemima to be specifiable by reference to the definite form of *some* sortal completed by reference to some distinguishing features.

At this point some strong objections will be entered. It will be said that, even though the definite form of a sortal in specificatory role is required for the very articulation of the propositions of existence implied by the use or occurrence of the definite form of a sortal in a 'picking out' role, and for the articulation of the true existential propositions presupposed in referring, this is by no means the end of the matter. A person may use e.g. 'The last man to be executed in public in England' in a specificatory role in answer to the question: 'To whom are you referring by "Lord Lovat"?' or 'Which man do you mean, which man are you referring to in your use of "That man"?'; 'Which apple are you referring to in your use of "That apple"?' ('The first apple in the second row'.) And in that the definite form of a sortal answers such questions surely: (i) reference is presupposed; (ii) in that reference is presupposed, existence is presupposed. So have we not a case for holding that at least in one use of the definite form of a sortal in specificatory role, existence is implicitly presupposed? This type of case, this use of the definite form of a sortal cannot be denied. However in that the definite form of a sortal occurs or is used in such a case it has the use of, occurs in the role of, *clarifying* the reference by specifying the reference and in this context reference *has been made* but *clarification* of what is being referred to is called for. This use, or occurrence, of the definite form of a sortal in a specificatory role i.e. *clarificatory* role rests on:

(i) the use (or occurrence) of the definite form of a sortal to specify reference or intended reference prior to any reference *being* made, for as I have argued, referring to something presupposes being able to specify it in *some* way. That is, that it is a presupposition of being able to refer to an object on an occasion that one is first able to specify it - issue some definite specification (in more customary terms issue some 'definite description') and the use of the definite form of a sortal in a *clarificatory* role or use rests on such a use in that it supposes reference has been made. Thus the sense in which it may be held that the use of the definite form of a sortal in specificatory role *presupposes* existence is *secondary*.

(ii) the use or occurrence of the definite form of a sortal in specificatory role in the context of an existential proposition just in that the 'clarificatory' use, occurrence, presupposes existence.

Now it will be objected: is it *still* not the case that whilst referring to something presupposes being able to specify it, does not the ability to specify something presuppose that that thing *exists*; that *there is* something to be specified? I have already argued against a parallel objection in relation to specifying via the indefinite form of a sortal. All I shall add here is that if we adopt the position that existence is presupposed we are in deep trouble. Affirmation of this thesis entails its own denial, for in order to state the thesis we *already* have to say what it is that is presupposed to exist and this *itself* involves sortal specification.

To briefly summarise then on the relations between (1) referring and existence; (2) referring and identifying; (3) identifying and existence. I have argued: (1) that referring presupposes existence; (2) that referring presupposes (rests on) identifying both in the sense of: (a) sortal picking out or discrimination; (b) sortal specification in that it presupposes the truth of a proposition to the effect that there is just one *F* such that ...; (3) that sortal identifying in the sense of sortal 'picking out', discrimination, *implies* some existential proposition to the effect that there is just one *F* such that ...; that sortal identification in the sense of sortal specification can be said (in general) neither to presuppose nor imply existence but is required both for: (A) the very formulation of propositions of existence implied by sortal identification in the sense of sortal 'picking out'; (B) the very formulation of propositions of existence whose truth is presupposed by an expression in referring (logical subject) role within the context of an subject-predicate proposition. We may note here that even if one were to accept Strawson's account of, analysis of, 'logical subject' in terms of 'identifying' in the sense of 'picking out', it would *still* be the case that sortals are required for the notion of 'occurring in the role of logical subject' as thus analysed. Hence on *either* account of 'occurring in the role of logical subject' sortals are required.

I. Explanations of the Subject-Predicate Distinction in the sense of Answers to Two Questions

I now turn to the question of an explanation of the subject-predicate distinction in the sense of an answer to the questions:

(1) What makes the subject-predicate distinction *possible*?
(2) What *underlies* the subject-predicate distinction?

I.1. My answer to the first question is by now clear; it is the presence and recognition of sortals for I have argued: (i) that at one level or aspect of language i.e. language viewed as a symbolic system, sortals are required for the introduction of both names and predicables (predicates) in so far as one is speaking of the names of particulars in the sense of 'objects'; (ii) that at another level or aspect of language i.e. speech, sortal identification is prior to referring and predicating and indeed, in the senses explained in Chapter II, prior to other forms of identification; (iii) that in regard to the question of an *explanation* of the distinction in the sense discussed earlier, sortals in a specificatory role are required for the very articulation of the propositions of existence which in the one case are presupposed and in the other case cannot be said to be presupposed by expressions occurring in, or being used in, subjective and predicative role respectively; (iv) that parallel considerations apply in relation to the formulation of a *criterion* for the subject-predicate distinction.

My answer to the *second* question, construed as the question: 'Is there any *kind* of proposition in a language which employs proper names and predicables (predicates), in a mode of discourse which introduces particulars in the sense of objects, the truth of which is necessary for the introduction of propositions of a subject-predicate kind in that language, mode of discourse?' is equally in terms of existential propositions which invoke sortals in the form:

[$(\exists x)F$]..., as opposed to Strawson's answer in *Individuals* and 'Particular and General' which I shall later argue: (a) cannot in fact be sustained; (b) is based on a confusion between (i) *necessarily* to be a particular is to be a particular of a certain sort or kind and (ii) to be a particular is *necessarily to be of a certain sort or kind*; (c) is misconceived.

It will have been noted that I have just introduced the thesis: *Necessarily*: to be a particular is to be of a certain sort or kind i.e. to be specifiable by reference to a sortal *or* by what I have earlier described as a 'matter' term, 'mass' term, precursed in a certain way or a Geachian 'substantival general' term precursed in a certain way. Or, more correctly, by a term, symbol, *formed from* a 'matter' term, 'substantival general' term (or symbol) which furnishes a principle of counting and distinction in its own right. And this brings me to the important modification of my thesis alluded to in various places above. To be consistent with what I have maintained in Chapter I viz., that sortals in the sense introduced are not reducible to principles of counting and distinction formed from 'matter' terms and *equally* that such principles of counting and distinction are not reducible to, nor can be said to rest on sortals, I *equally* have to allow that: (a) what makes the subject-predicate distinction possible is the presence and recognition of principles of counting and distinction formed from 'matter' terms; (b) what underlies the distinction are existential propositions articulated by reference to 'matter' terms in the form: [$(\exists x)M$]M_1x ... : 'There exists a (parcel, lump, etc.) of matter (metal, liquid, rock) such that it is gold, water, stone etc., which ...'. That is, I have to allow that we may introduce or have subject-predicate propositions in a language, form of discourse, which invokes *only* particulars of a certain kind, as opposed to of a certain sort. I have not discussed the relevant working out of this imposition in the above discussions but concentrated on the issue of sortals in the sense introduced since I have been concerned with the subject-predicate distinction in the general context in which we speak of *objects*, not in the context in which we speak of parcels of matter. As I remarked in Chapter I, it is in accordance with the normal philosophical use of 'particular', 'individual' that in that we so speak we are speaking of *objects* but we *may*, of course, speak of 'particulars', 'individuals' in the sense of *parcels of matter*. In that I have spoken of sortals as prior and requisite in regard to points (i)-(iv) above, I also have to say that in a language, system of discourse which is limited to speaking of parcels of matter or principles of counting and distinction formed from 'matter' terms or 'substantival general' terms, then in such a language, in such a system of discourse, *such* items are prior and requisite. What we may *not* do as, by implication, I have argued in Chapter I is to claim that *either*:

(A)(1) recognition and use of the subject-predicate distinction together with:

- (i) an explanation of this distinction in the sense discussed earlier;
- (ii) a criterion for the distinction; in a *system in which we speak of parcels of matter*, is dependent on, i.e. logically dependent on:

(B)(1) recognition and use of the subject-predicate distinction together with the attendant explanation and criterion in a *system in which we speak of objects introduced by reference to sortals*;

(C)(1) the truth of the existential propositions which may be said to underlie the subject-predicate distinction as manifested in a language or system of discourse in which the proper names of parcels of matter occur or are used in the role of logical subject;

rest on, is necessarily dependent on, the truth of some existential proposition invoking a sortal such that one could say that the *basic form* of existential proposition was an existential invoking a sortal;

or:

(A)(2) recognition and use, manifestation of, the subject-predicate distinction together with: (i) an explanation of this distinction in the sense discussed earlier; (ii) a criterion for the distinction in *a system in which we speak of objects introduced by reference to sortals*, is dependent on, i.e. logically dependent on:

(B)(2) recognition and use, manifestation of, the subject-predicate distinction together with the attendant explanation and criterion *in a system, mode of discourse in which we speak of parcels of matter*;

(C)(2) the truth of the existential propositions which may be said to underlie the subject-predicate distinction as used, manifested, in a language or system of discourse in which the proper names of objects occur or are used in the role of logical subject;

rest on, is necessarily dependent on, the truth of some existential proposition invoking a 'matter' term such that one could say that the *basic form* of existential proposition was an existential invoking a 'matter' term.

It may be commented that a difficulty arises for (C)(2) above which does not arise for (C)(1). It would, for example, be implausible to argue that a presupposition of 'Ullswater' occurring in subjective role in:

'Ullswater is seven miles long'

(where 'Ullswater' is the proper name of a stretch of water) is that there exists some lake, some *F* which is a stretch of water. And this is not simply for the reason that it is only contingent that a stretch of water is an *F*, but more importantly for the reason that such an argument, such a position, would demand that the introduction of propositions of existence mentioning a particular introduced by reference to principles of counting and distinction formed from a 'matter' term, *logically rested* on the introduction of a proposition of existence invoking a sortal - that we *could only* introduce the one in that we first introduced the other - which would be to *deny* that principles of counting and distinction formed from 'matter' terms were principles of counting and distinction in their own right.

The difficulty for (C)(2) is as follows. It might be said that 'Aristotle' in 'Aristotle is wise' (where 'Aristotle' is the proper name of a man) presupposes the truth of an existential proposition invoking a 'matter' term such as:

> 'There exists a parcel of matter, a parcel of flesh and bone'

or even:

> 'There exists an animated parcel of flesh and bone'

since it is not contingent that a man is a parcel of flesh and bone and an animated parcel of flesh and bone. However from the fact that it is not contingent that if X is a man then X is a parcel of flesh and bone, it does not follow that and indeed cannot follow, that the introduction of propositions of existence invoking the sortal 'man' logically rests on the introduction of propositions invoking 'matter' terms; that we could only introduce the former in that we may first introduce the latter, for this would be to deny that principles of counting and distinction furnished by sortals were principles of counting and distinction in their own right. To speak in crude terms: to hold that it is a logically necessary condition or presupposition of Aristotle being a man, that there must first be the truth of some existential proposition 'There exists a parcel of flesh and bone' or even 'There exists an animated parcel of flesh and bone', would render man a characteristic of such a parcel. But man is not a characteristic feature of a parcel of flesh and bone; rather being a man is what a certain parcel of flesh and bone animated in a certain way *is*. 'Aristotle', where this is the proper name of a man, in such a proposition as 'Aristotle is a man', presupposes the truth of an existential proposition invoking a 'matter' term in that, and only in that, Aristotle is a *man*, *not*, as it were *independently of* him being a man - otherwise *man* simply becomes a characteristic of a parcel of matter; 'man' becomes a predicate of a parcel of matter and loses its status as a principle of counting and distinction in its own right. Further, we may note that whilst from the fact that if X is a man it follows that X is an animated parcel of flesh and bone, we cannot say that *every* proposition of existence invoking a sortal entails a proposition of existence invoking a 'matter' term e.g. 'God exists'; yet as Aquinas argues there is no matter in God. It is thus *contingent* that propositions of existence containing sortals entail propositions of existence containing 'matter' terms. We may reasonably hold that propositions of existence containing 'material' sortals *entail* propositions of existence containing 'matter' terms but it does not follow from this that such propositions are presupposed in the sense of 'logically prior' - that we may only introduce propositions of existence containing sortals in that we may first introduce such propositions, such propositions as a type. To hold *this* does precisely suppose, reverting back to the specific case, that being a man *is* a characteristic of an animated parcel of flesh and bone and not what it is for a certain organised parcel of flesh and bone *to be*. In this case for an animated parcel of flesh and bone to exist, to be, *is* for it to be a man, or, more generally, an *F*. To say that a certain animated parcel of flesh and bone is a man is to say *what* it is, not to attribute a characteristic to it. In such a case we only have one such animated parcel in that we have one man, one *F*, and we can only count the number of animated parcels in that we can count the number of men, of *F*'s. What we may hold is that propositions of existence formed from 'matter' terms are *empirically* prior to propositions of existence formed from 'material' sortals but, for reasons given, we may not hold that such propositions are *logically* prior and no thesis about what is *empirically* prior or is empirically the case entails any logical thesis.

In that one is concerned with the question of what underlies the subject-predicate distinction, one is concerned with the question of what *types* of existential proposition must be true in order for the distinction to operate, to be manifested. One is concerned with the question of which *types* of existential proposition are *logically* basic, not with the question of which types of existential proposition are *empirically* basic. And it is my case that we have two basic types of existential proposition neither of which is reducible to the other: neither of which can be said to be logically more *fundamental* than the other. I shall later claim that one mistake, though not the only one, which Strawson makes in his claim that existential propositions formed from 'feature' universals together with the notion of 'placing' underlie the subject-predicate distinction is that a logical distinction rests on, or may be said to rest on, what is simply empirically the case. In that I hold that we have, that there are, *two* basic types of existential proposition and that neither is reducible to or can be said to rest on the other, I am committed to the thesis that *necessarily* to exist is to be specifiable by reference to some sortal or to be specifiable by reference to a symbol ('term') formed from a 'matter' term or 'substantival general' term viz., *necessarily* to exist is to be specifiable by reference to a symbol (a 'term') which furnishes a principle of counting and distinction in its own right.

I.2. It follows from my explanation of the subject-predicate distinction in the sense of an answer to the question: 'What makes this distinction possible?' in terms of the presence and recognition of sortals that neither sortals themselves nor their definite form may occur in or be used in the role of logical subject. It will be protested that in e.g.:

'Man is rational'
'God is eternal'

it certainly seems that we have a case of a sortal occurring or being used in subjective role. In these cases however 'Man' is ambiguous between being the name of human kind and being a sortal and 'God' is ambiguous between being the *name* (proper name) of a god and a sortal. In that 'Man' and 'God' are construed in the first way we have a potential subject-predicate proposition with Man and God as potential subjects. In that 'Man' and 'God' are sortals we do not have potential subject-predicate propositions for in such a case 'Man' is writable as 'Every man'; 'Some man' and in 'Every man is rational'; 'Some man is rational' we do not have subject-predicate propositions expressed. 'Every man is rational' comes out as a universally quantified proposition traditionally expressed as: 'For every x, if x is a man, x is rational': $(x)(Mx \supset Rx)$; on my account it comes out as 'For every F, if F is a man, then F is rational': $[(x)F](Mx \supset Rx)$, where 'F' is a 'higher' order sortal, 'M' is 'man' and 'R' is rational. This is in accordance with my thesis that we require a sortal in the formulation of the quantifier sense of 'is' and the fact that 'M' occurs in predicative position here does *not* for reasons given earlier, commit me to the thesis that a sortal *is* a predicate. It is a case of it being necessary to write a sortal as a predicate granted the language of predicate calculus. We could write 'Every man is rational' simply as '(x)man (x is rational)' *pace* Quine[64] and retain the point that we require a sortal in the formulation of the quantifier sense of 'is'. Again 'Some men are rational' comes out as the particular quantified proposition:

> 'For some x, if x is a man, x is rational'
> $(\exists x)(Mx.Rx)$

On my account:

> 'For some F, if F is a man, F is rational'
> $[(\exists x)F](Mx.Rx)$

or:

> $(\exists x)$ man (x is rational)

And likewise in the 'God' case where 'God' is a sortal e.g., 'God is eternal' - 'Every God is eternal'; 'Some God is eternal' would on my account come out as:

> $[(x)\ F](Gx \supset Ex)$

or more simply:

> (x) God (x is eternal)

or again, for 'Some God is eternal':

> $[(\exists x)F](Gx.Ex)$

where 'F' is a 'higher order sortal; 'G' = 'God'; 'E' = 'eternal', or more simply:

($\exists x$) God (x is eternal)

As regards the definite form of a sortal 'The F which ...'; 'The F who ...'; it will be objected that such forms may occur in a referring role and hence may occur as logical subject in a subject-predicate proposition. I do not deny that the definite form of a sortal expression may occur in subject or referential position but in so far as it is so able and so does then it is no longer a *sortal* but a *quasi* name. Let me try to explain this. The use of say 'The man who broke the bank at Monte Carlo' in a referential role presupposes the existence of the man who broke the bank at Monte Carlo and in that it presupposes the existence of the man who broke the bank then it presupposes the specificatory use of the definite form of the sortal 'The man who broke the bank at Monte Carlo'. In other words the use of the definite form of a sortal expression in referential, subjective, role presupposes its use as a *sortal*. We cannot say that the definite form of a sortal in referential role *remains* a sortal for in such a role it: (a) presupposes existence; (b) its role is precisely to refer and not to 'pick out' or 'specify', yet we cannot go as far as saying that in such a role it is a name, for occurrence in a certain role: (i) is not a *defining* characteristic of a class of symbol; (ii) hence not a *sufficient* condition for change of status. It, even so, can be maintained that in that the definite form of a sortal occurs in or is used in a referring, subjective, role, what we have is a *quasi* name; a case of the definite form of a sortal *posing as* a name. This can be argued from the considerations: (a) in that it so occurs or is so used we can rightly ask for a specification of the reference (as we can with names or other 'singular terms' which may be regarded as names or pseudo (*quasi*) names (e.g. 'That cat')); (b) in that it so occurs or is used, a name may be coherently substituted for it - which is not the case in that it is or remains the definite form of a sortal, any more than a name may coherently be substituted for a definite description.

Parallel consequences follow from my 'explanation' of the subject-predicate distinction in terms of the presence and recognition of principles of counting and distinction formed from 'matter' terms. 'Gold melts in *aqua regia*' (e.g.) cannot be analysed as a subject-predicate proposition where 'Gold' is not construed as the proper name of a metal. 'Gold dissolves in *aqua regia*' is to be analysed out as either: 'All gold dissolves in *aqua regia*' or 'Some gold dissolves in *aqua regia*' and these propositions would be given formal expression in a mode parallel to that expressing universal and particular quantification as given above. I give the following as an illustrative example. 'All gold dissolves in *aqua regia*' - $[(x)M](Gx \supset Ax)$, where 'M' is some 'higher order' matter term (e.g. 'metal'), 'G' = 'gold'; 'A' = 'dissolves in *aqua regia*'. (Or, more simply, following Quine, (x)gold (x dissolves in *aqua regia*)). An objector here will doubtless raise a parallel problem to that raised above concerning the definite form of a sortal and occurring in the role of logical subject; here too I would give a parallel answer to that given above. To take an illustrative example: 'The piece of gold in the corner of the room' in that it occurs or is used in a referring (logically subjective) role, is a *quasi name* for reasons analogous to those given in connection with e.g. 'The man who broke the bank at Monte Carlo'.

I.3. It will now be said that having been honest about the consequences of my explanation of the subject-predicate distinction in terms of the presence and recognition of sortals, it is now time to be equally honest about the consequences or alleged consequences of my thesis that *necessarily* to be is to be specifiable by reference to a sortal or to be specifiable by reference to a principle of counting and distinction formed from 'matter' terms, for it might be held that these consequences are indeed embarrassing. I shall attempt to illustrate such alleged embarrassing consequences by reference to the case as regards sortals; a parallel situation arises or may be alleged to arise in the case of 'matter' terms, but I claim that so far as I can cope with the former alleged embarrassing consequences, I can cope with the latter. The alleged embarrassing consequence is as follows. In so far as I hold that necessarily to exist is to be specifiable by reference to a sortal, that this constitutes existence, how can I permit or acknowledge such propositions of existence as 'Men exist', 'Trees exist', where 'exists' here is the 'redundant predicate' sense;[65] or again such propositions as 'There are men', 'There are trees' - in general, such propositions as '*F*'s exist', 'There are *F*'s' where '*F*' is a sortal? I have earlier (Section I.2) offered an account as to how I can admit and acknowledge such propositions but such an account, it will be said, lands me in the same embarrassing position as Aristotle[66] in his analysis of general propositions of existence, namely, turning what is apparently a contingent statement of fact into a necessary truth. For I have held that e.g. 'Man exists' is to be construed as 'Man is (e.g. a rational) animal'; 'Cats exist' as 'Cats are feline animals'; 'Trees exist' as 'Trees are plants (of a certain type)'. To give the Aristotelian example itself: 'Ice exists' comes out as 'Ice is solid'. So, it will be said, either I deny that general propositions of existence are possible or such general propositions are turned into necessary truths. Either of which positions is absurd; hence I am in a fundamental dilemma.

It would certainly be absurd to deny that general propositions of existence containing sortals are impossible and I have not advocated this view. I thus turn to the second horn of the dilemma. First, it is to be noted that I can maintain my thesis without embarrassment in relation to the quantifier sense of 'is'. 'There are *F*'s' comes out as 'There are F_1's which are *F*'s' or 'There are F_1's which are *G*', where '*G*' introduces a defining characteristic of *F*, and F_1 is a 'higher order' sortal. For example:

> 'There are men' - 'There are animals which are human';
> 'There are cats' - 'There are animals which are feline';
> 'There are books' - 'There are volumes of printed material which are (plus some defining characteristic of books)';

or again, to take the Aristotelian example:

'There is ice' - 'There is water (liquid) which is solid (frozen)'.

There is no embarrassment in these cases since in each case we have: (a) a perfectly intelligible, if somewhat sketchy, account; (b) *contingently* true propositions. Even though to be a man (e.g.) is to be a human animal is definitionally true, it is *contingent* that *there are* animals which are human. Equally, whilst to be ice is to be frozen water or more generally frozen liquid, that *there is* water (liquid) which is in such a state is *contingently* true. So the real embarrassment apparently lies in the 'redundant predicate' sense of 'is' - in such propositions as 'Men exist'; 'Ice exists' parsed out in Aristotelian fashion. I hold that ultimately there is no embarrassment and to claim that there *is*: (a) rests on a misunderstanding of Aristotle; (b) a dogma. To support (a).

In holding that e.g. 'Men exist', 'Ice exists' are writable as 'Men are rational animals', 'Ice is frozen liquid', no *existential claim* is either being made or is implicit, rather an account of what man's existence *consists in* is being offered or again what the existence of ice consists in is being offered. Equally in the singular case 'Socrates exists', parsed out as 'Socrates is a man', no singular existential claim is being made; rather an account of what it is for Socrates to exist is being offered. But to offer an account of what it is for something, general or particular, to exist is not itself to make any existential claim. The misunderstanding of Aristotle is, I think, twofold. (i) to think that in such propositions as 'Man exists', 'Ice exists', where 'exists' is a redundant predicate an existential claim is being made either explicitly or implicitly; (ii) that an analysis of a proposition of existence is itself an existential proposition. As regards (i) to maintain such a position is incompatible with 'exists' being a redundant predicate and, as I have just argued, with what Aristotle is seeking to do, and to hold that the predicative sense of 'exists' *implicitly* makes an existential claim would be to hold that the 'predicative' sense is reducible to or implies the quantifier sense. That the predicative sense is reducible to the quantifier sense is a thesis I have already argued against holding that we need the redundant predicative sense in order to formulate the quantifier sense. I have earlier acknowledged that there is a sense in which the redundant predicative sense implies the quantifier sense, rests on the quantifier sense, but since the quantifier sense for *its very formulation* requires the redundant predicative sense, we may hold that the latter sense is *ultimately* prior. As regards (ii) we may clearly not hold that an analysis of an existential proposition is itself an existential proposition, any more than we may hold that an analysis of a chemical compound is itself a chemical compound.

To turn to (b). The dogma is that *all* propositions of existence are contingently true. Where this is taken to read: 'All sentences *propounding* existence are contingently true', there is, as far as I can see, no inherent objection; but where this is taken to read: 'All existential propositions are contingently true' this is surely false. We have just been considering counter cases. Others have also thus argued.[67]

I thus claim that neither I myself, nor Aristotle, are committed to turning general propositions of existence into necessary truths and hence the second horn of the dilemma is avoided.

I.4. In the above discussions I do not claim to have dealt with all objections which might be raised against the consequences of my thesis that what makes the subject-predicate distinction *possible* is the presence and acknowledgement of *sortals* (in a language which invokes the notion of 'an object'), and the presence and acknowledgement of principles of counting and distinction formed from 'matter' terms (in a language which invokes the notion of 'a (material) particular'), or the consequences of my thesis concerning *what underlies* the subject-predicate distinction, but I do claim to have tackled apparently overriding objections. On the latter issue I shall have to face the position advocated by Strawson and this I shall certainly be doing in Chapter X. And there are further general difficulties concerning my thesis that *necessarily* every variable is determined - in relation to issues raised in this chapter - that we may *not*, *pace* Quine eliminate 'range indicators' e.g. write 'Every man is mortal' as '$(x)(x$ is a man $\supset x$ is mortal)', as opposed to: (x) man (x is mortal) or $[(x)A](Mx \supset M^1x)$. This matter will be taken up again in my discussion of Quine in Chapter XI. My next immediate task is to set out what I claim to be the *general consequences* of failing to acknowledge the category of the sortal as a category of symbol in its own right; the *general consequences* of treating sortals as 'predicates'.

Notes

1. 'Subject and Predicate', *Mind*, 1950, p. 463.
2. *Individuals*, (London: Methuen, 1959), p. 143.
3. My parentheses.
4. My parentheses.
5. My parentheses.
6. My parentheses.
7. *Individuals*, p. 143.
8. My parentheses.
9. My parentheses.
10. My parentheses.
11. *Ibid.*
12. My parentheses.
13. *Ibid.*, p. 144.
14. Geach, *Reference and Generality*, pp. 31-32.
15. *Individuals*, p. 144; my parentheses.
16. *Ibid.*, p. 145.
17. *Subject and Predicate in Logic and Grammar* (London: Methuen, 1974), p. 9.
18. *Ibid.*, p. 11.
19. *Ibid.*, p. 9.
20. *Ibid.*, p. 10.

21. *Reference and Generality*, p. 23.
22. *Individuals*, pp. 187-188.
23. *Ibid.*
24. In his famous paper 'On Referring', repr. in *Logico-Linguistic Papers* (London: Methuen, 1971), pp. 1-27.
25. *Individuals*, pp. 190ff.
26. *Ibid.*, p. 190.
27. *Ibid.*
28. *Mathematical Logic*, pp. 67-69.
29. *Individuals*, pp. 187-188.
30. *Ibid.*, p. 187.
31. *Ibid.*
32. *Ibid.*
33. *Reference and Generality*, p. 25.
34. cf. *Individuals*, pp. 186-189.
35. cf. Strawson's list at *ibid.*, p.139.
36. *Ibid.*
37. *Methods of Logic* (New York: Henry Holt and Co., 1950), p. 205.
38. *Ontological Relativity and Other Essays* (New York: Columbia University Press, 1969), p. 62.
39. *Philosophy of Logic* (Englewood Cliffs, N.J.: Prentice-Hall, 1970), pp. 27-28.
40. *Ontological Relativity*, p. 95.
41. P.F. Strawson, *Subject and Predicate in Logic and Grammar*, p. 12.
42. my parentheses.
43. *Ibid.*, pp. 12-13.
44. 'Singular Terms and Predication' in *Logico-Linguistic Papers*, p. 65.
45. *Ibid.*, p. 66.
46. *Ibid.*
47. My parentheses.
48. My parenthesis.
49. *Ibid.*, p. 67.
50. *Ibid.*
51. *Ibid.*, p. 69.
52. *Ibid.*, p. 67, n. 1.
53. *Ibid.*, p. 67.
54. *Ibid.*, p. 68.
55. *Ibid.*, p. 67.
56. *Subject and Predicate in Logic and Grammar*, pp. 32-33.
57. *Ibid.*, p. 33.
58. *Ibid.*
59. *Ibid.*, p. 32.
60. *See* Owen in 'Aristotle on the Snares of Ontology', *op. cit.*
61. Irving Copi, 'Essence and Accident', repr. in J.M.E. Moravcsik (ed.), *Aristotle - A Collection of Critical Essays* (London: Macmillan, 1968), pp. 151ff.
62. Michael Durrant, 'Essence and Accident', in *Mind*, Vol. LXXXIV, N.S. No. 336, October 1975, pp. 595-600.
63. *See* ch. IX.

64. *Mathematical Logic*, p.67; but this may be thought to present a problem if it is construed as committing one to the thesis that every variable is *essentially* determined (*see* p. 101 above) but there is no need to thus construe it.
65. *See* Aristotle: *Metaphysics* 1003b 27; 1054a 16-18; *De Interpretatione* 21a 21-3; G.E.L. Owen, 'Aristotle on the Snares of Ontology', *op. cit.* This is the two-place incomplete use of 'is' ('*x* is *P*'; '*x* is not *P*') where '*P*' takes as values some predicate, the predicate varying with the sort of object we assign the verb (*see also*: G.E.L. Owen, 'Plato on Not-Being', in Gregory Vlastos (ed.) *Plato: I Metaphysics and Epistemology* (Garden City, New York: Doubleday and Company, Inc., 1971), pp. 223-267; repr. in G.E.L. Owen, *Logic, Science and Dialectic*, *op. cit.*, pp. 104-137).
66. *See* G.E.L. Owen, 'Aristotle on the Snares of Ontology' (*op. cit.*) and Robin Attfield, 'How Things Exist: A Difficulty', *op. cit.*
67. e.g. Alvin Plantinga, *God and Other Minds* (Ithaca, New York and London: Cornell University Press, 1967), ch. 2, sec. IV; Anthony Kenny, 'God and Necessity', in Bernard Williams and Alan Montefiore (eds.), *British Analytical Philosophy* (London: Routledge and Kegan Paul, 1966), pp. 131-151.

Chapter IV

General Consequences of the Failure to Acknowledge the Category of the Sortal

A. The Attempted Reduction of Sortals to Predicates and its consequences

Failure to recognise the category of the sortal in the sense of acknowledging *only*, as principles of counting and distinction in their own right, such principles formed from 'matter' terms thus reducing sortals to predicates, I claim commits one to either:

(I) An impossible account or no account of what constitute an instance of a sortal; or again, an unintelligible account of what so constitutes;

or

(II) a doctrine of *relatively* basic names or singular symbols (as opposed to a doctrine of *absolutely* basic or *absolutely* simple names), the bearers of which may constitute an instance;

both of which consequences I hold to be objectionable.

A.1. How do such consequences follow? One who reduces sortals to predicates is necessarily faced with the question: 'What *instances* or may instance a sortal?', since predicates ('predicables') are instantiable. In relation to this question he must maintain either:

(1) Every particular name, *qua* being the name of an individual instancing a sortal, is *necessarily* a name relating to a sortal i.e. every particular name *necessarily* supposes for its introduction and use some sortal and *this* commits him to an impossible account or no account of what constitutes an instance of a sortal or, again, an unintelligible account as will later be argued;

or:

(2) Every particular name, *qua* being the name of an individual instancing a sortal, is *not necessarily* a name relating to a sortal i.e. every such

particular name does *not necessarily* suppose for its introduction and use some sortal.

Now (2) is compatible with:

(3) Every particular name, *qua* being the name of an individual instancing a sortal, is *necessarily* a name relating to an individual introduced by reference to a principle of counting and distinction formed from a 'matter' term.

Since (2) is compatible with (3) it permits (3) and hence permits the introduction of *relatively* simple basic names and correspondingly *relatively* simple basic individuals, as opposed to *absolutely* simple basic names and correspondingly *absolutely* simple basic individuals. Further, (3) not only permits such introduction, but one who is committed to (3) is committed to such introduction since: (a) it is an assumption in the present context that principles of counting and distinction in their own right are necessary for the introduction and continued use of proper names; that *absolutely* simple names are *out*; (b) it is also an assumption in the present context that every basic proper name is introduced by, and has its continued use by reference to, a principle of counting and distinction in its own right viz., a sortal or 'matter' term. On this account, what constitutes an instance of a sortal is an individual (or an individual name) introduced by, and having its continued identity determined by, a principle of counting and distinction supplied by a 'matter' term. Such are the *relatively* basic individuals and relatively basic names. Such a doctrine of 'relatively basic' names and individuals i.e. (II) in Section A above, is open to objection precisely in that it requires that proper names and individuals introduced by and having their identity determined by reference to sortals, rest on the introduction of such names and such individuals. This is the particular application of the general thesis objected to in Chapter II, namely, that principles of counting and distinction furnished by sortals are *not* such principles in their own right; that they are, in effect, reducible to 'predicates'. That (I) in Section A above is objectionable *of itself* is clear; but that one who reduces sortals to predicates is committed to it depends upon it being shown that (1) above so commits him; that this is so I shall argue later.

A.2. The position may be put at a 'metaphysical' level, in 'metaphysical' terms as follows. Failure to recognise the category of the sortal *universal* as a principle of counting and distinction in its own right, acknowledging *only* principles of counting and distinction formed from 'Feature' *universals* reduces sortal universals to *universals* introduced by predicates viz., 'Characterizing' *universals*. This has the consequence of committing one to *either*:

(I_1) An impossible account or no account of what constitutes an instance of a sortal universal; or again, an unintelligible account;[1]

or

(II_1) A doctrine of relatively simple (as opposed to absolutely simple) basic particulars introduced by reference to principles of counting and distinction formed from 'Feature' universals; *both* of which consequences I hold to be objectionable.

The reasoning for such consequences is parallel to that offered above; but in the light of what is to come it is of worth, and indeed helpful, to present the case again in appropriate 'metaphysical' terms.

One who reduces sortal universals to 'predicates' or 'characterizing' universals is faced with the question: 'What instances a sortal universal?' and in relation to this question he must maintain either:

(1_1) Every particular is *necessarily* a particular of a certain sort; *necessarily* an instance of a sortal universal;

or:

(2_2) Every particular is *not necessarily* a particular of a certain sort; *not necessarily* an instance of a sortal universal.

(1_1) commits him to (I_1), as will later be argued in a fashion parallel to the argument for (1) in the earlier version committing its propounder to (I). (2_2) is compatible with, permits:

(3_1) Every particular is necessarily a particular of a certain kind and thus permits one to introduce 'relatively simple', 'basic' particulars, subject to conditions parallel to those given at the corresponding stage of the earlier argument;

(3_1) also *commits* one to a doctrine of relatively simple basic particulars introduced by reference to 'matter' terms, since (3_1) is incompatible with:

(4_1) Every basic particular is not necessarily a particular of a certain kind or sort.

That (1_1) above commits one to (I_1) will be defended later as promised. (II_1) is objectionable precisely in that it requires that particulars of a *sort* rest upon the introduction of particulars of a *kind* and this, once again, is the particular application of the general thesis that principles of counting and distinction furnished by sortal universals are *not* such principles in their own right, but are reducible to predicates ('Characterizing') universals.

B. The consequences of failure to recognise the Sortal as a Principle of Counting and Distinction in its own right *and equally* those of the failure to recognise principles of Counting and Distinction in their own right formed from 'Matter' terms

B.1. Failure to recognise the category of the sortal as a principle of counting and distinction in its own right, *and equally* failure to recognise principles of counting and distinction formed from 'matter' terms as such, i.e. the *conjoint* failure to recognise *both* as such principles, reducing *both* to predicates, commits one to *either*:

(III) an impossible account or no account of what constitutes an instance of a sortal or 'matter' term precursed in a certain way; or again an unintelligible account;

or

(IV) a doctrine of *absolutely* simple, basic, names; *absolutely* simple, in that they presuppose, rest on, *no* other category of general symbol or 'term' *at all*; basic, in that other categories of singular symbol are reducible to them, but not conversely.

Why is this so? One who reduces sortals *and* 'matter' terms precursed in a certain way (e.g. 'stretch of water') to *predicates* is necessarily faced with the question: 'What instances a sortal or "matter" term (suitably precursed)?' and in relation to this question he must maintain *either*:

(i) Every particular name, singular symbol, is *necessarily* a name relating to a sortal or 'matter' term (suitably precursed) i.e. presupposes for its introduction and use some sortal or 'matter' term (suitably precursed);

or

(ii) Every particular name, singular symbol, is *not necessarily* a name, singular symbol relating to a sortal or 'matter' term (suitably precursed) i.e. does not necessarily presuppose for its introduction and use some sortal or 'matter' term (suitably precursed).

(i) commits one to an impossible account or no account, or again an unintelligible account of what constitutes an instance of a sortal or 'matter' term (suitably precursed) i.e. to (III) above, for reasons parallel to those to be offered in relation to (1) in my first discussion committing one to (I) and (1_1); in my second discussion committing one to (II_1). (ii) commits one to (IV), which is unintelligible in that on such a doctrine we should never, in principle, be able to determine the reference or bearer of such names, for these names are such that

(a) they do not necessarily presuppose *any* category of general symbol for their introduction or use - which entails (b) *necessarily* they do not so presuppose.

In corresponding fashion, in 'metaphysical' terms, failure to recognise sortal universals as principles of counting and distinction in their own right and *equally* failure to recognise 'Feature' *universals* (suitably precursed) as such principles i.e. the composite failure to recognise both, reducing both to *universals* introduced by predicates ('Characterizing' universals) commits one to *either*:

(III_1) an impossible account or no account, or again an unintelligible account of what constitutes an instance of a sortal universal or a 'Feature' universal (suitably precursed);

or

(IV_1) a doctrine of *absolutely* simple and basic individuals (particulars), 'pure' particulars: 'absolutely simple' in that they presuppose no universal at all for their introduction; 'basic' in the sense that other types of particular rest on them, are eliminable in favour of them.

The reasoning in favour of this is parallel to that given in the corresponding case above, but, once again in the light of what is to come, I shall give it. One who reduces sortal universals to universals introduced by predicates ('Characterizing' universals) is again faced with the question: 'What instances a sortal universal?'; and similarly one who reduces 'Feature' universals (suitably precursed) is likewise faced with the question: 'What instances a "Feature" universal (suitably precursed)?' In relation to this question he must maintain *either*:

(iii) Every particular is *necessarily* a particular of some sort or kind; an instance of some sortal universal or 'Feature' universal (suitably precursed);

or

(iv) Every particular is *not necessarily* a particular of a certain sort or kind; not necessarily an instance of some sortal universal or 'Feature' universal (suitably precursed);

(iii) commits him to (III_1) for reasons to be offered parallel to those to be offered in relation to the corresponding cases above; (iv) commits him to (IV_1) for reasons parallel to those offered in support of (ii) above, committing one to (IV) above.

C. Some Elucidations and Expansions

Having thus set out what I claim to be the general consequences of failure to recognise the category of the sortal, and also the consequences of failure to recognise principles of counting and distinction formed from 'matter' terms equally as such principles in their own right, treating *both* as predicates, it is now time to elucidate and expand on: (i) what I have claimed is an impossible or no account of what constitutes an instance of a sortal; (ii) what I have claimed is an unintelligible account. Equally with regard to what constitutes an instance of a 'matter' term. I shall, in fact, not explicitly engage in the latter task but shall concentrate on the 'sortal' case as this is my principal interest; it will be clear to the reader what parallel arguments may be adduced in the latter type of case.

C.1. In that someone holds: (1) every particular name, singular symbol, is *necessarily* a name relating to a sortal (as in A.1. (1) above); or (2) every particular is *necessarily* a particular of a certain sort (as in A.2. (1_1) above), then he is committed to holding that what instances a sortal is a name specified by reference to that sortal itself; or again that what instances a sortal universal is a particular specified by reference to that sortal universal itself. Let me consider the 'metaphysical' formulation since the point I wish to make is easier to illustrate in this mode. What will instance the sortal universal *cat* is the particular *cat*; what will instance the sortal universal *man* is the particular *man*. Sortal universals are *necessarily* required to specify what instantiates them or may instantiate them. But just this is impossible. No universal may of necessity be required to specify its own instances since it is *contingent* what instantiates any universal i.e. what instantiates a universal is *contingently* an instance of that universal. The thesis that a universal, of necessity, is required to specify what may instance it yields the paradoxical result that all propositions expressing what Strawson calls the 'instancing' relation would be necessarily true. To refer to the earlier examples. What can instantiate the sortal universal *cat* is a particular *cat* - say Jemima. Since a particular cat is necessarily a cat, 'Jemima is a cat' becomes a necessarily true proposition. It is in that the thesis that what instances a sortal has a necessary reference to the sortal itself entails the above paradoxical result that I claim this thesis is impossible.

It may be counter claimed here that in that a sortal universal of necessity is *used* to specify its own instances the above paradoxical result follows, but that all one need hold is that the sortal universal is only *mentioned* in the specification of its own instances, in answering the question: 'What constitutes an instance of a sortal?'. But in that case no *specification* of what constitutes an instance of a sortal is supplied; no answer to the question is provided. This is one ground on which I claim that we have *no* account of what constitutes an instance of a sortal. I produce another ground below.

At this point another counter claim will be entered. It will be said that the above paradoxical consequence may be avoided and indeed is to be avoided in the following way. What constitutes an instance of a sortal universal does not *have* to be specified by reference to *that* sortal universal itself. Indeed A.1.(1) above[2] is expressed in terms not of 'some *one* sortal' but of *some* sortal; and A.2.(1_1)[3] is expressed in terms of 'necessarily a particular of a certain sort', not in terms of 'necessarily a particular of *some* one sort' (and similarly in relation to B.1.(i)).[4] Hence what constitutes an instance of a sortal universal need not be specified by reference to that sortal universal itself; it *may* be specified by reference to another sortal universal and hence the paradox avoided. In relation to the earlier examples, what may instance the sortal universal *cat* is the (a) particular feline animal; what may instance the sortal universal *man* is the (a) particular rational animal. This move however will not suffice since to maintain it we either: (i) *indirectly* have to introduce the sortal universal *cat* and *man* to specify their instances in that we have to understand that a cat is at least a feline animal; that a man is at least a rational animal; or (ii) we cannot introduce the instances as instances of *those* sortal universals. And this holds *whatever* the sortal concerned. It is contingent of which sort a particular is an instance, but *whatever* the sort, on the thesis under consideration, in order for it to be an instance of *that* sort, it necessarily has a reference to that sort or it cannot be introduced as an instance of that sort. This is the further sense in which I speak of *no* account being offered of what constitutes an instance of a sortal or sortal universal.

C.2. I now turn to the sense in which I claim that one who reduces sortals to predicates presents an unintelligible account of what constitutes an instance of a sortal. Using the 'metaphysical' formula, what instances a sortal is a particular necessarily specified by reference to some sortal. So we have to introduce a sortal in order to specify what constitutes an instance. We may, however, *never* employ a sortal in order to specify what constitutes an instance of a sortal; for nothing can specify *its own* instances, otherwise that which does the specifying *itself* becomes an instance of *itself*, which is absurd. Consider a parallel case in the case of predicates. To hold that what can constitute and instance of a predicate (or 'predicable') is specifiable only by reference to a predicate itself (or 'predicable' itself) makes a predicate (or 'predicable') an instance of *itself*, which is absurd. In general, what *sort*, *type* of thing can constitute an instance of *X* can never be specified, introduced, by reference to *X* itself, without the 'type error' resulting that *X* itself becomes an instance of itself. Further, we may observe that, what sort of thing can constitute an instance of *X* can never be specified only by reference to *X* itself, otherwise there can be no radical distinction between: (i) names and predicables; (ii) particulars and universals; (iii) what *has* instances and what *does* the instancing; (iv) more generally, between particulars and principles of counting and distinction for particulars.

It will now be objected: 'Surely it is to be *expected* that what instances a sortal will be specified by reference to that sortal itself?' Indeed it could not be otherwise, for surely it is necessarily true concerning any instance of a sortal that it can only be specified by reference to that sortal itself? More generally, it is necessarily true concerning an instance of anything general, where 'being general' is a sortal or 'matter' term (precursed in a certain way), that it can only be specified by reference to the sortal or 'matter' term itself. To so object however commits the objector to an impossible doctrine of what constitutes an instance of a sortal (or 'matter' term), no doctrine, or an unintelligible doctrine, as I have above argued. Paradoxes may be said to be the result of a type error and in this case I suggest that the type error is to suppose that sortals may *themselves* be said to *have* instances. But if *this* is the case how do we explain the initial force of the objection raised and the apparent necessity in: 'Indeed it could not be otherwise for surely it is necessarily true concerning any instance of a sortal that it can only be specified by reference to that sortal itself?' The explanation, I hold, is that the truth of the matter is expressed by:

(1) *Necessarily* to be an instance, to be a particular, is to be specified (or specifiable) by reference to some sortal (or some 'matter' term, precursed in a way already given);

but this does not entail:

(2) To be an instance, to be a particular, is *necessarily* to be so specifiable.

The converse however is the case: 'A is necessarily P' does entail '*Necessarily* A is P', but this has not, and cannot, be appreciated by those who treat sortals (or 'matter' terms) themselves as predicates. The thesis I advocate is (1), and on this thesis sortals (and 'matter' term suitably precursed) may not be treated as predicates since the question of what constitutes an instance may not be intelligibly raised concerning them. Symbols, terms, universals, which furnish principles of counting and distinction in their own right for objects, for particulars, may not *themselves* be said to *have* instances but rather themselves *furnish us with the notion of 'an instance'*. And that they themselves may not said to have instances is shown by the paradoxes generated by possible answers to the question: 'What instances a sortal?', 'What instances a sortal universal?', 'What instances a "matter" term (suitably precursed) or "Feature" universal (suitably precursed)?'

C.3. The consideration that symbols, terms, universals, which furnish principles of counting and distinction in their own right may not themselves be said to have instances, but themselves furnish us with the notion of 'an instance', leads me on to some further general consequences of the failure to recognise the true nature of sortals (and also of 'matter' terms so precursed as to supply principles of counting and distinction in their own right) namely:

(1) the search for a basis for the introduction of 'particulars in general';
(2) the search for a basic notion of 'an instance';
(3) the search for an underlying basis for the subject-predicate distinction as manifested in a language in which the categories of name and predicable (predicate) are introduced or present, or in a mode of discourse in which particulars are introduced or spoken of, in the sense

of a search for some one type of *empirical* proposition which contains no reference to sortals and is formed independently of any reference to sortals.

for these (correlative) consequences *deny* that:

(a) *Necessarily* to be a particular is to be of a certain sort or kind; to be specifiable by reference to a sortal or 'matter' term suitably precursed, maintaining rather: to be a particular is *necessarily* to be of a certain sort or kind; is *necessarily* to be specifiable by reference to a sortal or 'matter' term suitably precursed.

(b) Such principles of counting and distinction do not furnish us with the notion of an 'instance'; rather they are to be reduced to predicates and the basic notion of 'an instance' is to be supplied by *another* item, notably, in the case of Strawson to be considered later, by reference to that of 'placing'.

D. Summary of the General Consequences of the failure to acknowledge the Category of the Sortal

To summarise. We have, I hold, the following general consequences of failure to acknowledge the category of the sortal:

Either:

I (a)(i) a doctrine of *relatively* simple, basic, proper names introduced by and having their use determined by principles of counting and distinction formed from 'matter' terms together with the thesis that such principles are prior to all other principles of counting and distinction. And, correspondingly, a doctrine of *relatively* simple, basic, individuals together with the thesis that 'Feature' universals precursed in a certain way to form principles of counting and distinction are prior to all other such principles;

(a)(ii) a doctrine of *absolutely* simple, basic, names of *necessity* having no reference to sortals (or 'matter' terms) for their introduction or use and, correspondingly, a doctrine of absolutely simple individuals, 'pure' particulars. To this we may add:

(a)(iii) a doctrine of *absolutely* simple, basic, terms, symbols, in favour of which all 'singular terms' may be eliminated;

or:

(b) a doctrine to the effect that sortals may *themselves* be said to have instances and that principles of counting and distinction formed from 'matter' terms may *themselves* be said to have instances.

II. The search for a basis for the introduction of *particulars in general*;

III. The search for a basic notion of 'an instance'.

IV. The search for an underlying basis of the subject-predicate distinction as manifested in a language in which the categories of name and predicable (predicate) are introduced or present, or in a mode of discourse in which particulars are introduced or spoken of, in the sense of a search for some one type of *empirical* proposition which contains no reference to sortals and is formed independently of any reference to sortals.

D.1. Having thus set out what I hold to be the general consequences of failure to recognise the category of the sortal, I shall, in the rest of this book:

(i) seek to establish that the various authors I have mentioned have failed to acknowledge the category of the sortal;

(ii) work out some particular difficulties for and consequences of their position;

(iii) seek to illustrate some of the general consequences given above in their particular instantiation as follows:

I (a)(ii) by reference to Russell;
I (a)(iii) by reference to Quine;
I (b) by reference to Strawson and (indirectly) to Frege;
II, III, IV by reference to Strawson.

It is *arguable* that Frege is committed to a version of I (a)(ii), in so far as he holds that a proper name does not of its *essence* have a 'sense' and that this is sufficient for him to be committed to I (a)(ii). To claim that Frege holds that proper names do *not*, of their essence, have a sense is to go against received authoritative opinion (e.g. Michael Dummett's definitive work on Frege)[5] but I think that such a contrary thesis is arguable and shall present some case for it. I shall however contend that even if it can be shown that for Frege 'ordinary' proper names do *not* of their essence have a sense, this of itself is not sufficient to show that he is committed, ultimately, to a version of I (a)(ii) *but* in that he is *not* so committed he *is* committed to a version of I (b) and thus in the final analysis may be used as a particular illustration of I (b). The noticeable absentee here is a particular illustration for I (a)(i). I suspect Geach may be committed to such a position in that he seems to be committed to the priority of 'substantival general' terms over 'count' terms - 'count' terms are included under 'substantival general' terms (cf. *Reference and Generality*, pp. 39-40) but I cannot get as far as actually *showing* Geach is so committed. The most one can get as far as showing is that Geach regards sortals ('count' terms) as a species of substantival general term from which it does not follow that he holds a doctrine of priority of such terms over 'count' terms, over sortals. I (a)(i) might be thought to have particular exemplification in Strawson but on close investigation this is not so. I (a)(i), as far as I can judge, remains without any clear particular exemplification in recent philosophy but the consideration that no recent philosopher has advocated it, or clearly advocated it, does not detract from the importance of highlighting it as a general consequence of the failure to acknowledge the nature of sortals. It is clearly a position which might prove highly tempting and arguably both Aquinas and Aristotle were so tempted - the latter in order to supply an adequate account of 'substantial change' in *Physics*[6] and subsequent writing.

Notes

1. The attention of the reader is drawn to the distinction between the use of (I_1) and (1_1).
2. p. 148.
3. p. 150.
4. p. 151.
5. Michael Dummett, *Frege: Philosophy of Language* (London: Duckworth, 1973).
6. A, ch. 7.

Chapter V

Frege: Sortals as 'Concepts'

A. That Frege fails to recognise both the definite and indefinite form of the Sortal as a *Distinct* and *Primary* Category of general symbol (expression)

I shall first argue that as Frege fails to recognise the definite form of the sortal ('The so-and-so') as presenting a distinct and primary category of singular symbol (expression), regarding such a form as a type of name, so also he fails to recognise the indefinite form 'an *F*' as presenting a distinct and primary category of general symbol (expression), regarding a 'concept word' or 'concept expression' as predicative and of its essence predicative. As his doctrine of singular symbols (expressions) does not cater for a class of such expressions or symbols which could serve in the role of identifying an object in either of the senses already introduced, equally his doctrine of 'concept words', 'concept expressions' does not cater for a class of general expressions (symbols) which could serve to identify an object in the sense of specifying what it is.

A.1. To turn then to the case of the definite form of the sortal. The only category of singular symbol which Frege recognises is the category of the (singular) *name*. He distinguishes between: (i) *actual* proper names of which 'Aristotle' is given as an example; and we may safely say other ordinary proper names such as 'Michael' or 'Joseph' would equally count as actual proper names; (ii) proper names which 'represent' actual proper names, i.e. such proper names as 'The celestial body most distant from the earth'; 'The planet Venus'; 'The number 4'. That Frege counts expressions: (a) formed from a demonstrative together with a substantive noun; (b) formed from the definite article plus a substantival noun or phrase; (c) formed from the definite article together with a substantival noun plus some completing relative clause, as illustrated by the following examples: 'That cat'; 'The angel'; 'The Lion of Judah'; 'The man who broke the bank at Monte Carlo'; 'The horse which won the Derby in 1989', as names, even though names which 'represent' proper names, is clear. For example, he writes in the *Grundlagen*:[1]

> To start with it will not do to call a general term the name of a thing. This leads straight to the illusion that number is a property of things. The business of the general term is precisely to signify a concept. Only when combined with the definite article or a demonstrative pronoun can it be counted as the proper name of a thing, but in that case it ceases to count as a general term. The name of a thing is a proper name.

and a little later he writes: 'With a concept the question is always whether anything, and if so what, falls under it. With a proper name such questions make no sense'. In 'On Concept and Object' he writes: 'The singular definite article always indicates an object, whereas the indefinite article accompanies a concept word'.[2]

Russell objected to Frege's account of expressions of the form: 'The so and so' as a type of name, radically distinguishing such 'definite descriptions' from names. My objection is that Frege treats the definite form of a sortal as a name failing to appreciate that we need such forms of expression for the very introduction and continued use of proper names themselves. It might be objected that my criticism of Frege here is misplaced i.e. that Frege does *indirectly* recognise the definite form of the sortal even though he regards such a form as a name. He does so indirectly recognise in that he regards such names as 'The celestial body most distant from the earth' as proper names which 'represent' actual proper names, represent actual proper names in a mode of representation or 'presentation'. Could not then the doctrine of 'representative proper name' be an attempt to recognise a category of expression or symbol which is required for the introduction and continued use of 'actual proper names'? After all, he does allow that 'The celestial body most distant from the earth' (i) would be a grammatically well-formed expression; (ii) can be said to represent a proper name; (iii) that every grammatically well-formed expression 'representing' a proper name always has a sense.[3]

Now whilst from this it can be claimed that it follows that: (i) 'representative' proper names of their essence have a sense or 'present' a sense; (ii) the sense of 'actual' proper names is given by such 'representative' proper names, the thesis that in the doctrine of 'representative' proper names we have a veiled recognition of the definite form of the sortal unfortunately cannot be substantiated since:

(a) the class of such names also includes such forms as 'That cat', 'This car' etc.;

(b) it would require that Frege holds that sense is *logically prior* to reference, whereas such a thesis is in doubt in spite of some evidence to the contrary, for he writes:

> In the case of an actual proper name such as 'Aristotle' opinion as to the sense may differ. It might for instance be taken to be the following: the pupil of Plato and teacher of Alexander the Great. Anybody who does this will attach another sense to the sentence 'Aristotle was born in Stagira' then will the man who takes as the sense of the name: the teacher of Alexander the Great who was born in Stagira. So *long as the reference remains the* same, such variations of sense may be tolerated ...[4]

We may thus determine the reference independent of the sense and as this is so we strictly have no need for 'representative' proper names as far as *logical* considerations go, from which it follows that even if it can be claimed that the doctrine of 'representative' proper names is an implicit recognition of the definite form of the sortal, Frege's *logical* position is that we have strictly no need for such symbols.

(c) He holds that: 'The sense of a proper name is grasped by everybody sufficiently familiar with the language or totality of designations to which it belongs'.[5] Thus sense is given by reference to a 'totality of designations' and what is important about 'representative' proper names is their designations, as opposed to their sense - their designations as opposed to their ability to specify the reference or designation of proper names. 'Representative proper names' are, after all, proper names.

(d) Even if it still can be held that in the doctrine of 'representative proper names' Frege implicitly recognises the definite form of the sortal (in traditional language, recognises 'definite descriptions'), can it be said that he does so recognise in that Fregean 'representative names' can be said to perform the function which the definite form of the sortal may be said to perform - viz., introduce and serve as criteria of continued identity for a proper name? The answer to this is unfortunately in the negative since in that 'representative proper names' have of their essence a sense, they have their sense only in the context of a proposition. The sense of, for example, 'The teacher of Plato' will be given by the set of true propositions in which this phrase occurs. Once this thesis is advocated we seem to be in inexorable difficulties, as I have earlier argued, and the sense of the definite form of a sortal *cannot* be given in this way for the definite form of a sortal is required to determine the identity of propositions concerning names, as I have also previously argued.

I thus conclude that we may not say that in his doctrine of 'representative proper names', Frege implicitly recognises the category of the definite form of the sortal. I therefore now turn to establish the claim that, in corresponding fashion, he fails to recognise the indefinite form of the sortal as presenting a distinct and primary category of general expression or symbol.

A.2. In 'On Concept and Object', he writes: 'The concept (as I understand the word) is predicative. On the other hand, a name of an object, a proper name is quite incapable of being used as a grammatical predicate';[6] and again: 'the indefinite article always accompanies the concept word'.[7] In the *Foundations*,[8] we have: 'As soon as a word is used with the indefinite article or in the plural form without any article it is a general term'. From this latter passage then the indefinite form of a sortal is to be regarded as a general term. Frege indeed introduces general terms to elucidate 'concept words' as is clear from the passage from *Foundations* cited above.[9]

The concept of 'general term', though from one aspect of Frege's writing in the *Foundations*[10] relates to single words, it concerns more than what falls under the grammatical category of common noun. A general term is earlier[11] explicated in terms of a general term or *notio communis* and hence includes expressions which fall under other grammatical categories e.g. expressions which are grammatically adjectives such as 'red'. The contrast Frege is introducing here is that between a *notio communis* and a *nomen proprium* and 'red' in terms of such a contrast is certainly a *notio communis*: it does not name a single object and that 'red' is regarded by Frege as a 'concept word' can be clearly seen from a well known passage in his *Foundations*.[12] Here, having commented that the concept to which a number is assigned does in general isolate in a definite manner what falls under it, he adds:

> Not all concepts possess this quality. We can, for example, divide up something falling under the concept 'red' into parts in a variety of ways, without the parts thereby ceasing to fall under the same concept 'red'. To a concept of this kind no finite number will belong.

Frege's point here is that if I have a red coat, that coat falls under the concept 'red'; 'That coat' by Frege's criterion will name or stand for an object which falls under the concept 'red', '(---- is red)', but '---- is red' applies in exactly the same sense to each and every part of the coat *however* we may divide it or cut it up. If it is true that the coat is red, then it is equally true that the sleeves, back, and any arbitrarily chosen pieces, are red; '---- is red' is true in exactly the same sense of the parts of the coat as of the whole. The implication Frege draws from this is that the concept 'red' is one to which no *finite* number will belong but to which an *infinite* number will belong. He thought that you can make a *start* at counting the number of red things but that you may not make an *end*. This is certainly open to objection as Geach has pressed, but for my present purposes I need not now consider the objection and its importance since what I am concerned to show here is that Frege's dictum: 'the indefinite article always accompanies the concept word'[13] is not held by him to be violated in this case. In that we may *start* counting red things we may speak of 'a red'. And from this it follows that we may not delineate the class of 'concept words' by reference to the class of common nouns exhibiting the features exhibited by the grammatical expression of the sortal, such that it might be claimed that Frege implicitly recognises the category of the sortal expression in its indefinite form, for a sortal expression in its indefinite form is most certainly to be contrasted with a 'concept word' such as 'red'.

It might still however be pressed that Frege implicitly recognises as a distinct class of expressions, symbols, the indefinite form of a sortal in that he does distinguish concepts of a kind to which a finite number belongs and concepts of a kind to which *no* finite number belongs. To attribute such a position to him commits one to the view that he makes a *radical* distinction between the two kinds of concepts or 'concept words'. However he makes no *radical* distinction between, for example, 'table', 'man', 'apple' on the one hand - between expressions, symbols, which furnish principles of counting in their own right on the one hand; and 'red' on the other. All such symbols are for him 'logically predicative', for they are indeed 'concepts' and 'concepts' are (essentially) 'predicative'.[14]

One final attempt may be presented to support the thesis that Frege, in his notion of 'concept words' to which finite number belongs, *does* recognise the category of the sortal in that he recognises, for example, 'planet' or 'god' - in general, '*F*' as the unit relative to a finite number. I claim however that such a thesis cannot be upheld. For Frege it is *not* such items as 'planet' or 'god' which are units relative to a finite number. It has to be admitted that there are certain passages which might (initially) suggest that such items are units relative to a finite number. For example:

(i) at the beginning of para. 54 of the *Foundations* he writes:

> It now becomes possible to give a satisfactory definition of the term 'unit'. E. Schröder writes on p.7. of his text book - already referred to: 'This generic name or concept will be called the denomination of the number formed by the method given and constitutes, in effect, what is meant by its unit.'
>
> Adopting this very apt suggestion I propose to call a concept the unit relative to the Number which belongs to it.

From the fact that Schröder has spoken of a 'generic name' as what is meant by 'unit' and Frege takes this over and proposes to call a 'concept' the unit relative to the number which belongs to it, one might be tempted to infer that a generic name, such as 'planet', or a specific name such as 'moon' or 'satellite' will form the unit relative to a finite number and hence, on the principle that being a generic name is a sufficient condition for being a sortal, that Frege implicitly recognises sortals.

(ii) at *Foundations*, para 51, p. 64e we have:

> With a concept the question is always whether anything, and if so what, falls under it .... We should not be deceived by the fact that language makes use of proper names, for instance, Moon, as general terms and *vice versa*; this does not affect the distinction between the two.

(iii) a final passage from what Frege says about existence with specific reference to God's existence. At *Foundations*, para. 53, p. 65 we have:

> In this respect existence is analogous to number. Affirmation of existence is in fact nothing but the denial of the number nought. Because existence is a property of concepts the ontological argument for the existence of God breaks down. But oneness is not a component characteristic of the concept 'God' any more than existence is ...

It might be commented here: 'Surely Frege is saying that for God to exist is for there to be one God; to deny that God exists is to deny that there is anything falling under the concept "God" i.e. to assign the number "0" to the concept. In this way we can ask whether there are *many*, *only one*, or no gods;

we can make a beginning and an end of counting the number of gods that there are, so "god" is a unit relative to a finite number.'

However to take such passages as indicating that for Frege such items as 'moon', 'planet' or 'God' *are* the units relative to a finite number and so to claim that Frege does, after all, recognise sortals as a distinct class of symbol, expression, is to misrepresent his thought and would place him in the position of contradicting himself on the point that the concept is 'essentially predicative'. That 'moon', 'planet', 'God' are distinctly *not* for Frege 'units relative to a finite number' can be clearly illustrated as follows:

(1) Taking up the extract (i) above from para. 54 of the *Foundations*, we must note that he continues:[15]

> We can then give a sense to the assertions made about the unit, that it is isolated from its environment and is indivisible. For it is the case that the concept to which the number is assigned does in general isolate in a definite manner what falls under it. The concept 'letters in the word "three"' isolates the '*t*' from the '*h*', the '*h*' from the '*r*' and so on. The concept 'syllables in the word "three"' picks out the word as a whole and as indivisible in the sense that no part of it falls any longer under that same concept.

From this passage we see that for Frege what can be a unit relative to a finite number is not, 'letter' or 'syllable', but 'letter in the word "three"' or 'syllable in the word "three"'. A similar point is made earlier,[16] where he writes:

> Here I will add that a concept does not cease to be a concept simply because only one single thing falls under it, which thing, accordingly, is completely determined by it. It is to concepts of just this kind (for example *satellite of the earth*)[17] that the number 1 belongs, which is a number in the same sense as 2 and 3.

and also the following extracts:

(2)

> In the proposition "Jupiter has four moons", the unit is "moon of Jupiter".[18]

(3)

> The concept has a power of collecting together far superior to the unifying power of synthetic apperception. By the means of the latter it would not be possible to form the inhabitants of Germany together into a whole: but we can certainly bring them all under the concept 'inhabitant of Germany' and number them.[19]

These passages, I hold, show conclusively that Frege did not recognise sortals as a distinct category of symbol; for him the principle of counting is not 'man' or 'planet' but a functional expression such as 'planet of ----'. All sortals are thus reduced to functional expressions which are, he has said, 'essentially predicative'.

B. Some consequences and difficulties which arise from Frege's failure to recognise the Sortal as a *Distinct* and *Primary* Category of general symbol (expression)

What particular consequences and difficulties may be said to stem from this position?

B.1. As regards his failure to recognise the definite form of the sortal, the reduction of such a form to a 'representative proper name', we have the consequences: (a) that he has not the means of introducing the bearers of proper names; (b) of specifying their reference; (c) of providing a criterion of identity for their continued use. It will not suffice here to say that Frege could introduce the bearers of proper names, or indeed determine their reference, simply by 'ostensive definition' since: (i) even in the case of ostensive definition one must, in theory, be able to say *what* individual one is pointing to as the bearer of the name; but the only means Frege has of answering this question is by reference to the doctrine of 'representative proper names', which, still being proper names, cannot perform this function; (ii) to appeal to ostensive definition would render his doctrine of 'representative proper names' otiose. Again it will not suffice to say that he could furnish a criterion for the continued use of a proper name by reference to a 'representative proper name' for the same reason: a proper name may *have* a sense but: (a) it is not itself a sense; (b) may not be regarded as identical with its sense.

We may even press a further consequence - namely that Frege is engaged in an infinite regress. For granted that on his view a name must, in some sense, have a sense then: (i) the sense of an ordinary proper name is given by a 'representative proper name'; (ii) the problem arises as to the sense of the 'representative proper name'. This will have to be given, it seems, by yet another proper name and so on *ad infinitum*. If it be denied that the sense of a 'representative proper name' is a further name, then what is the alternative or are the alternatives? If it is held that it is to be given by a definite description (or a set of definite descriptions), we are confounded since Frege treats 'definite descriptions' as proper names. If it is held that it is to be given by a set of predicates which uniquely hold true of a given object, then we have to be able to specify the object in the sense of being able to answer the question what sort or kind of thing it is. To be able to do this however demands that we do not treat the indefinite form of a sortal as a predicate, which Frege precisely does as I have argued above. If it now be said that we cannot raise the question: 'What is the sense of a "representative proper name"?' since a 'representative proper name' *gives* the sense of an ordinary proper name and we cannot ask for the sense of a sense, we are faced with the difficulty that a name cannot *give* a sense, though it may *have* a sense.

Thus on this further point Frege, it seems, is either faced with an infinite regress or a series of difficulties which, within his framework, it seems impossible for him to answer.

B.2. As regards his failure to recognise the indefinite form of the sortal. To hold that the indefinite form of the sortal is reducible to a functional expression or symbol entails that such sortals as 'man', 'planet, 'moon' are of their *essence* part of such expressions as, for example, 'moon of Venus'; 'planet of The Sun'; 'man of England'; 'God of Jacob'. And similarly with 'syllable' or 'cat'. On such a thesis it seems that we can never, in principle, say that we are speaking of the *same sort of thing* when we speak of, for example, 'moon of Venus' and 'moon of (the) Earth'; or again when we speak of 'man of England' and 'man of God'; or yet again when we speak of 'God of Jacob' and 'God of Paul'. On such a position it seems that we must dispense with speaking of 'the same sort of thing'. To this it might be replied that Frege need not go as far as rejecting speech in terms of the 'same sort of thing'. It may be said that he could contend that, in particular, 'moon of the Earth' and 'moon of Venus' signify different planets and that we are speaking of the same sort of thing when we speak of 'moon of the Earth' and 'moon of Venus' in that in both cases we are speaking of (different) *planets.* This might appear to suffice as regards this particular example but clearly will not suffice in relation to such an example as 'man of God' and 'man of England'. We are prevented from saying that we are speaking of the same sort of thing in that we speak of, for example, Ted Heath as a man of England and Basil Hume as a man of God, for there is no clear 'common denominator' here as there might seem to be in the former case. Yet both Ted Heath and Basil Hume are the same sort of thing - namely men. I have just commented: 'as there might seem to be in the former case'; this *is* a case of *seeming* only since 'planet' *itself* too is, for Frege, 'essentially predicative'. It is writable as 'planet of *x*' so we are prevented from saying that in speaking of 'moon of the Earth' and 'moon of Venus' we are speaking of the same sort of thing.

At this point an objector will retort: 'Alright: we dispense with speaking of sorts and kinds of thing and speak in terms of predicative functions only and what is so objectionable about this?' One problem which immediately arises here is how we may determine that we have the same function present in both: (i) 'moon of Venus'; 'moon of the Earth'; or again in (say) (ii) 'man of England' and 'man of God' such that we may introduce 'sameness' in relation to functions. We may *not* say here that we have the same function present in 'moon of Venus' and 'moon of the Earth', since what can stand as a value for the variable in the function 'moon of ----' is the proper name of a planet, yet this assumes that 'planet' itself is *not* a functional expression which on Frege's thesis it precisely is. Similarly we may not say that what may so stand is, for example, the proper name of a heavenly body (celestial body) for the very same reason. We meet the same problem in relation to (ii) above, even if we could furnish a 'common denominator' for 'England' and 'God' - not even the most fervent patriot could claim that England and God were the same *F*. Further, for reasons offered in Chapter I, we may not hold that identity of function is determined by the function itself. If it be said that we have the same function present in, for example, 'man of England' and 'man of God', in that in both cases we simply have the function 'man of ----' completed by different proper names, then on the thesis that we may dispense with speaking of 'sorts of things' (and/or 'kinds of thing') we are still faced with an intractable problem since, as I have previously argued, *necessarily* any proper name is the name of something or other - some *sort* of thing or some *kind* of thing. However we are to determine identity of function, sameness of function, we may *not*, if what I have been concerned to argue earlier is right, even *introduce* the notion of a 'function' without that of a sort or kind of thing. Yet just this is proposed in the thesis that we speak in terms of predicative functions only.

Parallel problems to those raised earlier concerning how we are to speak of the 'same sort of thing' arise in the 'letter' and 'syllable' cases too. On Frege's thesis, it may be said, we could never, in principle, speak of the *same letter* occurring in the word 'three' and the word 'two'; of, for example, the letter 't', the *same* letter, occurring in the two words (or indeed of the same letter occurring in the original two words *and* in the 'two' of 'two words'). To this it might be objected that Frege is *not* committed to going *so* far. That whilst on his position we cannot speak of *the same letter* occurring in the two words, he is not committed to saying that 'letter in the word '"two"' is of its essence or nature a letter in the word "two", but *only* to saying that 'letter in the word "two"' is a 'letter in a word'; the point being that we may only speak of the same 'letter in a word' and not of 'the same letter'. But such a thesis has the quite unacceptable consequence that necessarily any letter is part of a word; a letter has its identity by reference to some word. This is unacceptable for a number of reasons; for example, how on such a thesis is it possible to produce or invent new words? Again, we may hold that letters may occur independently of *any* word as in initials and in acronyms, e.g. E.E.C. If these examples are held to be question begging for fairly obvious reasons then more examples come readily to hand - as in 'Row A', 'Paragraph B', 'Section A', etc. In such instances the letters are not parts of the words 'Row', 'Paragraph', 'Section' or indeed part of any further word. Yet it seems clear that we may speak of the same letter e.g. 'A', being used in one context to designate a row of seats, in another context being used to designate a section of a chapter in a book or whatever.

There are further particular unacceptable consequences and difficulties deriving from such a thesis as Frege seems to advocate. If every letter is necessarily a letter in a word, how may *the* thesis itself be coherently stated? - for we cannot speak of individual letters independently of *some* word. And again on such a thesis how is it possible to spell out *the letters* of a word? Further, if 'letter' is of its essence 'letter in a word' and 'syllable' is of its essence 'syllable in a word', then we are in a parallel difficulty to the difficulty expressed in the earlier cases *vis-à-vis* how it is possible to speak of either 'the same letter' or 'the same syllable'. To be able to speak in such a way Frege would have to allow that, for example, 'word' itself was not a functional expression, which he may *not* do for 'word' itself meets his criteria for being a 'concept word' and is therefore 'essentially predicative'. There is of course an added difficulty concerning these later cases. In the former cases of 'moon' and 'satellite', what at first sight may replace the blank in 'moon of ----', 'satellite of ----' is the proper name of some planet, whereas 'word' cannot replace the blank in 'syllable of a ----' since 'word' itself is not a proper name, but precisely a 'concept word'.

I shall now enlist just one further difficulty which I consider to be particularly dire in its consequence. If 'man', for example, is of its essence a predicative symbol, of its essence *part* of a complete symbol such as 'man of God', 'man of the Earth', 'man of England' then not only can we not speak of the *same F*, the same man, but no child could be asked to count *the number* of men, for we should have different *X*'s in each of the above cases. This consequence may be more clearly illustrated by exercising a certain latitude with regard to Frege's position which I trust is not radically unfair to him. If 'man' is of its essence *part* of a complete symbol such as 'man in church A', 'man on station B', then not only may we not speak of counting the same things in the two places, but no child could be asked how *many* men there are *altogether* for 'men in two places' would furnish yet *another* concept. That is, a child could not be given the following simple sum:

> If there are 5 men in church A, 16 men at station B, 27 men who are wise and 100 men who are foolish, how many men are there altogether?

An odd sum no doubt, but certainly an intelligible one to which we may expect the answer: 148. Not only may one enlist such a case as a dire consequence of Frege's position but one may also put one's finger on what seems to be a central mistake and which, were it not realised to be a mistake, might be advocated in favour of reducing sortals to predicates, to functional expressions. From the fact that we may only count the number of *F's* granted a certain context, framework of reference; that is, we may not count or be asked to count the number of *F's* - men, cats, churches or cars *period* - all counting is done within certain 'parameters', *it does not follow* that 'man', 'cat', 'church' are *parts* of expressions mentioning a context or specifying a context such that 'man', for example, becomes of its nature or essence *part* of some such expression as 'man in room R', 'man on station A' or 'man of Hellenic wisdom' or 'man of God'. Indeed it necessarily does not so follow, otherwise we are precluded from speaking of 'counting within a context' but to be precluded from such a way of speaking, to make as it were, man or church as part of the context lands us in the dire and seemingly intractable difficulties already discussed.

C. Is Frege committed to a doctrine of absolutely simple names and correspondingly, absolutely simple particulars, as a consequence of his failure to acknowledge the Category of the Sortal?

It might now be suggested that Frege is committed to a version of I (a)(ii) as set out in the previous chapter (p. 156) in that he holds that a proper name does not of its essence have a sense.

C.1. It might be said that to even contemplate the thesis that Frege holds that an ordinary proper name does not of its essence have a sense goes against what Frege was chiefly concerned to advocate and a principal contribution he made to philosophy. Such a charge would be brought by no less an authority than Michael Dummett who explicitly contrasts Frege's position with those who have advocated a doctrine of 'pure' proper names (e.g. Mill and Russell in their respective ways) and is epitomised by Searle when he writes:

> Frege's most important single discovery in the philosophy of language was the distinction between sense and reference ... in addition to the name and the object it refers to viz., its reference, there is a third element, its sense (or as we might prefer to say in English: the meaning or descriptive content of the name in virtue of which and only in virtue of which it refers to its reference). The sense provides the 'mode of presentation' of the object and referring to a reference is always achieved by way of sense .... Frege regards his theory of sense and reference as applying to definite descriptions like 'the man in a blue shirt' but also to ordinary proper names like 'Chicago' and 'Winston Churchill'.[20]

In the light of such authoritative testimony, how can one even contemplate arguing in favour of the above thesis? If it be thought that to attribute such a position to Frege is so outrageous as not to merit serious consideration, it is to be noted that Quine claims that his own doctrine of 'referential position' derives from Frege, yet his own doctrine ultimately commits him to a doctrine of absolutely simple ('pure') singular symbols which have reference only in the sense that all singular terms are reducible to pronouns, as far as ordinary grammar goes, and to the variables of quantification in terms of logical theory.

The case for saying that it is not a necessary truth for Frege that his 'actual' proper names have a sense; that as a matter of logic there may be 'actual' proper names without a sense - and we might add that this is necessary if Quine's understanding of Frege is correct - will presently be set out.

The contrary thesis, the 'received' thesis, will be advocated in the light of, e.g., the following extracts from his writings:

> A proper name (word, sign, sign combination, expression) *expresses* its sense, *stands for*, or *designates* its reference. By means of a sign we express its sense and designate its reference.[21]

This extract of itself however does not permit one to conclude that a proper name of its essence has or presents a sense. That it does not so permit may be argued as follows:

(i) to hold that such a conclusion follows is at least apparently incompatible with Frege's doctrine of singular symbols (proper names) which 'represent' actual proper names. In that such singular expressions as 'The planet Venus'; 'The number 4'; 'The celestial body most distant from the earth' *represent* 'actual' proper names i.e. represent proper names in some 'mode of presentation', then proper names themselves, of their essence, have or present no sense. They only have or present a sense in or under some mode of presentation. (I shall be considering the charge that this is a total misunderstanding of Frege's position on proper names when I have attempted to set out the case against the contrary, the 'received', thesis).

(ii) Such a thesis is incompatible with (or apparently incompatible with) the following illuminating passage:[22]

> The sense of a proper name is grasped by everybody who is sufficiently familiar with the language or totality of designations to which it belongs ...

when taken together with the footnote already cited on p. 160 above. For the benefit of the reader I take the liberty of citing the footnote again to bring the point home:

> In the case of an actual proper name such as 'Aristotle' opinions as to sense may differ. It might, for instance, be taken to be the following: the pupil of Plato and teacher of Alexander the Great. Anybody who does this will attach another sense to the sentence 'Aristotle was born in Stagira' than will a man who takes as the sense of the name: the teacher of Alexander the Great who was born in Stagira. So long as the reference remains the same, such variations of sense may be tolerated...

This passage reveals: (a) that reference is not logically dependent on sense for we may speak of the reference remaining the same independent of any sense; (b) we may speak of the same name 'Aristotle' having different senses or variations of sense. The identity of the name 'Aristotle' therefore is not dependent on its sense or indeed on any sense, for if it were dependent on some sense, how could we speak of the same name 'Aristotle' having a *variety* of senses? It, that is, the name 'Aristotle' is not, to use Frege's earlier terms[23] dependent upon any 'mode of presentation'. We can speak of the same name independently of any mode of presentation. It is thus contingent that an 'actual' proper name presents or has a sense. (Once again the charge that it is a total misrepresentation of Frege to hold that we may speak of the same name independently of any mode of presentation; that there can be 'actual' proper names independently of any mode of presentations, will be considered later).

Independently of any such charge being brought as a general charge, it will be protested that this passage itself and the earlier passages invoked do not present a sufficient condition for holding that Frege's 'actual' proper names do not of their essence have a sense. It will be pressed that there are other passages which suggest that we may only speak of reference in that we may first introduce a sense. For example in the following we have:[24]

> The regular connexion between a sign, its sense, and its reference is of such a kind that to the sign there corresponds a definite sense and to that *in turn*[25] a definite reference, while to a given reference (an object) there does not only belong a single sign.

Here surely, it will be said, sense is logically prior to reference; the reference of a sign can only be determined by reference to its sense or again, we can only determine the reference of a sign in that we first understand its sense. If however we accept this passage with its apparent message, how are we to avoid apparent inconsistency with the earlier passage and in particular with the footnote? The following I hold to be a plausible and indeed correct explanation since it neatly coheres with Frege's earlier remarks in 'On Sense and Reference' viz., it is *logically contingent* that the reference of a sign can only be determined by reference to its sense, but it is *epistemologically necessary* that we can only determine the reference of a sign in that we first understand its sense or senses. That is, *logically* reference is prior to sense (otherwise Frege would not be able to speak of a name having the same reference independently of some sense; or of a name having the same reference but variations of sense), but *epistemologically* sense is prior. That is, the sense of a name is only required for our *understanding* of that name; we have to know (understand) the sense of a name in order to determine its reference; but the sense of a name is not required for, as we might say, the name 'of itself' - independently of our understanding of it. (Once again, a challenge to this on the grounds of a general misunderstanding will be considered later.)

Now to support this explanation; to tie this explanation in with earlier passages in 'On Sense and Reference' and in particular with Frege's doctrine of 'mode of presentation'. In his famous discussion of the notion of 'equality' at the beginning of the paper,[26] Frege introduces the idea of a 'mode of presentation' in addition to the idea of a 'mode of designation' in order to explain how it is possible for a true statement of the form:

'*a* is identical with *b*'

to contain any more information than a true statement of the form:

'*a* is identical with *a*'

If we speak in terms of 'mode of designation' only, then we either have to say that equality is a relation between that which the names 'a' and 'b' *designate* or is a relation between the *signs* 'a' and 'b'. Neither account of equality (identity), Frege holds, is satisfactory. If we say that equality (identity) is a relation between what is designated by the signs 'a' and 'b' then 'it would seem that $a = b$ would not differ from $a = a$ (i.e. provided that $a = b$ is true). A relation would thereby be expressed of a thing to itself and indeed one in which each thing stands to itself but to no other thing.' On the other hand if we say that equality (identity) is a relation between the *signs* 'a' and 'b' then this relation would only hold between the signs in so far as they designated something. *But* it is an *arbitrary* matter that a sign designates something - 'Nobody can be forbidden to use any arbitrarily producable event or object as a sign for something', hence the sentence $a = b$ would express no proper knowledge. On neither thesis therefore, within the context of 'mode of designation' only, can we explain how it is possible for a true statement '*a* is identical with *b*' to contain any more information than a true statement '*a* is identical with *a*'; yet clearly '*a* is identical with *b*' *does* contain more information than '*a* is identical with *a*'.

In order to offer an explanation of this, Frege introduces the notion of 'mode of presentation' ('sense'). It is introduced by way of an example:[27]

> Let *a*, *b*, *c* be the lines connecting the vertices of a triangle with the mid points of the opposite sides. The point of intersection of *a* and *b* is then the same as the point of intersection of *b* and *c*. So we have different designations for the same point and these names ('point of intersection of *a* and *b*', 'point of intersection of *b* and *c*') likewise indicate the mode of presentation; and hence the statement contains actual knowledge.

Thus the reference of the expressions 'the point of intersection of *a* and *b*' and 'the point of intersection of *b* and *c*' would be the same, but not their sense. And similarly the reference of 'The Evening Star' would be the same as that of 'The Morning Star', but not their sense. Hence statements of the form: '*a* is identical with *b*' can be said to contain more information than statements of the form '*a* is identical with *a*' in that even though the reference is the same the sense is different and the statement conveys that one and the same object has the features specified in the different senses of the two expressions.

In support of the current thesis I draw the reader's attention to fact that Frege's writing here lends support to the view that for him reference is logically prior to sense even though sense is epistemologically prior. The problem of how '*a* is identical with *b*' can be said to contain more information than '*a* is identical with *a*', in that it *is* a problem regarding *information*, is an epistemological problem and Frege sets it explicitly in that context. The objection to the thesis that identity is a relation between signs only in so far as they designate something is that on such a thesis $a = b$ would express no proper *knowledge*.[28] And the introduction of the notion of a 'mode of presentation' (or sense) allows us to speak of $a = b$ as a statement 'containing actual knowledge'.[29]

That 'reference' or 'mode of designation' is what is *logically* prior and hence, from a logical point of view, it is contingent that a proper name has a sense; that an 'actual' proper name is a proper name under some 'mode of presentation' may be illustrated as follows:

(a) We should note that Frege introduces his notion of a 'mode of presentation' by commenting:

> A difference (i.e. in cognitive value of $a = a$ and $a = b$)[30] can arise only if the difference between the signs *corresponds* to a difference in the *mode of presentation of that which is designated* ...[31]

thus implying that what is designated is logically independent of its mode or indeed any mode of designation;

(b) In the illustration given, he writes:

> The point of intersection of *a* and *b* is then *the same as* the point of intersection of *b* and *c*. So we have different designations for the same point ...

He does *not* write:

> The point of intersection of *a* and *b* is the *same point* as the intersection of *b* and *c*.

The point therefore does not gain its identity from 'the point of intersection of *a* and *b*' or 'the point of intersection of *b* and *c*'. Yet 'the point of intersection of *a* and *b*' and 'the point of intersection of *b* and *c*' are said to indicate the mode of presentation. Hence it is not a necessary truth that the point gains its identity from a mode of presentation. It is a contingent truth that the point is the point it is under some mode of presentation;

(c) We should further note that he also says:[32]

> In our example, accordingly the reference of the expression 'the point of intersection of *a* and *b*' and 'the point of intersection of *b* and *c*', would be the same but not their senses. The reference of 'evening star' would be the same as that of 'morning star' but not the sense.

thus implying that logically reference is independent of sense although in order to *know* the reference we have to know the sense. Again, in relation to this passage he writes:

> Comprehensive *knowledge*[33] of the reference would require us to be able to say immediately whether any sense belongs to it. To such *knowledge*[34] we can never attain.[35]

(d) We may also cite here what he writes later on,[36] where he rightly wishes to distinguish between reference, sense, and associated idea. He writes:

> The reference of a proper name is the object itself which we designate by its means; the idea, which we have in that case, is wholly subjective; in between lies the sense, which is indeed no longer subjective like the idea, but is yet not the object itself.

> Then he gives the famous analogy of someone observing the moon through a telescope.[37] The moon itself is compared to the reference; it is the object of observation mediated by the real image projected by the object glass in the telescope and the retinal image of the observer. The former he compares to the sense, the latter to the idea. Frege admits that the optical image in the telescope is indeed dependent on the standpoint of observation but nevertheless he rightly holds that it is still objective in that the same standpoint may be occupied by many observers.

From this illuminating passage we may maintain that what Frege holds is that the sense is the medium of reference; it is we who need the medium of sense to appreciate, understand, know the reference as it is we who (then) needed the telescope to observe the moon. In order to understand appreciate, know the reference of 'The moon' (proper name) we have to understand, appreciate, know the sense of the name 'The moon', but nevertheless it is not *logically* necessary that the name 'The moon' has a sense. The name 'The moon' has a reference independently of any sense as the actual moon exists independently of any method of observation involving a medium.

Having thus advanced arguments in favour of the thesis that for Frege: (i) reference is logically independent of sense but not epistemologically independent of sense; (ii) reference is logically prior to sense but that sense is epistemologically prior to reference, I shall now attempt to redeem the promissory note issued hitherto and consider the general charge that to claim thus is a gross misunderstanding of Frege's position.

C.2. The crux of the general charge that there has been a gross misunderstanding of Frege as alluded to on p. 169 and p. 171 above is that for him to be an 'actual' proper name is necessarily to be an 'actual' proper name *within* a 'mode of presentation', and *this*, it will be reiterated, is one of Frege's major contributions to the philosophy of language; indeed according to Searle in the extract earlier cited, the major contribution. Let me first offer some comments on Searle, then on Gareth Evans and finally on Dummett.

First, as I have earlier indicated, I have no dispute with Searle on the question of 'representative' proper names being of necessity names within a 'mode of presentation'. Secondly, Searle's sentence: 'The sense provides the "mode of presentation" of the object' and 'referring to a reference is always achieved by way of sense' is ambiguous as between: (a) whether the reference of itself is always achieved by way of sense; (b) whether our referring is always achieved by way of sense. On the account offered above, reference of itself is not necessarily achieved by way of sense, but our referring is always achieved by way of sense. If it be replied that no such contrast may be made and hence no such ambiguity may arise since there is no such X as 'referring of itself' independently of human subjects referring, the reply to this is that I have produced an argument for saying that Frege himself did not assimilate the two, even if the two may be assimilated. Thirdly, Searle has to admit that Frege presents 'sense' as a 'third element', i.e. an element in *addition* to reference and idea; 'sense' is not given as the context in which reference takes place and yet if 'sense' is to be the context in which reference takes place it would be necessary for Frege not to posit 'sense' as a third element, something in *addition* to reference, but to make it clear that 'sense' was an element which embraced reference. Finally, to hold that an 'actual' proper name is necessarily a proper name within a 'mode of presentation' does not cohere with the astronomical analogy. Indeed it may be urged that not only does it not cohere with this illuminating analogy but that the whole analogy is misleading on such an account. As I have earlier commented the point of the analogy certainly seems to be that *we* need sense in order to determine reference as the astronomer needs the optical image; but reference may be determined independently of sense - indeed the analogy demands that this be so. One cannot say that for the moon (e.g.) to be (sc. exist) is for the moon to be observable in some observational system otherwise the analogy is rendered unintelligible; equally we cannot say that for the reference (referent) to be (sc. exist) is for the reference to be in some system of sense, in some 'mode of presentation', without rendering the analogy totally misleading, indeed unintelligible.

The crux of the general charge alluded to on p. 171 is that there is no such distinction as the distinction there mentioned between: (i) the name 'of itself'; (ii) our understanding of the name. To be a name is to be understandable by reference to a sense. This charge too is based on the assumption that for Frege to be a name is necessarily, necessarily as a matter of logic, to have a sense. To counter this I refer the reader to arguments already given. As far as *Frege* is concerned to make such an assumption would go against the thesis that sense is the *medium* of reference. The issue here is not whether it is the case that to be an 'actual' proper name is to be understandable by reference to sense, but whether Frege can be said to have held this view. And arguments to the contrary have been advanced. In this context of 'understanding' it is appropriate to turn to Gareth Evans' appreciation of Frege on 'sense'.

In his chapter on Frege in *The Varieties of Reference*, Evans offers an account of Frege's notion of 'sense' in terms of the notion of 'a way of thinking of something' and advocates the principle that it is not possible to think of an object save in some particular way. He writes:

> Someone who hears and understands an utterance of a sentence containing a singular term, such as 'Aphla is over 5000 metres high', must at least think of the mountain - in having, precisely, this thought: that what the speaker is saying is true if and only if that mountain is over 5000 metres high. Now, it is not possible to think of a mountain save in some particular way. If you gave the way in which our subject was thinking of it, you would be giving what Frege calls the sense he attaches to the name 'Aphla'. Frege's idea was that it may be a property of a singular term as an element of a public language that, in

> order to understand utterances containing it, one must not only think of a particular object, its Meaning, but one must think of that object *in a particular way*;[38]

This appreciation of Frege is consistent with the argument that, for him, the reference of a singular term is logically prior to its sense for 'sense' relates to *understanding*. The consistency however is subject to the proviso that for Frege there can be terms, elements in a public language, which may be understood independently of sense and this proviso is met in some of his writings as earlier illustrated. Of course it would be *prima facie* absurd to suggest that Frege or anyone else could allow the existence of singular and other terms (symbols) in a language and claim they were not *understandable*. But Frege is not open to such a charge. There are two possibilities open to him, the second of which I shall be discussing below. The first possibility is that a singular term is understandable simply in that it refers; there is no more to its meaning than its reference. The second possibility is that a singular term is understandable in that its reference is *identifiable* but for its reference to be identifiable is not necessarily to give the sense of the term. I shall later contend that it is the *second* position which is attributable to Frege and in this way he is not committed to a doctrine of 'simple names' or a doctrine of 'pure' particulars. In relation to Evans' account here, in which 'sense' is introduced in terms of 'a way of thinking of something', one may observe that in that it is 'a way of thinking of something' we must, antecedently, be able to speak of the something in order for there *to be a way* of thinking about that thing and this we shall have to do independently of sense. Or again, if we agree with Evans that it is a necessary truth that it is not possible to think of an object save in a particular way - that all thinking is 'internal' to a system of thought - then whilst 'sense' is to be thus conceived, 'sense' does not exhaust 'thinking'.

Evans holds that his principle corresponds to Dummett's principle that it is not possible to have bare knowledge concerning some object that it is the referent of a term. This position is also consistent with the thesis advocated, for here we have a thesis about *knowledge* and the latter position has acknowledged that Frege at least holds that in order for us to know the reference sense is required. *We* need sense in order to know, to understand the reference; but from this it does not follow that for a singular term to refer, to have a reference, and for that reference to be knowable, sense is required.

Finally to turn to Dummett. He writes,[39] to the effect that it is Frege's thesis that even proper names (ordinary proper names as opposed to more complex proper names, near 'definite descriptions' such as 'The evening star') have a sense to which there is more than just having a certain referent. The problem is: 'Is this a necessary truth as a matter of logic for Frege?'. I have advanced arguments to the contrary and not even Dummett can hold that ordinary proper names have of their essence a sense such that we may only use a proper name to refer in the context of a sense. This may be illustrated by reference to his allusion to Frege's metaphor of a route. He writes:[40] 'Frege used the metaphor of a route from the name to the referent: names with different senses but the *same referent*[41] correspond to different routes leading to the same destination'. In this very example we must be able to speak of 'the referent' independently of the sense. The destination itself cannot be specified, described, by reference to sense since: (a) the destination itself is not a route; (b) neither does it have its identity determined by reference to the route or routes - rather the route (or routes) has its (their) identity determined by reference to it. Similarly in the case of the example of 'Afla' and 'Ateb' which Dummett cites[42] and indirectly referred to by Evans above. One cannot hold that the explorers in the Evans' example may only refer to the mountain via the senses which each of them gives to the names 'Afla' and 'Ateb' respectively. In order to give their respective senses to the respective names *and* in order for it to be discovered that Afla and Ateb are one and the same mountain, there must be some independent description available and this independent description, present in the geographical discovery, is not present in either of the senses. Here it might be said that Frege need not hold either: (i) that the destination in the route example is specifiable, describable, in any other way than being the reference of a singular term (which of course is no way); or (ii) that the mountain in the explorers example is specifiable, describable, in any other way than being the reference of a further singular term. But in *that* case how may he speak of 'the same mountain' in the explorers case or 'the same destination' in the earlier case? Further, if he were to hold the positions in either (i) or (ii), then it certainly would seem that he *is* committed to a doctrine of 'pure' names - absolutely simple names as I have described them in I (a)(ii)[43] - proper names which require for their introduction and use *no* prior category of symbol. Yet he is clearly *not* so committed as can be seen from other examples as well as those just discussed. In the astronomical analogy, the background of the analogy clearly takes the name 'The Moon' to be the proper name of a *moon*. In the famous and by now (as Dummett comments) 'hackneyed' examples of 'The morning star' and 'The evening star', both have the same reference in that they refer to the same *planet*, though 'planet' does not enter the account of the sense. In the geometrical example 'The point of intersection of *a* and *b*', 'The point of intersection of *c* and *d*', both designate the same *point*, which is specifiable, describable independently of the sense of these phrases.

Thus, whilst we may firmly agree with Dummett that:

> If someone is to be able to employ a proper name in sentences or to understand its use in the utterances of others so as to be able to judge of their truth and falsity, he must know more about the expression than just that it is a proper name: he must know some correct means for recognising an object as the bearer of that name.[44]

and that Frege concurs with this, it does not follow that for Frege a proper name of *logical* necessity has a sense. All that follows for him, it can be argued, is that there must be some correct means of recognising the object as the bearer of the name. Such correct means is not necessarily the sense but, by reference to the examples given, as it were the *foundations* of sense; reflecting the point made earlier that in that 'sense' is 'a way of thinking about an object' or, to invoke the route analogy, a means to an end, then: (a) we must first be able to specify, describe, the object in order for sense to be a way of thinking about *it*; (b) we must first be able to describe, specify, the destination in order for sense to be a means of arriving at *it*. If it now be counteracted that the correct means of identifying or recognising the object is itself the sense or a sense of the name which designates the object then Frege's notion of 'sense' has itself undergone a radical change of sense.

Hence it may be argued that even though a proper name for Frege does not necessarily have a sense, as a matter of logical necessity, but only as a matter of epistemological necessity, it cannot be argued that this itself is a sufficient condition for holding that he is committed to a version of (I) (a)(ii)[45] viz., that 'actual' proper names are ultimately such 'pure' or 'absolutely simple' names.

C.3. If the above arguments can be refuted and it be held that, after all, for Frege an 'actual' proper name does of its essence have a sense, then he is committed to a version of I (b)[46] as follows. In that he holds that proper names of their essence have a sense and sortals in any expression of that sense are reducible to predicables, then he is committed to I (b) just in that he treats sortals as predicables and hence may be said to have instances. One may also note here the following paradox for Frege in that he treats 'actual' proper names having of their essence a sense - namely, that in that he reduces sortals to predicables he is unable to give a coherent account of the notion of a predicable or predicative function itself.

This may be seen as follows. At the particular level, in that to be an 'actual' proper name is to be an actual proper name within a mode of presentation then in that he does not recognise sortals but treats them as predicables and a predicable is, of its nature, a 'propositional fragment', he has to say and is willing to say, that the sense of a proper name is given by the set of propositions in which *it* occurs. He thus cannot give an account of the sense without introducing the name itself. At the most general level, in that 'actual' proper names have a sense, then the proper name variable '*x*' necessarily has a sense in that he has to admit that what may stand as a value for the variable '*x*' in a predicative function will be a proper name which necessarily has a sense. Yet the notion of 'a sense' is to be given in terms of the notion of a predicative function. Since Frege treats sortals themselves as essentially predicative, he may only introduce the notion of a 'predicative function' by reference to the notion of a 'predicative function' itself - which is absurd since, as we have seen, a predicable is *essentially* incomplete. We have no means of saying what can stand in the argument place of a predicative function without invoking the notion of a 'predicative function' itself, yet the notion of a 'predicative function' itself cannot permit us to do this.

Again, if the above arguments stand, then Frege is still in severe difficulties. For on the above arguments, whilst 'actual' proper names do not have of their essence a *sense*, they do of their essence have a mode of identification invoking such 'concept words' as 'planet', 'point', 'satellite'. In that Frege does not recognise sortals and treats, e.g. 'planet' etc., as predicative functional expressions, as '---- is a planet' he is committed to I (b) allowing that sortals may be said to have instances. Parallel to the paradox instanced above we may elicit a paradox in relation to the thesis that whilst 'actual' proper names do not have of their essence a sense, they do have, of their essence, a mode of identification invoking such 'concept words' as 'planet', 'point', 'satellite'. At the most general level, in that 'actual' proper names do, of their essence, have a mode of identification then we have to admit that what may stand as a value for the variable '*x*' in a predicative function will be a proper name under some mode of identification. Yet the notion of a 'mode of identification' is to be given in terms of the notion of a 'predicative function'. As Frege treats sortals as essentially predicative he may *only* introduce the notion of a 'predicative function' by reference to this notion itself which is absurd for, as we have seen, a predicative function, a predicable, is of its essence incomplete.

I thus conclude that either way Frege is committed to a version of I (b).

Notes

1. *The Foundations of Arithmetic*, trans. by J.L. Austin (Oxford: Basil Blackwell, 1950), para. 51, p. 63e (hereafter *Foundations*).
2. In 'On Concept and Object', repr. in Peter Geach and Max Black (eds.), *Translations from the Philosophical Writings of Gottlob Frege* (Oxford: Basil Blackwell, 1952), p. 45.
3. In 'On Sense and Reference', repr. in *Translations from the Philosophical Writings of Gottlob Frege*, *op. cit.*, p. 58.
4. *Ibid.*, footnote to p. 58; my italics.
5. *Ibid.*, pp. 57-58.
6. 'On Concept and Object', *op.cit.*, p. 43.
7. *Ibid.*, p. 45.
8. *Foundations*, para. 51, p. 64e.
9. *Ibid.*, para. 51, p.63 e.
10. *Ibid.*, para. 51, pp. 63-64e.
11. *Ibid.*, para. 50, p. 62e.
12. *Ibid.*, para. 54, p. 66e.
13. 'On Concept and Object', *op. cit.*, p. 45.
14. *Ibid.*, p. 43.
15. *Foundations*, para. 54, p. 66e.
16. *Ibid.*, para. 51, p. 63 e.
17. My italics.
18. *Foundations*, para. 54, p. 66e.
19. *Ibid.*, para. 48, p. 61e.
20. John Searle in his 'Introduction' to *The Philosophy of Language* (London: Oxford University Press, 1971), pp. 2-3.
21. Frege, 'On Sense and Reference', *op. cit.*, p. 61.
22. *Ibid.*, pp. 57-58.
23. *Ibid.*, p. 57.
24. *Ibid.*, p. 58.

25. My italics.
26. 'On Sense and Reference', *op. cit.*, pp. 56-57.
27. *Ibid.*, p. 57.
28. *Ibid.*; my italics.
29. *Ibid.*
30. My parentheses.
31. *Ibid.*; my italics.
32. *Ibid.*
33. My italics.
34. My italics.
35. *Ibid.*, p. 58.
36. *Ibid.*, p. 60.
37. *Ibid.*
38. Evans, *Varieties of Reference*, p. 16.
39. Michael Dummett, *Frege: Philosophy of Language* (London: Duckworth, 1973), p. 97.
40. *Ibid.*, p. 96.
41. My italics.
42. *Ibid.*, p. 97.
43. *See* ch. IV above, p. 156.
44. Dummett, *op. cit.*, p. 100.
45. *See* note 43 above.
46. *See* ch. IV above, p. 156.

Chapter VI

Russell: Sortals as 'Descriptions'

A. Russell's basic Categories of Symbol

I shall first seek to establish that Russell fails to recognise the category of the sortal.

Russell recognises as basic categories of symbol, *complete* symbols and *incomplete* symbols, and within these categories differentiates between simple complete symbols and complex complete symbols: simple incomplete symbols and complex incomplete symbols. Simple complete symbols are *names* - what he later designates as logically proper names: complex complete symbols would be many worded names: simple incomplete symbols are predicates (one place predicates); complex incomplete symbols are complex predicates and 'descriptions' (definite and indefinite).

Now it can in no way be held that Russell, in his doctrine of complete simple symbols - his doctrine of *names*, either explicitly or implicitly recognised the category of sortal since:

(1) Russellian proper names do not have the form of a singular form of a sortal: Russellian proper names are paradigmatically demonstratives. He writes:[1]

> A name, in the narrow logical sense of a word whose meaning is a particular, can only be applied to a particular with which the speaker is acquainted, because you cannot name anything you are not acquainted with ...

and later on the same page:

> The only words one does use as names in the logical sense are words like 'this' and 'that'. One can use 'this' as a name to stand for a particular with which one is acquainted at the moment. We say, 'This is white'. If you agree that 'This is white' meaning the 'this' that you see, you are using 'this' as a proper name.

No sortal is introduced in connection with the demonstrative and neither can such an introduction be implied: indeed necessarily not so, since on this suggestion the demonstrative would cease to be a complete simple symbol.

(2) Russellian proper names do not present us with a category of singular symbols which *could* serve in the role of identifying, in the sense of picking out an object. For Russell, proper names (logically proper

names) simply stand for objects: they of their nature or essence have no sense. Or rather, for them to have of their essence a sense or meaning is for them to stand for an object. There is no other sense in which they have of their essence a sense. This sense *is* their designation - which amounts to saying that in that they have of their essence a sense, they only have of their essence a designation, *not* a sense. In his *Introduction to Mathematical Philosophy*, Russell writes:[2]

> The distinction between a name and all other symbols may be explained as follows:
>
> A name is a simple symbol whose meaning is something that can only occur as subject i.e. something of the kind that, in Chapter XIII., we defined as an "individual" or a "particular".[3] And a "simple symbol" is one which has no parts that are symbols. Thus "Scott" is a simple symbol, because, though is has parts (namely, separate letters) these parts are not symbols.

And he continues:[4]

> We have, then, two things to compare: (1) a *name*, which is a simple symbol, *directly designating an individual*,[5] which is its meaning, and having this meaning in its own right, independent of the meaning of all other words; (2) a *description* which consists of several words, whose meanings are already fixed, and from which results whatever is to be taken as the "meaning" of the description.

Proper names then symbolise and have a sense in that they 'directly designate' or 'name' individuals or 'particulars' which are their meaning. It is of their essence that they designate only. Russell's doctrine of 'direct designation' precludes him from holding that such names have of their essence a sense, for to hold that such names have of their nature a sense would entail that they could only *indirectly* designate an individual, i.e. in Russell's terms designate an individual via a 'description' which could give their sense. But, as is well known, for Russell, names which can only so indirectly name are not logically proper names, but merely veiled descriptions. Ordinary proper names which *do* of their essence have a sense and can only designate in that they have a sense, are for Russell, veiled or truncated definite descriptions. In his 'Lectures on the Philosophy of Logical Atomism',[6] he writes:

> You see, therefore, that this proposition 'Romulus existed' or 'Romulus did not exist' does introduce a propositional function, because the name 'Romulus' is not really a name but a sort of truncated description. It stands for a person who did such-and-such things, who killed Remus, founded Rome, and so on. It is short for that description; if you like, it is short for 'the person who was called "Romulus"'. If it were really a name, the question of existence could not arise, because a name has got to name something or it is not a name, and if there is no such person as Romulus there cannot be a name for that person who is not there, so that this single word 'Romulus' is really a sort of truncated or telescoped description, and if you think of it as a name you will get logical errors.[7]

Genuine, or logically proper names, Russell holds are therefore such expressions as 'this' and 'that', which directly designate particulars - in Russell's case sense data - which are the meaning of such logically proper names - even though *that* sense data are the meanings of such logically proper names is itself a contingent matter.

In that Russellian proper names do serve to directly designate particulars and do not of their essence either have or 'present' a sense, they do not present us with a category of singular symbols which *could* serve in the role of identifying; in the sense of 'picking out' objects: that is, they could not, and of their nature could not, perform the role of singular sortal expressions (symbols).

(3) Russell's 'proper names' cannot be said to *implicitly* require the category of the sortal; that the category of sortals is recognised by him as *implicit* in their introduction. This is particularly clear from the passage I cited earlier from his *Introduction to Mathematical Philosophy*, where he writes that a name has its meaning in its own right, *independently of the meaning of all other words*, and hence, *a fortiori*, independently of sortals. Indeed I am maintaining at various stages of this book, that Russell emphatically does not hold the thesis that demonstratives are dependent for their reference upon sortals.

B. That no case can be made for Russell's *implicit* recognition of Sortals in his Theory of Descriptions

Even though absolutely no case can be made out for the thesis that in his doctrine of proper names Russell, either explicitly or implicitly recognises the category of the sortal, it might well be pressed that he at least implicitly recognises the category of singular symbols which could serve in the role of picking out (identifying) an object - and hence the category of the definite singular sortal - in his doctrine of singular complex symbols, that is, in his doctrine of 'definite descriptions'. However, where such symbols have of their essence a sense, they explicitly do not have a sense in their own right: definite descriptions for Russell are *incomplete* symbols and are thus precluded from playing an identifying role.

He writes:[8] 'One very important point about the definition of "a so-and-so" applies equally to "the so-and-so": the definition to be sought is a definition of propositions in which this phrase occurs, not a definition of the phrase itself in isolation.' If a definite description is of its essence part of a proposition, it may only have a sense *qua* being such a part. In this way it is an incomplete symbol; it has, to refer back to an earlier passage cited,[9] no 'meaning in its own right'. Thus Russell holds[10] that 'when a (definite) description occurs in a proposition, there is no constituent of that proposition corresponding to that description as a whole. In the true analysis of the proposition, the description is broken up and disappears'. In other words, definite description are essentially fragments within propositions, they only have sense within the context of a proposition, and the propositions in which they occur are analysable out in terms of proper names (i.e. logically proper names) and predicates. In crude, definite descriptions are eliminable in favour of predicates. Thus, to cite the classic example,[11] Russell writes:

> We may say that "the author of *Waverley*" means "the value of *x* for which '*x* wrote *Waverley*' is true". Thus the proposition "the author of *Waverley* was Scotch", for example, involves (i.e. is analysable out as):[12]
>
> (1) "*x* wrote *Waverley*" is not always false;
> (2) "if *x* and *y* wrote *Waverley*, *x* and *y* are identical" is always true;
> (3) "if *x* wrote *Waverley*, *x* was Scotch" is always true.
>
> These three propositions, translated into ordinary language, state:
>
> (1) at least one person wrote *Waverley*;
> (2) at most one person wrote *Waverley*;
> (3) whoever wrote *Waverley* was Scotch.
>
> All these three are implied by "the author of *Waverley* was Scotch". Conversely, the three together (but no two of them) imply that the author of *Waverley* was Scotch. Hence the three together may be taken as defining what is meant by the proposition "the author of *Waverley* was Scotch".

It will now be commented that whilst I have illustrated that for Russell definite descriptions are incomplete symbols, it has yet to be shown that definite descriptions *qua* incomplete symbols are precluded from playing an identifying role. The reason is a follows:[13] No symbol which is of its essence incomplete, i.e. is of its essence solely a fragment within a proposition can serve to identify an object since the identification of objects is presupposed by the idea of a proposition in the following sense. Unless we were able to identify objects independently of the propositions in which the names of such objects occur, we could not introduce the idea of propositions which give us true or false predications of objects;[14] the idea of propositions *about* objects. And further, unless we were able to make such independent identification, we could not introduce the idea of the *same* proposition. For at least a *necessary* condition of two sentences expressing the *same* proposition, is that the *same* object is referred to by the particular subject term; a sufficient condition for two sentences *not* expressing the same proposition, is that a *different* object is referred to. But a condition of speaking of 'the same object' is that a category of expression (symbol) be acknowledged which furnishes a criteria of identity for objects in its own right, which no symbol which is of its essence only a part of a proposition could do. Therefore in that an expression (a symbol) is of its essence part of a proposition it is precluded from playing an identifying role: it presupposes, rests upon, a category of expression (symbol) which could perform this role in that the introduction of the idea of a proposition which is true or false of something or other rests upon the introduction of such a category of expression (or symbol). We can put the matter, perhaps more crudely, in another way. In that, crudely speaking, definite descriptions are eliminated in favour of predicates and predicates are of their essence, *parts* of propositions, predicates could not, of their essence, serve in such an identifying role for precisely the reason that being true or false of something or other, presupposes that one can independently identify, by a category of expression (symbol) which is not essentially predicative, what it is that predicates are true or false of: or that concerning which the predication as a whole is true or false of.

At this point it is pertinent to comment on Russell's 'translation' into ordinary language of his theory of descriptions, for it might be thought from that translation that Russell at least *in theory* has the idea of singular expressions (symbols) which could perform the role of identifying in that he recognises the existence of sortal expressions. Russell analyses the definite description, in 'informal' terms, in terms of 'at least one *person* wrote *Waverley*', etc. But to claim that he thereby recognises the category of the sortal, and hence has a means for the introduction of a category of singular expressions (symbols) which could play an identifying role, is actually without foundation. In this way his 'translation' is highly misleading. 'Person' itself for Russell is an 'indefinite description' and indefinite descriptions themselves are, of their essence, constituents of propositions. They too have no meaning 'in their own right', which is the case as concerns sortals. The proposition in which such indefinite descriptions occur are in turn analysable out in favour of propositions containing logically proper names and predicates only. For example, in 'On Denoting',[15] Russell writes:

> Suppose now we wish to interpret the proposition, 'I met a man'. If this is true, I met some definite man; but that is not what I affirm. What I affirm is, according to the theory I advocate:
>
> '"I met *x* and *x* was human" is not always false'
>
> Generally, defining the class of men in the class of objects having the predicate *human*,[16] we say that:
>
> 'C (a man)' means '"C (*x*) and *x* is human" is not always false'.
>
> This leaves 'a man', by itself, wholly destitute of meaning, but gives a meaning to every proposition in whose verbal expression 'a man' occurs.

To claim that Russell, theoretically, has a doctrine of singular expressions (symbols) of the kind required to perform an identifying role precisely necessitates that he does not construe all such phrases as 'The man who broke the bank at Monte Carlo' or 'The horse which won the Derby last year' as 'definite descriptions' i.e. expressions which signify in that they are reducible to propositional fragments which, *qua* being propositional fragments, are true or false of something or other, i.e. *describe* something or other, since description presupposes identification. It is indeed to be noted that Russell makes no clear, let alone radical distinction, between 'definite descriptions' which contain sortals and those which do not, but contain only what Strawson characterizes as 'characterizing universals'.[17] And this is only to be expected granted his reduction of sortals to fragments of propositions in his theory of indefinite descriptions. For example in giving examples of indefinite descriptions and definite description he writes:[18]

> Ambiguous: A man, a dog, a pig, a cabinet minister;
> Definite:
> The man with the iron mask
> The last person who came into this room
> The only Englishman who ever occupied the Papal See
> The number of the inhabitants of London
> The sum of 43 and 34

Of the first group (indefinite descriptions), whereas the first three examples are sortals, the fourth is certainly not. One can only count the number of cabinet ministers in that one can count the number of *A*'s (in this case men or persons) who *are* cabinet ministers, i.e. who satisfy the predicate (description) '---- is a cabinet minister'. Of the second group, whilst the first two examples contain sortals, the next two do not. One can only count the number of Englishmen in that one can count the number of men who are English; and again one can only count the number of the inhabitants of London in that one can count the number of people who inhabit London.

This criticism of Russell does not commit me to holding that such expressions as 'The inhabitant of London' could *never* serve to identify; only that its identifying role is secondary. In only serves to identify in that it is writable as, for example, 'The person who inhabits London',[19] yet for Russell 'The inhabitants of London' is never writable as 'The person who inhabits London' - or rather, in so far as it is, such a formulation is misleading, as we have seen with the earlier example of 'The author of *Waverley*'. Upon analysis it comes out as: 'There is at the least and the most one *x* who is human and who inhabits London'. But precisely in that: (i) for Russell 'The inhabitant of London' is not writable, 'in the last analysis' as 'the person who inhabits London'; (ii) for him 'the person who inhabits London' is writable as 'there is at the least and at the most one *x* who is human and who inhabits London', he cannot be said in his doctrine of definite descriptions to recognise a category of singular symbols which could serve in the role of identifying, i.e. serving to pick out a particular. Correlatively, in that he reduces all sortals, i.e. general form of the sortal, to indefinite descriptions, he fails to recognise a category of general expression (symbol) which could perform the role of identifying a particular in the sense of specifying what it is.[20] It has indeed been commented that Russell fails to recognise the identifying force of definite descriptions;[21] I hope now to have produced some explanation of why Russell could not recognise such a force: his failure to so recognise is unfortunately a necessary failure.

Having thus sought to establish that Russell fails to recognise the category of the sortal, I shall now turn to the task of working out particular difficulties for, and consequences of, his position.

C. Particular Difficulties in Russell's position

These difficulties will be centred round the Russellian doctrine of proper names. As we have seen a proper name for Russell is a simple complete symbol which directly designates an individual, which is its meaning and the name has this meaning in its own right 'independently of the meaning of all other words'.[22] But in that such a name has such a meaning, 'independently of the meaning of all other words', then the question arises as to: (1) how we determine this meaning, i.e. how we determine the designation of such proper names in general and their reference on any given occasion: (2) how we determine that we have the same Russellian proper name on two occasions of use? As regards (1), we: (a) can furnish no answer; (b) necessarily can furnish no answer. We can furnish no answer since (i) to furnish an answer in terms of a further Russellian proper name is clearly impossible: one cannot determine the designation or reference of a proper name by reference to a further proper name, and Russell does not, in any case, introduce a hierarchy of proper names and even if he were to he would find himself in an infinite regress. (ii) to furnish an answer in terms of definite descriptions is equally impossible since the analysis of a definite description implicitly contains of its essence reference to a Russellian proper name. *Necessarily* we can furnish no answer since: (a) (i) and (ii) exhaust all the possibilities; (b) a proper name for Russell is *defined* by him partially in terms of having its meaning 'independently of the meaning of all other words'.

As regards (2), Russell is holding the thesis that necessarily a proper name (in his sense) has no nominal essence, even in the sense advocated by Geach;[23] and in that he is so holding he is holding that it is a feature, and necessarily a feature of proper names, when properly understood, (or of 'real' proper names), that there are no criteria of identity for such symbols. The absurdity of such a thesis is clear viz. we could never be said to have two occurrences of the same proper name or be said to use the same proper name on two occasions of use. As Professor Anscombe has written, commenting on the mistake that there is no 'nominal essence' of individuals:

> Locke said that if you take a proper name, 'A', you can only discover whether A is, say, a man or again a cassiowary, by looking to see if A has the properties of a man or a cassiowary. This presupposes that, having grasped the assignment of the proper name 'A', you can know when to use it again, without its being already determined whether 'A' is the proper name of, say, a man or a cassiowary: as if there were such a thing as *being the same* without being the *same such-and-such*. This is clearly false.[24]

That is, in order for us to speak of two occurrences of the *same* proper name or be said to use the *same* proper name on two occasions, it must be the case that it *is* already determined that '*A*' is the proper name of some *F*, (where '*F*' is either a sortal expression in the sense I originally introduced or a sortal expression in Strawson's extended sense, such as 'flake of snow': 'bit of stone'), if '*A*' is to be the proper name of an individual, as opposed to a proper name of a kind, such as 'gold', 'silver'. I shall not here consider the latter since Russell is clearly concerned with the proper names of individuals. We reject Anscombe's thesis, and, I should hold, that of Aristotle, on pain of being able to speak of 'the same proper name'.

Here, however, Russell may reply that it is only a contingent feature, and a dependent one at that, that names may be used on more than one occasion to refer to the same object: or, to put the matter in terms more accordant with his way of thinking, that we speak of two occurrences of the same proper name. A *logically* proper name, a 'real', 'genuine' proper name, may only be used *once*; may only occur once. Proper names are *unique* in their designation and reference. Hence the above objection is otiose; and the second problem I initially introduced (2) is no problem. Necessarily we may not supply a criterion as to identity for proper names, in the strictest sense of 'proper names' and one is under no obligation, from a logical standpoint, to take ordinary proper names as paradigm. That Russell would reply in such a way can be seen from a passage in his 'Lectures on the Philosophy of Logical Atomism' where he considers what sort of language a logically perfect language would be.[25] In a logically perfect language the words in a proposition would correspond one by one with the components of the corresponding fact, with the exception of logical operators. There would be one word and no more for every simple object and every thing that is not simple would be expressed by a combination which would be a function of such simples. Such a language would be analytic, having syntax only and no vocabulary. If you add a vocabulary, such a language, whilst being logically perfect, 'would not only be intolerably prolix, but, as regards its vocabulary, would be very largely private to one speaker ... all the names that it would use would be private to that speaker'.[26] And Russell goes on to assume that, 'we have constructed a logically perfect language',[27] as a background to his analysis of facts.

Russell then, we might reasonably hold, could reject (2) as a problem for his theory, but he is still left with (1). As regards (2), we might object that a sign which was logically private to a given speaker would be no *symbol* and hence no *name* at all, on the basis of Wittgenstein's well known and often rehearsed objections to a 'private language', but this is not the line of attack I shall develop against Russell. My line of attack is much more specific and will get us to the point that in order to propound (2) he *has* to answer (1), which I have contended that necessarily he cannot. In order to claim that (2) is no problem, that Anscombe's point is otiose, he has to introduce the doctrine of names which logically may only occur once, or only be used on one occasion. My case is that in order to speak of a name which may only, as a matter of logic, occur *once*, or only by used on one occasion, one *still* has to have a principle for determining the reference or designation of the name, even on that single occasion, otherwise one cannot speak of *the* reference or designation of such a name. That is, in order to speak of the supposed reference or designation of such a name, we have (a) to allow (b) to furnish an answer to the question '*What* does the (logically proper) name refer to or designate?'. Yet in order to introduce the idea of some *one* object which the supposed proper name designates or refers to, Russell (a) has to allow the existence, in the language in which such logically proper names are introduced, of symbols (expressions) which supply principles of counting in their own right viz. sortals; (b) has to acknowledge that *necessarily* there is associated with such a logically proper name *some one or other* sortal expression (symbol), even though it is *contingent which one*, in order that the reference or designation of the name may be determined: in order that we may determine that *that* name is the name of *that* individual (object). Yet this is just what Russell, of necessity, is forced to deny in his definitive doctrine that a name has it meaning in its own right, 'independently of the use of all words'. My thesis precisely is that even the 'logically proper names' which Russell regards as the only true or 'real' names, cannot have meaning independently of *all* other words, viz. not independently of sortals, precisely for the reason that even these proper names are held to designate (or refer), in each case, some *one* individual; yet in order to introduce the idea of *one* individual we need expressions (symbols) which furnish us with principles of counting in their own right. An expression may be unique in its reference or designation in that it may refer to or designate one *F*, which has certain uniquely distinguishing characteristics, but it may *not* be unique in its reference, its designation, in that it refers to *no F*: to purport to be unique at the expense of *not* being subsumable under some principle of counting and distinction which is a principle of counting and distinction in its own right, is to purport that which is *impossible*, not to purport that which is in some way a superior kind of uniqueness. For to be unique, is to be a unique *something or other*, a unique *F*, where '*F*' is such a principle.

My third difficulty is that given the doctrine of logically proper names which Russell advocates, it is not possible for him to offer a coherent doctrine of ordinary proper names. Leaving the problem aside of construing ordinary proper names as 'veiled definite descriptions' and for present purposes accepting his construction, Russell in so construing ordinary proper names, is faced with the intractable problem that the analysis itself invokes the notion of a logically proper name which is itself impossible in that we can never, in principle, determine the reference or designation of such names. And my fourth difficulty naturally follows on from this - namely that for the same reason, Russell's doctrine of 'definite descriptions' and 'indefinite descriptions', is faced with an intractable problem. Indeed I hold that it is the impossible doctrine of Russellian logically proper names which is the principal objection to Russell's renowned theory of descriptions, *not* the confusion of meaning and truth which Strawson[28] lays at his door, since for Strawson's case to be plausible one has to: (i) hold that Russell's concern with meaning was *not* with meaning in the sense of what a sentence or proposition or even expression *says*, but rather with what *use* a sentence or expression has - which is not the case;[29] (ii) hold that the meaning of a sentence or expression is given in its use: that one subscribes to the analysis of meaning in terms of use: which I have argued is viciously circular.[30]

My fifth, and final difficulty, is related to Russell's doctrine of logically proper names - namely his doctrine of an 'essentially undetermined variable *x*'. It is my contention that what such an 'essentially undetermined variable *x*' may be said to range over, is in the last analysis such logically proper names, and in that logically proper names are what constitute values for such a variable, Russell is faced with (the again) intractable difficulty of determining that we have *one* instance of such a variable: determining that we have *one* value for the variable. My contention just mentioned may be challenged. It may be countered that Russell is not committed to holding that what strictly can replace the *x* in, say, the function: '*x* wrote *Waverley*' is a logically proper name, since what he holds is the doctrine that '*x*' is an 'essentially undetermined variable' in that '*x*' can be replaced by *any proper name whatsoever*; it is not restricted to replacements by, say, the proper names of men or persons; when completed by the proper name of an entity other than a man/person, what we have is not a piece of nonsense, but simply a false proposition. Now certainly this interpretation is initially possible in that it is consistent with Russell's theory of types, but on further investigation it turns out to be implausible. For this answer yields the result that any *ordinary* proper name, irrespective of what it is the name of, may replace '*x*' and hence be a 'logically proper name' which *ex hypothesi* cannot be the case, since *ordinary* proper names are not names at all for Russell, but 'truncated' or 'veiled' definite descriptions.

D. Consequences of Russell's failure to acknowledge the Category of the Sortal

The consequences which directly arise from the difficulties I have set out above are:

(1) Russell propounds an incoherent doctrine of names: granted my arguments above there can be no such logically proper names;

(2) He is prevented from introducing the notion of an 'ordinary' proper name since he can provide no criterion for discrimination and continued identity of such names;

(3) He can provide no analysis of 'ordinary' proper names in the way in which he advocates: this analysis is *radically* deficient, even granted the thesis that 'ordinary' proper names *are* and may satisfactorily be treated as 'truncated descriptions';

(4) His theory of descriptions, both definite and indefinite, fails: is radically deficient, and cannot be coherently advocated;

(5) His doctrine of 'an essentially and wholly undetermined variable "*x*"' is radically deficient and cannot be coherently advocated.

In short, Russell's whole logical apparatus is radically undermined by his failure to acknowledge the category or the sortal: that the 'modern' version of Russell viz. the logical apparatus of Quine is also undermined by such failure I shall be arguing later.

E. That Russell is committed to a doctrine of *absolutely* simple, basic Names and concomitantly to a doctrine of *absolutely* simple or 'pure' particulars

I said at the end of Chapter IV that I would seek to illustrate thesis I (a)(ii) by reference to Russell - namely that Russell is a philosopher who posits a doctrine of *absolutely* simple basic names, of necessity having no reference to sortals or 'matter' terms for their introduction or use and concomitantly, posits a doctrine of *absolutely* simple or 'pure' individuals.

That Russell is committed to such a doctrine of names and concomitantly such a doctrine of *absolutely* simple, 'pure' individuals, I claim follows from: (A) his thesis that *definitionally* logically proper names *directly* designate individuals, which are their meanings, *and* that this meaning is a meaning such names have in their own right, independently of the meaning of all other words. As this is *definitionally* the case, namely, it is *necessarily true* that logically proper names *directly* designate individuals, then such names *are absolutely* simple basic names, *necessarily* they have no reference to sortals, or indeed *any* other symbols for their introduction or use. Concomitantly, the doctrine of individuals introduced by reference to such names, are *absolutely* simple or 'pure' individuals: *necessarily,* for Russell, to be an individual, is to be of *no* sort or kind. It might be said that this is somewhat unfair to Russell, since what he is holding is *not*: (1) Necessarily to be an individual is to be of no kind or sort; but *either* (2) to be an individual is *necessarily* to be of *no* kind or sort or; (3) to be an individual is *not necessarily* to be of any kind or sort. In so far as Russell might be said to hold (2), he is committed to (1) since (2) entails (1). And I hold Russell may not *simply* hold (3), for this is incompatible with it being definitionally the case that logically proper names have their meaning in their own right independently of the meaning of all other words. For this latter thesis demands, at least, that to be an individual is *necessarily not* to be of any sort or kind, and 'A is not necessarily P' does *not* entail 'A is necessarily not P' whereas 'A is necessarily not P' entails 'Necessarily: A is not P'. Hence: '*Necessarily*: to be an individual is not to be of any sort or kind', which is equivalent to: 'Necessarily to be an individual is to be of no sort or kind'.

That Russell is committed, in parallel fashion follows from (B) his doctrine of 'an essentially and wholly undetermined variable "*x*"'. In Chapter I, I distinguished between the thesis: (1) that every variable is *essentially* determined; (2) that it is essential that every variable is determined. I claimed that Russell was right to reject (1), but wrong to reject (2). Yet Russell I claim as well as rejecting (1) also rejects (2) since his doctrine of 'an essentially undetermined variable "*x*"' is ambiguous and as such forces him to think that (2) entails (1). And in that he rejects (2) his *is* committed to a doctrine of absolutely simple basic names, and correspondingly to absolutely simple particulars, 'pure' particulars, since in rejecting (2) he is rejecting the thesis that what may stand as a value for his 'essentially and wholly undetermined variable "*x*"' in the function, say, '*x* is wise' is the proper name of *any F whatsoever*. And it is a necessary truth that what may stand as a value for his 'essentially and wholly undetermined variable "*x*"', is *not* the proper name of *any F* whatsoever, on the grounds that were it solely a *contingent* truth, then the variable '*x*' would *not* be 'essentially and wholly undetermined'. Hence all that may so stand is an *absolutely* simple basic name and, since the meaning of such absolutely simple basic names is an absolutely simple, 'basic', pure particular, Russell is committed to the doctrine specified in I (a)(ii) earlier and furnishes a prime illustration of it.

In spite of the above arguments, someone may *still* claim that I have been unfair to Russell in citing him as an example of a philosopher committed to *absolutely* simple basic names and concomitantly *absolutely* simple or pure individuals on the following grounds. If this were so how can Russell: (i) admit that what is designated by a logically proper name ('this', 'that') is a sensum or sense datum; (ii) that it is a contingent fact, and not a necessary truth, that what are designated by logically proper names are sense data? Granted Russell's position *does* commit him to 'absolutely simple', 'pure particulars', how can he consistently hold that we may: (1) in some sense specify what they are; (2) claim that it is only contingent that they are thus specified? To hold such a position surely commits Russell to holding, to invoke his terminology, that whilst it is a necessary truth that his 'individuals', 'particulars' fall under *some* description, it is contingent that they fall under the description they do in fact fall under. In other words, do not the very considerations here indeed commit Russell to holding the very thesis I am accusing him of necessarily denying? My answer to this problem is that it is Russell who is in the serious difficulty here, not me. For him to hold that it is a necessary truth that his 'individuals', 'particulars' fall under *some* description, even though it is contingent which one, is to contradict a crucial axiom of his system, namely his doctrine that his proper names have a meaning in their own right and do not presuppose for their introduction, or imply in their intelligible use, the meaning of *any other* word whatsoever. On the basis under consideration they would presuppose or imply just that - namely some description. If a supporter of Russell then advances the thesis that he is not committed to the doctrine of which I accuse him, on the grounds that he (Russell) is really holding that not only is it *contingent* that his particulars fall under the particular description that they do fall under, but it is *also* contingent that they fall under *any* description, then this is insufficient for Russell's doctrine that it is definitionally the case that logically proper names have their meaning in their own right independently of the meaning of all other words - independently of descriptions, by parity with the argument considered in (3) earlier. In crude terms, Russell cannot hold that it is simply contingent that his particulars, 'individuals' fall under some description: he *has* to hold that it is necessary that his particulars, individuals fall under *no* description, for descriptions *themselves* are analysed out in terms of logically proper names and predicates.

I thus finally conclude that Russell is so committed and does present us with an illustration of I (a)(ii) in Ch. IV above. Russell cannot hold the thesis that it is contingent that his particulars fall under *any* description: he has to hold that it is necessary that his particulars fall under no description, for descriptions are themselves analysed out (in part) in terms of logically proper names.

Notes

1. In his 'Lectures on the Philosophy of Logical Atomism', repr. in R.C. Marsh (ed.), *Logic and Knowledge* (London: Allen and Unwin, 1956), p. 201.
2. *Introduction to Mathematical Philosophy* (London: Allen and Unwin, 1930), p. 173.
3. He writes: (*ibid.*, p. 142) 'We shall define "proper names" as those terms which can only occur as *subjects* in propositions... We shall further define "individuals" or "particulars" as the objects that can be named by proper names.' (NB: by 'term' here Russell means 'meaning').
4. *Ibid.*, p. 174.
5. My italics.
6. Repr. in Marsh, *op. cit.*, p. 243.
7. *See also ibid.*, p. 200.
8. *Introduction to Mathematical Philosophy*, p. 172.
9. *Ibid.*, p. 174.
10. Marsh, *op. cit.*, pp. 247-248.
11. *Introduction to Mathematical Philosophy*, pp. 176-177.
12. My insertion in parentheses.
13. This argument is a brief recapitulation of an argument already offered earlier in ch. I, sec. B.1, pp. 25-28.
14. cf. Geach in *Reference and Generality*, ch. 2.
15. 'On Denoting', repr. in Marsh, *op. cit.*, p. 43.
16. Russell's italics.
17. Strawson, *Individuals*, pp. 168-172.
18. 'Lectures on the Philosophy of Logical Atomism', repr. in Marsh, *op. cit.*, p. 243.
19. cf. my earlier discussion of Frege in ch. V above.
20. In the light of a possible objection here, it has to be pointed out that Russell does not consider 'matter' terms (Quine's 'mass' terms): but even if he had done so we have no good reason to suppose that Russell would do other than reduce *these* also to predicates.
21. cf. Kripke in 'Identity and Necessity', *op. cit.*
22. Russell, *Introduction to Mathematical Philosophy*, p. 174.
23. *Reference and Generality*, ch. 2.
24. Anscombe and Geach, *Three Philosophers*, p. 8.
25. Repr. in Marsh, *op. cit.*, pp. 197-198.
26. *Ibid.*, p. 198.
27. *Ibid.*
28. In 'On Referring', *op. cit.*
29. cf. Gareth Evans, *The Varieties of Reference*, p. 52, n. 14; p. 67.
30. Attfield and Durrant, 'The Irreducibility of "Meaning"', *op. cit.*

Chapter VII

Geach: Sortals, Substantival General Terms and General Names

A. Geach's recognition of the category of Substantival General Term does not entail recognition of the category of the Sortal as either basic or distinctive

In his discussion of the kind of general terms which can ever occur as names,[1] Geach distinguishes between those general terms concerning which it makes sense to prefix the words 'the same' and those concerning which it does not make such sense. Further it is only concerning the former that the question: 'How many?' can be raised. 'Letter' (in the typographical sense) is given as an example of the former: 'sea' as an example of the latter. The ground of the distinction between general terms was, Geach holds, recognised by both Frege and Aquinas. The former come under the heading of 'concepts' which 'sharply delimit' what they apply to i.e. Frege's 'concept words' properly so called:[2] the latter do not. As I have earlier commented,[3] Geach rightly criticises Frege's account of 'red' as a concept which determines no finite number: contrary to Frege, Geach rightly holds that 'the same red thing' supplies no 'criterion of identity'. What supplies a criterion as to identity is a general term which is prefixable by the words 'the same'. But from the fact that Geach recognises such general terms, and in so doing, implicitly in one sense recognises the category of the sortal since he designates such terms, following Aquinas, as *substantival* general terms; and a sufficient condition, though not a necessary condition of a general term being substantival, is that it takes a number on its own account - supplies a principle of counting and distinction for particulars in its own right - *it does not follow* that Geach recognises the category of the sortal either:

(a) as a category of general term which is *basic*: rather they are a subclass of his substantival general terms - which gives rise at least to the suspicion that for him the basic class of substantival general terms are 'matter', 'mass' terms and hence that all sortals (in my first and original sense) are reducible to matter terms precursed in a certain way, viz., by 'a bit of'; 'a lump of' etc., though for reasons offered I do not think this charge can be fully directed against Geach;[4]

or:

(b) as an independent distinctive category of symbol in the way in which I have sought to claim it is.

For it is possible to claim that Geach either:

(i) reduces substantival general terms (and hence sortals) to their occurrence as names or predicables (predicates) in propositions;

or:

(ii) actually reduces substantival general terms (and hence sortals) to a certain sort of predicable, following Frege.

B. Geach's reduction of Substantival General Terms to Names or Predicables (Predicates) and its difficulties

B.1. The evidence for (i) is indirect. It is that Geach certainly speaks of a certain kind of general term - his substantival general terms - which can occur as names or again as predicables and produces criteria for determining when such substantival general terms are occurring as names.[5] It may be said that it does not follow from this that Geach is committed to the thesis that such substantival general terms (and hence sortals) are actually *reducible* to their occurrence as names or predicables. But whilst this does not formally follow, I can find nothing in his writing which would *prevent* him from having to hold that it does so follow. It might be suggested that he could maintain that such general substantival terms could retain their identity and independence; they are, after all, not categories of expression or symbol which of their essence or nature may be regarded as categories of expression or symbol which can either stand for an object (or objects) or be combined with another expression (symbol) to form a proposition, even though they may *occur* as such expressions or symbols. But if this line of thought is adopted then one is faced with the difficulties: (a) that Geach actually assimilates his substantival general terms to Frege's 'concept words'[6] (and hence sortals to 'concept words'); and Frege's 'concept words' are 'essentially predicative'; (b) he writes:[7] 'to link this up with what I have been saying, we need only observe that for Frege a concept was what language represented by a general term'. In other words, that the correct logical form of substantival general terms (and hence sortals) is the form of the concept word '---- is *F*'. And Geach certainly seems to concur with this. I thus hold that the only alternative Geach may offer to the charge (i) that he reduces substantival general terms to occurrences as names or predicables (or predicates) is (ii) that he actually *reduces* substantival general terms (and hence sortals) to a certain sort of predicable, following Frege. *Both* these theses, I hold, are open to severe difficulties and disastrous consequences. I shall now proceed with the difficulties I claim in relation to (i) and then turn to (ii), though I shall have occasion to mention (ii) in dealing with (i).

B.2. The difficulties that arise for (i) are both general and specific in relation to Geach's own writing. One general difficulty is that of speaking of a general term which furnishes a criterion as to identity, or a general term which furnishes a principle of counting in its own right, now having the *use* of a predicable, now having the *use* of a general name. In that *X* is such a term, it can be held that it cannot be said to have the *use* of a name, occur as a name or have the *use* of a predicable, occur as a predicate for to say that, for example, a sortal is a general term which supplies a principle of counting in its own right is to *give its use*: and similarly as regards substantival general terms supplying criteria as to identity; and hence neither can be said to *have* a use, either as a name or a predicable. In that a sortal, for example, is a general term defined by reference to its use: in that a substantival general term is a general term which perform a certain function, then they themselves cannot be said to *have* a use or function as a category of term or symbol. One cannot have the use of a use or a function (in this sense) of a function.

As a potential way out of this difficulty, it might be urged that Geach's point could be put in terms not involving 'general terms' but simply 'common nouns' - viz. that certain common nouns now have the use of names, occur as names: now have the use of supplying criteria as to identity. But to so put the point is to be open to the objection: (a) that Geach is not merely engaging in an exercise in English or Latin grammar; (b) although on p. 40 of *Reference and Generality* he is speaking of common nouns, if substantival, being used in acts of naming, he earlier speaks of substantival *general terms* and goes on to speak of substantival general terms occurring in propositions 'now as a name, now as a predicable'. Whilst we may perhaps speak of common nouns having the use of names etc., there is indeed objection to general terms which furnish criteria as to identity (and general terms which supply principles of counting in their own right) having such a use or so occurring for reasons offered above.

I said there was also a specific problem for Geach. Although he speaks of substantival general terms occurring in propositions now as a name, now as a predicable, there is some difficulty for such talk even within his own logical vocabulary. For names and predicables (rightly) are held by Geach to be classes of expression which may occur *in* certain roles within the context of a proposition (e.g. in the role of logical subject), it is therefore difficult to see how he himself can speak in terms of substantival general terms occurring in propositions, now as a name, now as a predicable - viz. how names and predicables may themselves be regarded as *roles* within the context of a proposition.

B.3. Having faced these difficulties for even *speaking* in terms of substantival general terms as occurring now as a name, now as a predicable, within the context of a proposition, I now have to face the challenge: *What are the particular difficulties and alleged disastrous consequences of so speaking?* The severe difficulties, I claim, are that in that such general terms do occur as either names or predicables, then they can no longer retain their status as general terms which supply criteria as to identity or which supply principles of counting in their own right. For in that they occur on an occasion as a predicable, in that they are on occasion construable and writable as a predicable, then following Geach's account of 'predicables' they 'apply to' or are 'true or false of objects'; but in that they are true or false of objects, they cannot serve as a criteria as to the identity of objects or as principles of counting for objects. That which is true or false of an object, cannot at the same time, also be a criterion as to the identity of an object or a principle of countability. Indeed such a function is necessarily excluded. Geach may reply to this by saying that whilst it is true that *F* cannot be both: (i) a predicable; (ii) such a principle on an occasion, *F* may nevertheless: (a) occur as a predicable on one occasion; (b) yet be such a principle on another. But this clearly won't do since on the thesis being considered to be such a principle for Geach is for such a substantival general term to occur in a proposition and to occur in a proposition *is* to occur either as a predicable or name. Of course to fully make my point here I have to argue that it is equally impossible for Geach's substantival general terms to always occur as general names, within the context of a proposition and retain their status either as criteria as to identity or principles of counting. This I shall now do in general terms and then specifically illustrate the point by detailed reference to Geach's writing, thus taking up my earlier promise[8] that I would illustrate my general thesis that sortals may not be reduced to general names. My case is that in that a sortal or substantival general term occurs as a general name, then following Geach's own account of: (i) proper names, (ii) general names, both proper names and general names signify in that they either: (a) stand for a simple object; (b) stand for several objects. Thus, in that a sortal or substantival general term occurs as a general name, then it signifies in that it stands of several, i.e. a number of objects: but in that it simply stands for a number of objects, its cannot, at the same time, *also* furnish a criterion as to the identity of those objects or a principle of counting in regard to those objects. Geach may once again reply that, agreed, in that such a general term occurs in the role of standing for a number of objects, it cannot at the same time perform its function of being a criterion as to identity or principle of counting as to those objects: he may even argue that this is of necessity excluded - but, we may surmise, he would reply, that this does not prevent him saying that such a substantival general term might nevertheless occur in such a role on another occasion. But this will not do, since for Geach, on the thesis being considered, the only way that such a substantival general term may occur in the context of a proposition *is* as a general name or predicate. These exhaust the possibilities for such general terms occurring in the context of a proposition. And thus I claim we are faced with a disastrous consequence, in that Geach does reduce such general terms to occurrences as predicates or names within the context of propositions. The effect of reducing *all* such terms to occurrences as name or predicables is that they may no longer retain their status as criteria as to identity for individuals or principles of counting for individuals; yet, as I have earlier argued, we need such general terms in such status in order to introduce the notion of 'name' and 'predicable' in the first place.

I thus now turn to Geach's treatment of substantival general terms occurring as general names with a view to explicating the prime difficulties involved.

B.4. I have already produced a quite general argument for the thesis that general names rest upon the introduction of symbols (expressions), sortals. I now turn to the details of Geach's case that sortals, or more generally, substantival general terms, can occur as names and are in effect, on one Geachian thesis, reducible to such an occurrence. That he regards, for example, 'cat' occurring as a general name and hence reducible to a general name is clear from his writing.[9] He presses the case that just as 'Jemima' can stand as a logical subject, so can 'cat': indeed the use of 'Jemima' as logical subject, he claims rests on the use of 'cat' as such a subject.[10] But then the problem arises as to what 'cat' stands for. He writes:

> Not for Jemima, nor for any definite cat rather than any other; the story may have been true of Jemima, but as told it was not a story about Jemima or about any definite cat rather than any other. Certainly not for an indefinite cat; there is no such animal. In an act of naming "cat" potentially stands for any cat you choose .... I think, therefore, we must suppose "cat" in the story to refer to any and every cat, equally and impartially. This result will no longer be suspect if we remember that a proposition making a predication "*F*( )" about *a* cat must have the same truth value as would belong to a disjunction "*F*(Jemima) or *F*(Ahab) or *F*(Tibbles) or *F*(Mehitabel) ..." (and so on for the complete list of cats); and this disjunction is symmetrical as regards the names of cats occurring in it.[11]

However, *that* Geach requires for the introduction of 'cat', an expression (symbol) which does *not* signify in that it stands for a number of objects, even a disjunctive set in which it does not signify in that it is a general name, but an expression (symbol) which signifies in that it supplies a principle of counting can easily be illustrated. For he writes, as cited above, 'In an act of naming, "cat" potentially stands for *any cat*[12] you choose'.[13] The second occurrence of 'cat' here, i.e. in 'any cat you choose', cannot be an occurrence in a naming role, i.e. an occurrence as a name, for otherwise it could not serve to indicate or specify the range of choice, which is primarily its function. If 'cat' in 'any cat you choose' were simply the occurrence of a substantival general term as a name, or simply a name, then we should be faced with the further question of determining the range of application or reference of the general name. Yet Geach acknowledges (and rightly) no such further question.

Let me now take another illustration of the difficulty for Geach's thesis.

He has written, as cited above: 'I think therefore we must suppose "cat" in the story to refer to any and every cat equally and impartially'. Once again 'cat' in 'any and every cat' cannot be regarded as a general name having significance in that it stands for a number of objects; for on such an hypothesis, it could not perform the role, in conjunction with 'any' or 'every', of identifying/indicating the range of reference for 'cat' in the story. A similar point may be made in connection with Geach's requirement 'and so on for the complete list of cats': 'cats' here has the role of making clear of the extent of the list, which it could not do if it were simply occurring as a general name or simply were a name. For if 'cat' here were simply so occurring or being, then we should have to introduce a further expression (symbol) which could perform the role of indicating, specifying the range of application (or reference) for the general name 'cat'. Yet Geach has no theoretical means for introducing such a further and yet more fundamental category of expression, symbol, since for him 'animal' or 'creature' would still be regarded as general terms, occurring as, or actually being general names and exactly the same problems would arise at this level too.

Thus I claim Geach's treatment of sortals - his substantival general terms - as names leads him to the difficulty of not coherently being able to make his point.

B.5. Having thus presented objections and even indicated the disastrous consequences that may be said to arise from the treatment of substantival general terms as occurrences as either predicables (predicates) or names, I now turn to the thesis of the explicit reduction of such general terms to predicables; thesis (ii) in A. above. The *difficulties* introduced in such a reduction will have already become clear to the reader:

(1) To treat such substantival general terms - in particular to treat sortals as predicables, reduces one to saying that such sortals as 'man' or 'cat' are reducible to '---- is a man', '---- is a cat', which 'apply to' or are 'true of' objects, but in that such terms 'apply to' or are 'true of' they cannot serve as criteria as to identity of objects or as principles of counting for objects. For, as I earlier argued, that which is true or false of an object cannot serve as a criterion of identity or principle of counting for an object. Indeed such a function is necessarily excluded. We have to be able to say what it is that the predicable applies to or is true or false of and no predicable can perform this function. If then all such general terms are reducible to predicables, we would never be able to say - and in theory never be able to say - what our predications (in Geach's sense of predications) were *about*. And if we could never in theory say this then we could never be able to assess the truth value of any predication.

(2) Further, the thesis that such terms are actually reducible to predicables raises a parallel difficulty discussed earlier; namely, that on such an account one cannot say, and necessarily cannot say, that such terms are able to retain their status as criteria as to identity or principles of counting.

C. Further criticisms of Geach's thesis that the introduction of Proper Names merely presupposes the introduction of Names

I have already criticised, in part, one interpretation of Geach's thesis namely that the introduction of proper names merely presupposes the introduction of names viz. that expressions whose introduction is necessary for us to speak of the *same* proper name may themselves be regarded as names. It is however of worth to repeat that argument here: (a) for its own sake; (b) to consider an objection.[14]

C.1. Geach claims that a general term cannot be used as a name (which is a sufficient condition for such a term being able to occur as a logical subject [cf. *Reference and Generality*, Chapter 2, Section 23]) unless it is substantival. As we have already seen the condition for substantivality is whether it makes sense to prefix the words 'the same' to a general term. This criterion is offered on the principle that 'when the same name is used in two acts of naming, we can always ask whether the same thing is named'.[15] It is my contention that whilst it is true that when the same name is used in two acts of naming we can indeed ask whether the same thing is named, it does not follow from this that such general terms as can be substituted for 'thing', in 'the same thing named', can itself be occurring as a name and hence in Geach's sense of 'name' as a logical subject.

For example, if one says:

(1) Socrates loved Xanthippe
(2) Socrates detested the Sophists

here we have the same name viz., 'Socrates' being used in two acts of naming (in Geach's sense) and we can, indeed we have to, ask whether the same thing is being named in order to begin to decide whether these propositions are true or false; but one must not confuse an answer to the question: 'Is the same thing being named by these two occurrences of the name "Socrates"?', which might take as its answer: 'Yes, the same name "Socrates" names one and the same man viz., the philosopher who lived in 5th century Athens, who was the teacher of Plato etc.', with a name i.e. one must not confuse the specification of what is named with a name. An expression of the form 'the same ----', where the blank takes a general term, is indeed needed for *A* to have the use of a proper name but neither from this nor the parallel consideration above does it follow that such a general term can occur as a name. It seems that Geach has moved from: (i) stating a necessary requirement for the continued use of a name to; (ii) maintaining the thesis that any general term which can supply a criterion of identity for the continued use of a name *A* can itself occur as a name, which is to maintain, without further reasoning being offered, that a condition for the use of a name is itself a name, or at the very least can occur as a name. Without further reasoning it is not at all clear why this should follow and indeed it only follows on a confusion. Further, to maintain the thesis that a condition for the use of a name is itself a name, engenders a regress since for that 'name' we can also ask whether the same thing is named on two different occasions of the same 'name'. To be fair to Geach however he has both a prior argument and a further argument.

The 'Prior Argument' (*ibid.* Section 23):

Geach would contend that such general terms can occur as names, since such an expression occurring in a sentence expressing a proposition of a subject-predicate type can be linked up with an act of naming - in Geach's example 'Smith nearly went to prison', 'Smith' occurs as a name since 'Smith' links up with an act of naming in the situation in which I say 'Smith!' pointing to a man and affirming that he was the Smith who nearly went to prison.[16] I fear there is an important snag in this argument which, when expressed, will not permit Geach to make the move he wants to. 'Smith' being a name in 'Smith nearly went to prison' crucially depends upon the link up mentioned and the snag comes precisely in Geach's description of the link - namely in my affirmation that the Smith pointed to is one and the same Smith as the Smith who nearly went to prison. By writing the all-crucial link as: 'The Smith pointed to is one and the same Smith as the Smith who nearly went to prison', Geach unfortunately leads us astray, for on his own principles expressed elsewhere[17] he should have written:

> 'The man pointed to, who bears the name "Smith" is one and the same man as the man who nearly went to prison'

since Geach subscribes to Frege's principle that proper names cannot be precursed by the definite article, or by a number, or by 'the same'. Now since in order to state the link properly we need to introduce a term which can furnish us with: (a) a criterion of identity; (b) a criterion of continued identity - namely 'man' in this case - that term itself cannot function as a name, without the above mentioned confusion and regress being respectively involved and generated. Yet Geach, inadvertently I suspect, allows that it can function as a name precisely by writing what can, on his own admittance, paradigmatically occur as a name in its place viz., the proper name 'Smith'. One might further argue against Geach's contention that a general term is an expression which could always be used outside the context of a sentence in a simple act of naming; this distinctly goes against Frege's position. However it is not necessary for me to argue against this here, since Geach does not claim that it is a sufficient condition of an expression being a name in a sentence which has the role of a subject-predicate proposition that that expression could, in other contexts, be used for a simple act of naming; it is the link that is all crucial here.

The 'Further Argument' (*ibid.* Section 34):

He writes:

> So far, as I have said, we have not found any names other than proper names to be used as logical subjects of propositions *regardless of whether the things named are present or absent;* proper names are at any rate the only obvious examples. However, as I shall try to show, it is plausible to suggest that general terms (substantival ones, that is) also admit of such use as logical subjects. Indeed, although the use of proper names in that capacity is much more easily recognised, it is arguable that such use depends on the possibility of general terms also being logical subjects.[18]

What is his case for this? He begins his argument by commenting:

> For every proper name there is a corresponding use of a common noun preceded by "the same" to express what requirements as to identity the proper name conveys: "Cleopatra's Needle" - "the same (bit of) stone"; "Jemima" - "the same cat" .... In all these cases we may say that the proper name conveys a *nominal essence*: thus, "cat" expresses the nominal essence of the thing we call "Jemima".[19]

With this I am not in dispute; it brings out the point that a condition of employing proper names for objects is that such objects must be of a certain sort or kind and that for a proper name to have a continued use we must be prepared to say that it names the same X on different occasions of use. However the heart of Geach's argument is contained in the following paragraph:

> I might tell a story involving Jemima and the river Thames without using either of these proper names; I might refer to Jemima as "a cat" and the Thames as "a river" when I first mentioned them, and thereafter speak of "the cat" and "the river" - sc. "the same cat" and "the same river". A hearer unacquainted even by hearsay with Jemima and the Thames would lose absolutely nothing by this suppression of the proper names. So if "cat" in this story telling did not retain the use of a logical subject, how could "Jemima" have such a use? How could we make out that "Jemima" has what it takes to be a logical subject, but "cat" has not?[20]

Now, whilst in the context of a story 'a hearer unacquainted even by hearsay' with Jemima and the Thames would lose absolutely nothing by the suppression of the proper names, since in a story the question of *which* cat or *which* river would not arise, such a question might well do so and could in theory always do so in a non-story telling context. Even in a non-story telling context however (it might be urged), nothing is lost by the mere suppression of the proper names. Agreed, nothing is lost by the mere suppression of the proper names, but something *is* lost if all one introduces in lieu of such proper names is 'a cat', 'a river' (sc. 'the same cat', 'the same river'), for in a non-story telling situation if we are faced with a purportedly true proposition: 'A cat fell into a river', where 'a cat' is interpretable as 'some cat' in the definite sense of 'some'; 'a river' as 'some river' in the definite sense of 'some', then the questions 'Which cat?'; 'Which river?' certainly arise and seek an answer either in terms of a definite description which can perform the role of uniquely identifying the cat or river introduced, or the proper names of the cat or river so introduced. What can putatively stand as a logical subject is either a definite description of a kind which can perform an identifying role or a proper name. All Geach's argument indicates is that definite descriptions of such a kind can stand as logical subjects, not that a certain kind of general term can. His case (here) for saying that some general terms can relies on his story-telling situation, where the question 'Which one?' need not necessarily arise, though of course it could do so if his story about a cat and a river formed part of a larger story or a collection of stories in which a number of cats or rivers were introduced. It seems that Geach's case would only hold if someone introduced the proposition 'Some cat fell into some river', where the sense of 'some' introduced was the indefinite sense - 'Some cat or other fell into some river or other'. Here, *ex hypothesi* the question 'Which one?' cannot arise and I am sure that Professor Geach would not wish to say that either 'cat' or 'river' was occurring as a logical subject in such a proposition.

In sum, whilst Geach's point that proper names convey a nominal essence is perfectly correct, he has not shown that it follows from this that such nominal essences themselves must be able to stand as logical subjects - be names.

C.2. The objection earlier alluded to is as follows: it will be objected that at least the initial argument given above rests on the principle that Geach would treat terms which can supply a criterion of identity for the continued use of a name *A*, as a sort of proper name: which is false. In this context he regards such terms as general names. My argument only shows that Geach cannot treat such terms as proper names: it does not, of itself, establish that it is absurd to hold 'That a condition of the use of a name is itself a name'. However, regarding such expressions as can form a criterion of identity for the continued use of a name as general names, does not alter the force of my contention. To regard such expressions (symbols) as general names, or occurring as general names, raises all the problems introduced earlier and in effect shows that it is absurd to hold that a condition of the use of a name as a logical subject is itself a name. Such a position refuses to acknowledge that such expressions *qua* supplying principles of counting in their own right are required for the introduction of *both* proper names *and* general names; where general names are treated, as in Geach's writings, as general terms (expressions) which signify in that they stand for a number of objects. Geach's thesis that proper names merely suppose the introduction of *names* has the consequence that he cannot introduce names.

D. The disastrous Consequences of reducing Substantival General Terms and Sortals to Names or Predicables (Predicates)

Having set out the difficulties for and indicated what I hold to be the disastrous consequences for the theses:

(i) that substantival general terms are reducible to names or predicables in that they are reducible to their occurrences as names or predicates in propositions;
(ii) that substantival general terms and hence sortals are actually reducible to predicables.

I now turn to an explicit statement of these consequences.

D.1. *For the thesis that such substantival general terms are reducible to occurrences as general names*

This thesis has the consequences:

(1) That we cannot introduce proper names, in the ordinary sense, for on such a thesis the only category of expression (symbol) at our disposal for the introduction of a proper name would be one which signified in that it stood for several objects, without any means of indicating the nature of the objects so stood for and hence without any means of indicating the bearer of any individual proper name. It is of no avail here to reply on such a thesis we could still introduce the notion of a *logically* proper name, in Russell's sense (the demonstrative 'this', 'that' or the modern equivalent), since I have already produced reasons for rejecting the thesis that demonstratives may be regarded as a sort of name or may be introduced or used external to the context of a sortal[21] and in any case Geach himself explicitly rejects the thesis that demonstratives are a type of name.[22] Further, not only could we not introduce ordinary proper names, but, even if, *per impossible* we were able to introduce them: (i) we would not be able to use them on one occasion - say we were using this name on this occasion, since we would have no means of specifying the reference; (ii) we would not be able to use them on a number of occasions since we would have no means of determining that we had the *same* reference on a number of occasions.

(2) That we cannot introduce general names since, as we have seen, a condition for the introduction of general names is that we have at our disposal a substantival general term which is not itself reducible to occurring as or being a general name. Equally, even if *per impossible* we were able to introduce general names, we would not be able to use them on a single occasion since we would have no means of specifying the several objects they purportedly name and we would not be able to

use them on a number of occasions since we would have no means of determining that they referred to the same number of objects on those occasions.

(3) That we cannot introduce the notion of a general name variable, as Geach later requires,[23] since we have in principle no means of indicating the reference range of such variables: we have no means of indicating what range of values may be substituted for the variables. It will not do here to say that what may be substituted for a general name variable is simply any general name whatsoever, for any general name is the general name of a set of objects or potential set of objects: it stands for each such object equally and impartially and we have, in principle, no means of indicating its range or potential range of objects. In that it is held that the sense of a general name is given in its extension, we have no means of specifying its potential extension; and in that it is held that the sense of a general name variable is given by the range of general names (arguments) which may replace it to yield a true or false proposition, we have no means of determining the range of general names which may so replace it. If we reply any general name whatsoever may replace it we are once again faced with the problem just mentioned above.

(4) That sortals, substantival general terms are *themselves* treated a variables,[24] whereas I have argued that as substantival general terms and in particular sortals are required for the introduction of: (i) proper names; (ii) proper name variables; (iii) general names; (iv) general name variables, one is reduced to reducing sortals to categories of symbol/expression which themselves require sortals for their intelligible introduction and hence reduced to absurdity.

(5) That we may not introduce the notion of the 'nominal essence' of an individual in Geach's sense of nominal essence, for in that in, e.g. 'Jemima is a cat', 'cat' gives us the nominal essence of Jemima, then it distinctly may not be construed as occurring as a general name, for a general name signifies in that it stands for a number of objects, yet merely standing for a number of objects is not a sufficient condition, and necessarily not a sufficient condition of being able to say, specify, what Jemima is and hence give the nominal essence of Jemima.

(6) That sortals and substantival general terms, reduced to occurring as names, can no longer retain their status as general terms which supply criteria as to identity or principles of counting; and necessarily *cannot* do so, and hence we have no means of introducing general terms which may supply us with such criteria or principles unless it can be argued that this may be done in terms of predicables, which I shall argue it cannot.

D.2. *For the thesis that such substantival general terms and hence sortals are reducible to occurrences as predicables*

This thesis has the consequences:

(1) That such substantival general terms can no longer retain their status as general terms which supply criteria as to identity or principles of counting and that it is a matter of necessity that they cannot do so. From which it follows that we have no means of introducing general terms which may supply us with such criteria or principles, for this may not be done by the means of names as argued above.

(2) That we may not introduce the notion of 'name' or 'predicable' since the introduction of these notions is dependent on such substantival general terms *not* being reducible to their occurrences as predicables: i.e. dependent on such general terms retaining their status as criteria as to identity and principles of counting.

(3) We would in theory never be able to say what our predications were *about*, and hence we would never be able to assess the truth value of any predication.

(4) That, in the last analysis, one would not be able to give a coherent account of the sense of a proper name. For on such an account what gives the sense of a proper name is a substantival general term occurring as a predicable or set of predicables; but if what gives the sense of a proper name is simply that which applies to or is true or false of the bearer of the name, on an occasion, how may we say that such gives the sense of *the* name? We have already had to determine that we have a true or false predication concerning the bearer of the name and yet that which is true or false of the bearer of the name on an occasion is supposed to give this! In other words, the sense of the proper name cannot be given by any set of true predications concerning that name, for understanding that any set of predications concerning the bearer of the name are true, depends upon our first understanding the sense of the name which hence may thus not *itself* be given in terms of any set of predications concerning the bearer of the name and hence in terms of a predicable or any set of predicables.

It may be objected that this is only a severe consequence for Geach in that he holds that it is necessary that proper names have a sense. If it is merely contingent that proper names have a sense then this consequence is otiose. But I claim that it is integral to Geach's thesis that it is *necessary* that proper names have a sense even though they do not necessarily have the sense they have. One might be tempted to the view that for Geach it is only an *accident* that proper names have a sense, in that he often speaks solely in terms of proper names standing for or naming objects - for example, he writes:[25] 'The object named by a name may be called its bearer' and it might be suggested that, for him, it is an accident to our use of a proper name that we claim the ability to identify the object so named (or that it is accidental to the occurrence of a proper name that *some* identity of the object so named is required). But such a suggestion appears clearly inconsistent with his implicit criticism of Russell;[26] and that he *does* hold that proper names of their essence have a sense can be independently substantiated by his thesis that we may only speak of the same proper name in that we can speak of *the same A*, where '*A*' is a term that can be prefixed by 'the same' and thus form a criterion as to identity for the names. I thus claim that the consequence under discussion is applicable to Geach: for him it would entail that we may not introduce the notion of a 'proper name'.

D.3. *For the thesis that such substantival general terms and hence sortals were actually reducible to predicables*

This thesis entails all the consequences set out in relation to D.2. above and it would be tedious to repeat them.

I thus maintain that Geach's failure to recognise the category of the sortal results in a fundamental disaster for his own theory of naming and predication.

Notes

1. *Reference and Generality* (Ithaca, New York: Cornell University Press, 1962), pp. 38ff.
2. cf. ch. V above.
3. *Ibid.*
4. *See* ch. IV above, pp. 157-158.
5. *Reference and Generality*, ch. 2, secs. 32-35, pp. 40ff.
6. *Ibid.*, pp. 38-39.
7. *Ibid.*p. 38.
8. *See* Introduction, p.1, p. 9 and ch. I, p. 28.
9. *See* particularly: *Reference and Generality*, ch. 2, sec. 34, p. 45.
10. *Ibid.*, p. 45.
11. *Ibid.*, pp. 45-46.
12. My italics.
13. *Ibid.*, p. 46.
14. The text of secs. C.1. and C.2. originally appeared in my paper 'General Terms and Logical Subjects', in *Canadian Journal of Philosophy*, Vol. III, No. 4, June 1974, pp. 525-538 and is reprinted here by kind permission of the editor.
15. *Reference and Generality*, p. 38.
16. I have rephrased Geach's argument in the first person to avoid excessive verbiage.
17. cf. his article on Aquinas in *Three Philosophers*, *op. cit.*, p. 109.
18. *Reference and Generality*, p. 43.
19. *Ibid.*, pp. 43-44.
20. *Ibid.*, p. 45.
21. *See* ch. II, sec. G, pp. 53-55.
22. *Reference and Generality*, p. 27.
23. *Ibid.*, pp. 159ff.

24. *Ibid.*
25. *Ibid.*, p. 29.
26. *Ibid.*, p. 27 and *see also* p. 44.

Chapter VIII

Strawson: Sortals - Failure to Recognise Their True Nature; His Dual Position

A. The Claim that Strawson recognises the true nature of the Sortal

Of the writers I have mentioned it will be pressed that Strawson recognises the true nature of the sortal in that he:

(i) explicitly introduces the notion of a sortal and sortal universal;[1]
(ii) explicitly introduces sortal universals as terms which supply principles of counting and distinction for particulars, and do so in their own right, presupposing no antecedent principle;
(iii) contrasts sortal universals with other types of universal which do not supply such principles (e.g. 'characterizing universals');
(iv) actually analyses the subject-predicate distinction by reference to the notion of 'identifying a particular'[2] and I have earlier contended that it is the prime function of the definite form of the sortal to identify.[3]

Even granted these points however, I shall contend that he nevertheless fails to recognise the sortal as a logically independent category of expression. Rather, I shall argue, he treats sortal expressions both (i) as a type of name or (ii) as predicative in that at the most basic level of predication they essentially play a predicative role. This dual position, I hold, leads to considerable embarrassment for his theory. In that he treats sortals as a type of name he is faced with the problem that names presuppose for their introduction and use some sortal or other,[4] and that in so far as names may at all be held to serve to identify they may only so serve via sortals.[5] Hence to treat sortals as a type of name entails: (a) that he cannot account for the introduction and continued use of names; (b) that he cannot give recognition to the concept of 'identifying' and hence the prime recognition to the concept as his theory requires.[6] It might be objected that (a) and (b) only hold if sortal expressions are treated as proper names; they do not hold if sortals are treated as general names. This objection itself however only holds if (c) it can be shown that Strawson himself holds sortal expressions to be general names; (d) if sortals themselves may be regarded as general names. As regards (c), such a contention is incompatible with his doctrine of sortal universals, for general names (or rather what they stand for) cannot be instanced; and further it is certainly arguable, granted

Strawson's doctrine of 'terms', that a sortal expression (e.g. 'man') is ultimately to be taken as an individual name designating a single sortal universal.[7] As regards (d), the thesis that sortals may be regarded as general names fails in the final analysis in the light of my arguments against Geach.[8] In that Strawson treats sortals as predicates (as predicative), being true or false of a subject, or 'applying to' a subject, predicates rest for their introduction upon a class of symbol which may serve to introduce subjects of predication, viz. upon sortals, as I have already argued.[9] Hence treating sortals as themselves predicates entails (e) that he cannot account for the introduction and continued use of predicates;[10] (f) that he cannot give due recognition to the concept of 'identifying' and hence the prime recognition to the concept as his theory requires.

Before I turn to substantiate my claims: (1) that Strawson does indeed treat sortals as a kind of name; (2) that he also treats them as predicative, as predicates, it is necessary to explain that the move from treating of *expressions* in (1), to treating of *roles* in (2), is not of my doing, but is one which is implicit in Strawson's own writings.[11]

B. That Strawson regards predicate expressions as a kind of *Name*, Sortal expressions as predicates and hence as a kind of *Name*

That Strawson does so treat sortal expressions as a kind of name can be seen from the fact that he regards all predicate expressions as a kind of name and he clearly treats sortal expressions (sortals), at least in the basic case, as predicates.[12] I turn first to his account of the 'grammatical criterion' for the subject-predicate distinction.

B.1. At *Individuals*, pp. 148-149 Strawson holds that whilst a 'subject expression' *merely* introduces its term into a remark having, in non-problematic cases, the grammatical form of a noun substantive, more particularly, a 'singular grammatically substantival expression',[13] a 'predicate expression' does not

> *merely* introduce its term, or introduce it with *merely* such an indication of style as is given by the case ending of the noun. It introduces its term in a quite distinctive and important style viz. the assertive or propositional style.[14]

Earlier Strawson has said that a predicate expression 'contains at least one finite form of a verb, in the indicative mood, which does not, within the limits of the B-expression (i.e. a predicate expression)[15] form part of a complete sentence or clause'.[16] From this it seems clear that by *predicate expression* is meant e.g. 'is wise', 'smokes', 'is a man', but that such expressions are to be regarded either (a) a name or (b) a combination of a name, together with some other type of expression, viz. 'is' in the case of 'is wise' or 'is a man' and distinctly *not* as expressions which are *fundamentally incomplete*, functional expressions of the form: '---- is wise', '---- smokes', '---- is a man', in the Frege, Geach, Quine sense. My reasons for holding this are as follows:

(1) Strawson writes:[17] 'A general requirement of both A- and B-expressions (subject and predicate expressions)[18] is that an expression of either kind should be capable of yielding an assertive sentence when *combined with*[19] some suitable expression of the other kind.' If Strawson here were truly recognising the category of predicable expression as opposed to that of a *name* he could not speak of *combination*, but rather, as regards 'predicate' expressions, of *completion*. We should note that he has explicitly written in the passage cited above that a predicate expression, whilst it contains at least one finite form of a verb 'does not ... form part of a complete sentence or clause'. This denies that a predicate expression is essentially a 'propositional fragment' or 'predicable', but is to be regarded as a sort of name or a name in combination with another sort of expression viz. the copula 'is'. Admittedly, Strawson speaks of the expression 'is wise' as demanding a certain kind of *completion*,[20] namely, completion into a proposition or propositional clause. However, in his notion of 'predicate expression' Strawson cannot be taken as introducing the notion of a Geachian predicable or Fregean 'concept' expression, since he has no good reason for speaking of *completion* as opposed to *extension* or of a form of expression which may, without violence to syntax, be *combined with* a 'subject expression' to form a proposition.

(2) As I have pointed out 'combined with' is precisely his later mode of talk and 'combined with' and 'completed by' are arguably incompatible notions. 'Combined with' relates to *elements* such that a proposition is viewed as a *combination of elements*: 'completed by' relates to *predicables, functions*, such that a proposition is viewed *not* as a *combination* of elements but as *formed from* a function together with a name. Unfortunately Strawson, on the one hand, wants to keep *out* of the Frege, Geach, Quine camp[21] and I hold *must* do so if his 'Category' criterion is going to be an explanation of the subject-predicate distinction;[22] and yet, on the other hand, his reference to propositional and non-propositional 'styles of introduction' seem to put him *in* it and indeed he wishes to gain benefit from this. For example, he writes: 'The latter expression (i.e. "is wise")[23] looks fragmentary just because it suggests a particular kind of completion; the former expression (i.e. "Socrates")[24] looks non-fragmentary just because it carries no such suggestion.'[25] The problem is that 'is wise' does not and cannot look or be regarded as fragmentary on Strawson's account since as I have argued, and will further argue, 'wise' is one of two *elements* which combine to form a complete whole (a proposition), not a *fragment* from which a whole (a proposition), in a given case, is formed. Indeed, as we shall later see, the 'is' in 'is wise' is separated off from 'wise', 'Socrates is wise' being construed as a relational proposition in a very special sense of 'relation'; and this latter construction gives rise to some embarrassment when Strawson claims

that in certain simple or paradigm cases there is an 'affinity' between the 'Grammatical' and 'Categorical' criteria for the subject-predicate distinction.

B.2. Subject expressions and predicate expressions for Strawson, introduce *terms* into *remarks* or *propositions*. I shall now continue my case by arguing that his doctrine of *terms* demands that predicate expressions be regarded as *names*. We must first ask why Strawson holds it necessary to introduce his notion of 'term'. His case is as follows. He writes:[26]

> We have seen that Geach's definitions,[27] in relying upon 'stands for' and 'about', are so far from differentiating A-expressions and B-expressions, that they may fairly be taken to mention a feature common to both.[28] It will be useful to have another way of speaking of this feature common to both 'Socrates' and 'is wise' in the remark 'Socrates is wise', and common to both 'Raleigh' and 'smokes' in the remark 'Raleigh smokes'.

He later continues:

> Let us say that anything which is introduced, or can be introduced, into a remark by an expression is a *term*. This piece of terminology has an obvious connexion with some of our earlier lists. By conjoining certain items from the functional distinctions of list I[29] we obtained such phrases as 'referring to something' and 'predicating something of it' .... These phrases yielded the distinctions of list III[30] between *term referred to* and *term predicated*. Now we can say that terms referred to and terms predicated are alike *introduced*.[31]

This doctrine of 'terms' is, I hold, open to severe objection as I shall attempt to show in a moment (B.4.): that it supposes or indeed demands that predicate expressions are a type of name may be seen as follows. The verbal expressions for terms introduced into a remark are regarded as names. 'Socrates' introduces the term *Socrates* into the remark 'Socrates is wise'; by the means of 'is wise' the *term wisdom* is introduced into the remark 'Socrates is wise'.[32] In order for 'is wise' to perform this function, 'wise' must be regarded as a name. If it were to be regarded as a predicable i.e. as '---- is wise' then: (a) Strawson could not speak of 'is wise' introducing a term into a remark; (b) he could not speak of e.g. *wisdom* being the term predicated: what would be predicated would be *is wise*. Treating 'is wise' as a predicative functional expression ('---- is wise') precisely does *not* commit one to the separate existence of *wisdom*; on such a treatment wisdom only exists in that some *F*'s are wise. Similarly with Strawson's case of 'Raleigh smokes', where the appropriate verbal expression for the term introduced by the predicate 'smokes' is 'smoking' and the term itself is *smoking*. By a parallel argument to that just offered we may contend that Strawson must treat 'smoking' as a name i.e. in this case that 'smokes' is writable as 'smoking'. In this case too, Strawson's doctrine of *terms* demands that predicate expressions ultimately i.e. with regard to what they may introduce, be regarded as *names* and in this case too he is committed to the independent existence of smoking, which he would *not* be committed to were he to treat 'smokes' as a predicable.

B.3. Having considered evidence for my thesis that Strawson treats predicate expressions as *names* by reference to the 'Grammatical' criterion and his doctrine of 'terms', I now turn to the 'Category' criterion to further substantiate this thesis. The best evidence we have here appears in his prologue to the statement of the 'Category' criterion i.e. in his doctrine of: (1) terms as a principle of collection of other terms; (2) non-relational assertive ties. That his doctrine of terms as principles of collection for other terms commits him to treating predicate expressions as names can be seen as follows. He writes:[33] 'A sortal universal supplies a principle for distinguishing and counting individual particulars which it collects' and again: 'Characterizing universals ... whilst they supply principles of grouping, even of counting particulars, supply such principles only for particulars already distinguished, or distinguishable, in accordance with some antecedent principle or method'. However in order for a sortal universal to supply such principles or a characterizing universal in its way to provide such principles, expressions for sortal and characterizing universals must be regarded as *names*, otherwise:

(a) he could not write later[34] that in e.g. 'Fido is a dog, an animal, a terrier', one and the same particular is sortally or instantially tied to a number of different sortal universals. *Dog*, *animal*, *terrier*, are the sortal universals named or designated by 'dog', 'animal' and 'terrier' respectively. Were Strawson to be treating 'dog', 'animal' and 'terrier' as predicative expressions (predicables) then he could not hold that it is *dog* or *animal* which is a principle of collection but rather that it is what is designated by '---- is a dog' or '---- is an animal' which is a principle of collection for particulars;

(b) he could not write[35] that in 'Socrates is wise', 'Socrates fights', 'Socrates dies' one and the same particular may be tied by a characterizing tie to many characterizing universals or that one and the same characterizing universal may be tied by the characterizing tie to many different particulars: Socrates, Plato, Aristotle are all wise, all die. *Wisdom*, *fighting*, *death* are the characterizing universals named or designated by 'wise', 'fights', 'dies', respectively. Were Strawson to be treating 'wise', 'fights', or 'dies' as predicative expressions (predicables), and not as names, then he could not hold that it is *wisdom*, *fighting* or *death* which is a (subsidiary) principle of collection, but rather that it is what is designated by or stood for by '---- is wise', '---- fights', '---- dies', which is such a principle of collection for particulars.

I now turn to the doctrine of non-relational assertive ties. He writes[36] that any term, particular or general, must be capable of being assertively tied to some other term or terms so as to yield a significant result, a proposition; and it is convenient to have and we do have, he holds, names for different kinds of assertive tying, based partly on differences in the types or categories of terms, partly on differences in purpose or context of assertion. Names for assert*ive* tying are introduced as follows: '... we may say of a speaker that he *characterizes* an object as such-and-such, or *instances* something as a so-and-so, or *attributes* something to something else',[37] from which we may conclude that the appropriate names alluded to are *characterizing*, *instancing*, *attributing*. Names for assert*ed* ties are introduced by reference to such forms as: '---- is an instance of ----'; '---- is characterized by ----'; '---- has the relation of ---- to ----': hence we get the names 'instantial tie', 'characterizing tie', 'attributive tie'. This doctrine of different kinds of assertive and asserted tying clearly commits Strawson to treating predicate expressions as *names* since what can fill the second gap in e.g. '---- is an instance of ----' is a name - or rather, to speak more precisely, a universal designated by a name. What can fill the gap must be an expression which is *complete* and has a sense independently of the form '---- is an instance of ----'. Similarly with '---- is characterized by ----'. What can fill the second gap in these formulations could never be a predicable expression without (a) absurdity resulting and (b) an incompatibility within Strawson's own position resulting. To introduce a predicable expression here to fill the appropriate gap we should have e.g. '---- is an instance of (---- is a man)' or '---- is characterized by (---- is white)'. These are absurdities in that functional predicable expressions themselves are what are instanced and hence they themselves cannot form part of a propositional formulation potentially asserting that something *is* instanced or characterized. For the above formulations to be intelligible the functional expressions would have to be mentioned, not used, i.e. we should have to re-write the form of words as '---- is an instance of "(---- is a man)"' but whatever the possibilities of such re-writing it is not a feasible re-writing of Strawson's account. What is instanced for Strawson is distinctly not a linguistic item such as a functional expression. Further, it is not possible to hold, granted Strawson's framework, that we may fill the appropriate gaps with predicable expressions since the use of predicable expressions does *not* commit us to positing the existence of sortal or characterizing universals whereas Strawson *is* so ontologically committed.

B.4. I shall now press the thesis that Strawson even requires sortal expressions to be single names by illustration from his doctrine of 'terms' in relation to his 'category' criterion and thus that his doctrine of 'terms' is open to severe objection. The defence of the thesis that Strawson's doctrine of 'terms' is open to severe, indeed quite devastating, objections is relevant to my general discussion of Strawson since, if it is successful, then the basic framework in which he plans to set out his explanation of the subject-predicate distinction is shown to be undermined at a crucial point.

As concerns Strawson's framework of speech functions and linguistic parts as introduced earlier,[38] we may ask: 'Need we introduce anything in addition to such speech functions and linguistic parts - viz. "terms"?'. I hold that we not only need *not*, but also that we *may* not. From the fact that every activity (function), in list A_I[39] represents the activity as having an object e.g. naming something, referring to something and that every activity (function) in list B_1[40] represents the activity as directed to the same object referred to by it, it only follows that we have to introduce *terms* as a common feature of both if we hold that the activity expressions represent the activity as having *just one* object; but from the fact that an activity has *an* object it does not follow that it has *just one* object which we then introduce or specify as a term. Further, from the fact that the third and fourth expressions in list I[41] represent the B-activity as having another object, predicating *something* of the first object, 'ascribing *something* to it', it does not follow that there is *just one* thing predicated i.e. a *term*, an additional entity. As there is no one or single thing named by a name and no one or single thing referred to by a referring expression, it is misguided to posit terms as non-linguistic entities in the sense of that one single thing named or referred to in any subject-predicate proposition or 'remark'. From the fact that we might hold that each name names or stands for a single object, it does not follow that there is, let alone must be, just one object named or referred to every time we use a name or a 'referring expression'. Equally, even though we may speak of 'predicating something of' the first object, or 'ascribing something to it', it does not follow that we predicate *just one* thing. Strawson illegitimately moves from: 'predicating *something* of' to 'predicating *just one thing* of', as he earlier illegitimately moved from: 'the activity has *an* object' to 'the activity has *just one* object'. What we predicate, using Strawson's framework, is some particular form of words or, alternatively, we 'ascribe a characteristic'; but there is no *one* thing which is predicated or which is attributed, a characteristic. We can no more ask *in general* 'What is predicated?', and get as our answer 'a term' than we can ask *in general* 'What is named or referred to?' and get as our answer 'a term'. I think Aristotle was making this point at *Metaphysics* Z3, 1029a 3ff. where he makes it clear that various things can be the subject of predication viz. matter, shape, and a combination of the two but there is no one single thing which is unchangeable and the subject of *every* predication. Yet a sortal universal, is for Strawson a term, that single X, one may hold, which is *named* by a sortal expression on every occasion of use of the expression.

B.5. We must finally note here that in that Strawson treats sortal expressions as a kind of *name* then he cannot account for the introduction and continued use of names or predicates. Furthermore, he cannot give due recognition to the concept of '*identifying*', which, as I have argued, precisely requires that sortals are *not* to be treated as a kind of name (*see* ch. I above, secs. B.1 and B.2).

C. That Strawson also treats Sortal expressions as *Predicative*

C.1. Having produced what I claim is sufficient evidence for the thesis that Strawson is committed to treating predicables, or more loosely, predicate expressions, as names, I now turn to support my case that he also treats sortal expressions as predicative in that at the most fundamental level of predication they essentially play a predicative role. On page 169,[42] 'Fido is a dog' is given as a case in which a particular may be said to be sortally or instantially tied to a sortal universal and the non-relational tie of *instancing* is given as a paradigm case of a subject-predicate proposition. On page 171,[43] in drawing up the formal criterion for the subject-predicate distinction in terms of the 'category' criterion, he writes: 'Suppose now, on the strength of these analogies and disanalogies we adopt the following ruling: the *primary*[44] sense of "*y* is predicated of *x*" is "*x* is asserted to be non-relationally tied to *y* either as an instance of *y* or as characterized by *y*"'. Thus '*x* is an instance of *y*' gives us the primary sense of '*y* is predicated of *x*', yet '*y*' is a sortal universal. It is difficult, indeed impossible, to see how Strawson may on the one hand give the primary sense of '*y* is predicated of *x*' by reference to *x* being an instance of *y*, where *y* is a sortal universal and yet *deny* that sortals, at the most fundamental level, in the primary sense, do not essentially play a predicative role. Indeed the very notion of a 'predicate' itself is introduced and analysed by reference to the notions of 'being an instance of' or 'being characterized by'. In his final formal statement of the 'category' criterion he writes:[45]

> In this way, by taking as the *fundamental*[46] case of *y* being predicated of *x*, the case in which *x* (a particular) is asserted either to be an instance of, or to be characterized by, *y* (a universal), and by proceeding thus to develop other cases by analogy or extension, we can build up a sense of 'to predicate' for which it is true that universals can both be simply predicated and have things predicated of them (i.e. be subjects) ...

yet in that sortals play an essentially predicative role, they may not play an essentially *identifying* role, since, as I have argued (in ch. II above, sec. D), predicating *presupposes* identifying. Furthermore, in that sortals play an essentially predicative role they may not account for the introduction or continued use of predicates.

C.2. That Strawson requires sortals to be both names and predicates may also be seen from his exposition of the 'type criterion'.[47] As I have argued earlier,[48] Strawson may only maintain his position that we may quantify into the predicate place in that what may enter the predicate place is a (single) name: thus sortals may only occupy predicative position in that they are construable as singular names (parallel to 'prettiness'); yet sortals are also required to be construed as predicative in that they may be said to be instanced in relation to some individual parallel to the way in which *pretty* is able to stand in a characterizing relation to some individual (*Betty* in the example).

D. A Summary Statement of Strawson's 'Dual Position'

That Strawson is unable: (i) to account for the introduction and continued use of names and predicates; (ii) to give recognition to the concept of 'identifying' are not the only consequences which flow from his dual position. Requiring that sortals be both names and predicates (perform a predicative role), leads to incoherence in any theory which adopts such a requirement since no symbol, term, can *both* stand for an object, or indeed a number of objects and have instances or be instantiable. Such incoherence I have attempted to illustrate earlier precisely by reference to his 'type' criterion.[49]

Further, taking one strand of the dual position, namely the treatment of sortals as names, leads him to being committed to a 'two name' or at least 'two term', theory of predication, with all the difficulties such a theory brings in its train. Such difficulties have been admirably exposed and developed by Geach[50] and I shall not pursue this consequence here. Taking the other strand, namely the treatment of sortals as predicates, leads either to an impossible (i.e. incoherent) account or no account of sortal instantiation, as I shall attempt to show in the next chapter.

E. Some *General* Difficulties for Strawson's 'Category' criterion for the Subject-Predicate Distinction

Before proceeding to substantiate my claim, arising out of his 'dual' position, that Strawson's treatment of sortals as predicates leads either to an impossible i.e. incoherent account or no account of sortal instantiation, it is important for me to present an account of what I hold to be crucial *general* difficulties which may be offered in relation to his formulation of the 'Category' criterion' for the subject-predicate distinction, since the 'Category' criterion is one of the three criteria[51] Strawson offers in terms of which he will explain the subject-predicate distinction.

The crucial *general* difficulties, I hold are as follows:

1. The framework in which the 'Category' criterion is set introduces us, I claim, to an unintelligible notion of '*remark*' and subsequent notion of '*proposition*';
2. This framework also introduces us to an objectionable notion of '*term*' i.e. a notion of 'term' which is illegitimately introduced (*see* sec. B.2.);
3. His formal account of '*to predicate*', which treats sortals as basically predicative, entails that he cannot give the all crucial recognition to the concept of 'identifying' (*see* sec. C.1.);
4. His treatment of sortals as *names* equally entails that he cannot give recognition to that concept and hence cannot subsequently analyse the notion of 'logical subject' in terms of 'identifying' (*see* sec. B.5.).

My reasons for holding 2, 3 and 4 have been given above. My reasons for holding 1 are as follows. Subject expressions and predicate expressions for Strawson introduce 'terms' into 'remarks' or 'propositions'.[52] As concerns 'remark', Strawson writes:[53]

> Let us say that the expression 'Socrates' ('Raleigh') serves to *introduce* the particular person, Socrates (Raleigh) into the remark, and that the expression

> 'is wise' ('smokes') serves to *introduce* the quality, wisdom (the habit of smoking) into the remark.

I claim this is quite unintelligible - how can a remark, which is a form of words or at the very most a form of speech, a *linguistic* entity by any account, contain as one of its elements a person and as another element a virtue or habit? There certainly appears to be a serious 'category error' in this position and yet Strawson, by his very own words, is prevented from saying e.g. that 'Socrates' introduces the name "Socrates" into the remark or that 'is wise' introduces the predicate or description "wise" into the remark. In that his doctrine of 'proposition' is subsequent to his doctrine of 'remark' the same objection holds.

Notes

1. *Individuals* (London: Methuen, 1959), pp. 168ff.
2. *Ibid.*, pp. 181-186.
3. In ch. II above, secs. A-D.
4. As I have argued in ch. I above, secs. B.1 and B.2.
5. *See* ch. II above, secs. A-C.
6. I have argued that other sorts of identifying either reduces to or are a quasi form of sortal identification in ch. II, secs. J and K.
7. *See* my discussion later in this chapter.
8. *See* my Introduction and ch. VII.
9. ch. I, sec. B.1.
10. He cannot account for the introduction of predicates in that if we cannot specify what predicates apply to or are true or false of, then the notion of a predicate applying to or being true or false of a subject loses its application. He cannot account for the continued use of predicates (that we have the same predicate on more than one occasion) precisely for the same reason. This is not to be thought to commit me to the thesis that predicate identification, i.e. identification of predicates, necessarily rests on sortal identification, for what I have earlier argued is that *individual identification* rests on sortal identification.
11. As Geach in his review of *Subject and Predicate in Logic and Grammar* has aptly pointed out (*see* my Introduction, n. 43).
12. *See* sec. C.1.
13. Strawson does not hold this to be a sufficient condition for a expression being a subject expression. (cf. *Individuals*, pp. 147-148).
14. *Ibid.*, p. 149.
15. My parentheses.
16. *Ibid.*, p. 147. Equally he does not hold this to be a sufficient condition for an expression being a predicate expression.
17. *Ibid.*, pp. 158-159.
18. My parentheses.
19. My italics.
20. *Individuals*, p. 153.
21. Quine has the Frege/Geach notion of 'concept word'/'predicable'.
22. *See* later, sec. B.3 of the present chapter and earlier in my discussion of Strawson's 'Type' criterion, ch. III, sec. F.2.

23. My parentheses.
24. My parentheses.
25. *Individuals*, p. 153.
26. *Ibid.*, p. 146.
27. Of the subject-predicate distinction (cf. *ibid.*, pp. 143-146).
28. Strawson's attack on Geach may well be challenged. I have done so in ch. III, sec. A.
29. cf. *Individuals*, p. 139.
30. *Ibid.*, p. 141.
31. *Ibid.*, p. 146.
32. *Ibid.*
33. *Ibid.*, p. 168.
34. *Ibid.*, p. 169.
35. *Ibid.*
36. *Ibid.*, p. 167.
37. *Ibid.*
38. *Ibid.*, pp. 139ff.
39. *Ibid.*, p. 139.
40. *Ibid.*
41. *Ibid.*
42. *Ibid.*
43. *Ibid.*
44. My italics.
45. *Individuals*, p. 172.
46. My italics.
47. In 'Singular Terms and Predication', *op. cit.*, pp. 65ff.
48. ch. III, sec. F.2.
49. *See* n. 48 above.
50. *Reference and Generality*, ch. 1 and *also* in 'A History of the Corruptions of Logic', repr. in *Logic Matters* (Oxford: Basil Blackwell, 1972), pp. 44ff, esp. pp. 47ff.
51. In *Individuals;* the others being the 'Grammatical' criterion (*ibid.*, pp. 139ff.) and the 'Mediating' criterion (*ibid.*, pp. 180ff.; *see* especially *ibid.*, pp. 187-188). In relation to the 'Category' criterion the 'Grammatical' criterion is distinctly secondary *(see ibid.*, p. 166). I am not rehearsing general difficulties for *this* criterion here for this reason, though I do hold that there are difficulties as is clear from this chapter and furthermore I hold that there are difficulties in precisely relating this criterion to the 'Category' criterion. For example, according to the *Grammatical* criterion, 'Socrates is wise' is formed from the verb form 'is wise' *plus* the name (or noun substantive); according to the *Category* criterion such a proposition is analysable as a relational proposition of the form: '---- is characterized by ----'; strictly it is formed by completing the two-place function. By the 'Grammatical' criterion the predicate expression is 'is wise'; by the 'Category' criterion the appropriate predicate expression is 'wise' and the predicate is *wisdom*.
52. *Individuals*, p. 146.
53. *Ibid.*

Chapter IX

Strawson: Sortals and Sortal Instantiation

A. That for Strawson one *can* raise the Question: 'What Instances a Sortal Universal?'; that Strawson is *not* committed to either a Doctrine of 'Pure' or 'Relatively Pure' Basic Particulars

In my last chapter[1] I claimed that in that Strawson treats sortals as predicates this leads to an impossible or no account of sortal instantiation. *That* treating sortals as predicates, in general, presents us with an impossible/unintelligible account or no account of what constitutes an instance of a sortal I have sought to establish earlier.[2] *That* Strawson is so committed I shall attempt to illustrate in Section B of this chapter.

First, however, I shall take up my promise in Chapter IV (sec. D.1.) that I would illustrate one general consequence of the failure to recognise the category of the sortal - namely the doctrine that sortals *themselves* may be said to have instances and that principles of counting and distinction formed from 'matter' terms *themselves* may be said to have instances - by reference to Strawson. A precursus to this is that one establishes that it is not a *misunderstanding* of Strawson that one can raise the question: 'What instances a sortal universal?'. It is to this issue that I immediately turn.

A.1. It might be held that it is abundantly clear from Strawson's writings[3] that he does require that sortals have instances, for according to him 'Fido is a dog' (e.g.) presents us with a paradigm case of a subject-predicate proposition and in such a proposition Fido is an instance of the sortal universal *dog*. In such a proposition Fido is sortally or instantially tied to *dog* (and a number of different sortal universals). In the light of this, how may someone hold that it is a *misunderstanding* of Strawson's position to think that one may raise the question: 'What instances a sortal universal?'.

The following line of thought may be advocated. The thesis that one may *not* raise the question is what lies behind Strawson's doctrine of the 'non-relational' nature of the various kinds of 'assertive tie' which he has introduced.[4] Such a line of thought however will not suffice since the point of *that* doctrine is that the various kinds of tie are not *themselves* to be regarded as *terms* i.e. as separately existing entities designated by names, otherwise he would find himself in an infinite regress; he would find himself subject to a version of the Academy's notorious 'Third Man' argument. The point of the 'non-relational' tie doctrine is that 'is an instance of' (e.g.) is to be written as '---- is an instance of ----' i.e. as a function containing two empty places that can be filled out by names or the equivalent of names. For the tie '---- is an instance of ----' to exist, to be in operation, is for *a* to be an instance of *B*, where '*a*' is a particular and '*B*' is a universal. However from the fact that the non-relational tie can have no separate existence it does not follow that the entities named by the expressions which can fill the gaps cannot have separate existence: indeed they can, since such expressions are precisely names. Hence the question: 'What can instance the instantial tie?'; 'What are the *a*'s which can form instances of a sortal universal?' can still be raised.

Even so, it might be said, we might *still* take Strawson as holding that the question: 'What does the instancing of a sortal universal?' *is* an impossible question to raise on the grounds of his explicit rejection of any doctrine of 'pure particulars'. An implication of his rejection of that doctrine is that for there *to be* particulars *is* for there *to be* entities falling under some sortal universal or 'general term',[5] whereas for there to be universals is not for there to be entities or 'terms' which are instanced. If we adopt this line of approach and construe his rejection of 'pure particulars' as reading that *necessarily* to be a particular is, in at least the primary sense, to be an instance of a sortal universal such that one cannot raise the above question, then:

(i) Strawson cannot consistently claim that the relation of 'is an instance of' is writable as '---- is an instance of ----';

(ii) he cannot consistently hold that the relation of 'instancing' gives us a paradigm or fundamental case of a subject-predicate proposition.

Let me now illustrate and defend these theses. (i) In that *necessarily* to be a particular is to be an instance of a sortal universal then we cannot independently specify what is to fill the first gap, the first argument place in the function '---- is an instance of ----'; yet we *must* be able to do just this if we are to treat 'is an instance of' in such a way and indeed it is a requirement of Strawson's thesis that one can write the relation of 'instancing' as '---- is an instance of ----'.[6] (ii) He cannot consistently hold both:

(A) *Necessarily* to be a particular is (primarily) to be an instance of a sortal universal such that we cannot raise the question 'What instances sortal universals?';

(B) that '---- is an instance of ----' or the instantial or sortal tie provided when this formula is appropriately completed, yields us with a fundamental case of '*y* being predicated of *x*', for whilst according to (A) we cannot raise the question: 'What instances a sortal universal?', we can *always* raise the question: 'What instances a predicate?', even though nothing may in fact do so.

Thus, I hold, Strawson must drop either (A) or (B) to remain consistent. To drop (B) renders his attempt to explain the notion of 'to predicate' by reference to the 'Category' criterion impossible. In that he insists on treating sortals as predicative and the sortal or instantial tie as an analysis or explanation of the subject-predicate distinction, central to his 'Category' criterion, he *has* to allow the question: 'What instances a sortal universal?'.

I thus conclude that the line of thought which would advocate that Strawson, if properly understood, cannot permit the question: 'What instances a sortal universal?' is to be rejected.

A.2. I thus turn to the question of whether Strawson is committed to a doctrine of 'pure particulars'. To hold that he *is* so committed would appear to be a non-starter on the grounds that he explicitly rejects the notion of 'pure particulars'. I have already alluded to this. In 'Particular and General', Section V,[7] he writes:

> Next, I want to make some general, and still propaedeutic, remarks about the notion of an individual or particular.
>
> > (1) The idea of an individual is the idea of an individual instance of something general. There is no such thing as a pure particular.

Further, to maintain a doctrine of 'pure particulars' would entail that we necessarily have no need for sortal universals; sortal universals would not simply be reduced to 'characterizing universals' but would be eliminable in that there would be no need for universals which furnish principles of distinction and counting for particulars. Yet Strawson puts a premium on such principles.

In the light of such considerations what possible grounds might one have for attributing such a doctrine to Strawson - for holding that what instances sortal universals are such particulars? Such possible grounds might be advocated as follows: that whereas there is certainly a sense in which we may affirm, with Strawson, that the idea of an individual is the idea of an individual instance *of* something in general, the truth of the matter is to be expressed by:

(1) *Necessarily*: every individual instance is an individual instance of something general;

whilst what Strawson works with is not *this* but:

(2) Every individual instance is *necessarily* an individual instance of something general;

and he fails to appreciate (1) or that (2) entails (1) but not conversely. He fails to appreciate (1) in that *were* he to appreciate (1) then he could not raise the question of what instantiates sortal universals. Working with (2) and failing to acknowledge (1), and that (2) entails (1), permits him to seek for criteria of distinctness and criteria of identity in general for the idea of a particular instance, for (2) reads:

> Concerning any individual instance it is necessarily true of *it* that *it* is an instance of something general.

That is, it permits him to seek for criteria of distinctness and identity for the notion of a particular instance *independently* of any general term viz., any sortal, but it only follows that he is committed to a doctrine of 'pure instances', let alone a doctrine of 'pure individuals' or 'pure particulars', in that he is *not* committed to the doctrine that one may specify what instances a sortal by reference to a sortal itself, viz., in that he is *not* committed to the doctrine that one may write (2) as:

> Concerning any individual instance i.e. any particular, it is necessarily true of *it* that *it* is an individual instance of *F* (where *F* is a sortal).

That this doctrine is impossible on any interpretation has already been argued in Chapter IV above, secs. C.1 and C.2; that Strawson is committed to such a doctrine will be argued for in Section B below. In the light of this latter I claim that we cannot foist on him a doctrine of 'pure individuals' - a doctrine of 'absolutely simple' particulars.

Further, it is of worth to note that even if one *could* argue that he is committed to a doctrine of 'pure instances', one could not thereby argue that he is committed to a doctrine of 'pure particulars' without gross unfairness since Strawson's concern, both in the relevant part of *Individuals* and in 'Particular and General', is to seek for a *basis* for the introduction of particulars. He is concerned to produce an account, an analysis, of the notion of 'an instance' at a fundamental level upon which the notion of 'a particular' introduced by reference to sortals or feature universals 'modified' in ways mentioned earlier rests. He writes at the beginning of Section VI of 'Particular and General':

> In the rest of this paper I shall try to do two things. First, I shall try to show how, in the case of *certain kinds* of particulars (particular instances of *certain kinds* of general things), the notion of a particular may be seen as something logically complex in relation to other notions (a kind of compound of these notions).[8]

and the 'other notions' alluded to here are the notions which he claims produce an analysis of the notion of a 'particular instance'.

Thus at the most we may only attribute a doctrine of 'pure' (particular) *instances* to Strawson and *this only* on the supposition that he does *not* subscribe to the position, doctrine, that one may specify what instances a sortal universal by reference to a sortal universal itself. That he does so subscribe, will, as I have said, be argued for in Section B below.

A.3. I thus now turn to the issue of whether Strawson can be said to be committed to a doctrine of 'relatively pure', 'relatively basic' particulars in the sense introduced in Chapter IV above viz., to a doctrine of such particulars introduced by reference to 'matter' terms - in his vocabulary 'feature' universals

- precursed in a certain way viz., by 'bit of', 'piece of' such that it is such particulars which instance sortal universals. The possible evidence for such a thesis relies on a certain interpretation of a passage in *Individuals* which, whilst initially having some plausibility, turns out, upon detailed investigation, to be implausible and such a doctrine is, in any case, incompatible with the doctrine I shall contend Strawson really is committed to - namely, that what instances sortals is specified by reference to sortals themselves.

In an important section of *Individuals*[9] and in a corresponding and more developed section in 'Particular and General'[10] (which will be the subject of more detailed discussion in my next chapter), Strawson raises the question of whether there is any familiar kind of universal and any familiar kind of statement introducing such universals, which exhibit the character of not containing or mentioning sortal or characterizing universals and hence which may provide a basis for the introduction of particulars which are instances of sortal universals and which may stand in a characterizing tie to characterizing universals. We are provided with an answer in terms of *feature universals* or *feature concepts* at the level of appropriate universals and *feature-placing* statements at the level of statement. Such 'feature' universals or 'feature concepts' are, e.g., *snow*, *coal*, *gold*, *water*, and the corresponding statements, such statements as:

> 'Now it is raining'
> 'Snow is falling'
> 'There is gold here'
> 'There is coal here'
> 'There is water here'

Strawson rightly holds that the universal terms here do not function as characterizing universals and neither are they sortals. The latter point has already been argued for earlier;[11] 'gold', 'water' etc., may not as such supply a principle of counting in their own right; indeed not a principle of counting or distinction at all. Strawson does contend however that whilst such universals do not provide 'a principle for distinguishing, enumerating, and identifying particulars of a sort', they can easily be modified so as to yield several such principles - we can distinguish, count, and re-identify *veins* or *grains* of gold, *lumps* or *dumps* of coal, *falls*, *drifts* or *expanses* of snow. Such phrases as 'lump of coal' introduce sortal universals, but 'coal', 'snow' *simpliciter* do not.

The thesis that Strawson is committed to a doctrine of 'relatively basic', 'relatively simple' particulars introduced by 'lump of coal' etc., is a *possible* thesis to advance here owing to a lack of clarity, ambiguity, in 'introduce', when he writes that such universals as 'lump of coal' *introduce* sortal universals. Does he mean (i) that such universals are the *basis* for the introduction of sortal universals as previously given earlier in the book or (ii) that such universals are *themselves* sortal universals? (i) is a possible interpretation here since: (a) he writes that no feature universal of itself provides a principle for distinguishing etc., *particulars of a sort*, but that modified they do provide such a principle for distinguishing etc., (implied: *particulars of a sort*), and it is possible to construe this as implying that such universals, as introducing such principles, provide a *basis* for distinguishing etc., *particulars of a sort*; (b) he has not earlier put forward such universals as *lump of coal*, in general '*lump of* ----', where '----' is replaceable by a 'feature' universal, as cases of *sortal* universals. On interpretation (i) the introduction of particulars of a *sort* would rest upon, or could be said to rest upon, particulars introduced by such principles, such universals, as *lump of coal* etc. i.e. by what I have earlier described as 'matter' terms precursed in a certain way and thus what could be said to instance sortal universals would be such particulars.

Whilst such an interpretation as (i) is possible, it must, on further investigation, be rejected in favour of (ii) - that '*bit of coal*' etc., are *themselves* sortals since:

(a) following the earlier account of 'sortal', what Strawson *means* by a sortal is a universal which of itself provides a principle for distinguishing, enumerating and re-identifying particulars;

(b) he explicitly contrasts '*lump of coal*', '*fall of snow*' etc., with *coal*, *snow*, etc., *simpliciter*. Sortals introduce particulars into discourse, equally 'lump of coal' etc., introduce particulars into discourse; 'lump of coal' etc., therefore may not be regarded as a *basis* for the introduction of particulars introduced by reference to sortals;

(c) he unambiguously says that 'lump of gold' *is* a sortal viz., a little later on in the discussion at *Individuals*,[12] he writes:

> For whilst particulars like pools of water, lumps of gold etc., do certainly belong to the class of basic particulars, they can scarcely be said to constitute a fair or reasonable selection from that class. *The sortal universals of which they are instances*[13] (*pool of water, lump of gold*) are alike in this: that their names incorporate, as a part, the names of kinds of stuff (*water, gold*) which seem supremely, or even uniquely, well adapted to be introduced as universal terms into feature-placing sentences.

Further, to hold that Strawson is committed to a doctrine of such 'relatively pure', 'relatively basic' particulars, would in effect commit him to denying that sortals furnished principles of counting and distinction in their own right for particulars. It would, in effect, reduce sortal universals to characterizing ones and thus not permit him to introduce his own original distinction.

B. The Case for saying that Strawson is committed to either an *unintelligible* and/or *impossible* Account or *no* Account of what constitutes an Instance of a Sortal

I shall now advance my case for saying that Strawson is committed to either (a) an impossible account and an unintelligible account or (b) an unintelligible account of what may constitute an instance of a sortal universal. Or, again, that as an alternative to these, he is committed to offering no account. In more simple terms he is committed to an unintelligible or no account of what constitutes an instance of a sortal.

B.1. Strawson lays himself open to the charge of presenting an impossible account (in the sense introduced in Chapter IV above)[14] in that he holds that an instance of a sortal, say, *man* or *apple* is a *particular man* or a *particular apple*. That he so holds may be illustrated by reference to his attempt to provide an explanation of the notion of an 'individual instance' in the cases of: (i) material names; (ii) substance names in the relevant passages of *Individuals* and 'Particular and General'. As concerns 'material names' it is difficult, nay impossible, to see how he can provide an explanation of the notion of an 'individual instance' in the case of 'material names' such as e.g. 'snow', 'water', which introduce 'feature' universals if he *does not* take it that an individual instance of e.g. *snow* is what is designated by 'This patch of snow'. Indeed he writes: 'Suppose we compare a feature-placing *sentence* ("There is snow here") with a phrase ("This (patch of) snow") in the use of which an individual instance of the feature is mentioned ...'.[15] The whole attempt to furnish an analysis of the notion of an 'individual instance of Φ' (where Φ is a feature universal) in terms of a feature universal together with the notion of 'placing' supposes that an individual instance of Φ is just this.

It may now be said that this point does not, of itself, show that he regards an instance of a sortal universal as having a necessary reference to the sortal itself; all the above shows is that in the case of an instance of a feature universal this is so. However, as I have argued above Strawson regards e.g. *lump* of coal, *dump* of coal, *expanse* of snow, etc., as *sortal* universals and an instance of such sortals would be the particular *lump* of coal etc.; what is mentioned by 'This (lump of) coal'. Thus even in this type of case he certainly appears to be committed to holding that an instance of a sortal is introduced by reference to that sortal itself.

The position is equally the case in regard to 'substance names' as discussed in 'Particular and General'.[16] Here his whole attempt to furnish an analysis of the notion of 'an individual instance of Ψ', where Ψ is a 'substance name' (introducing, one may take it, a sortal universal), in terms of a feature universal together with the notion of 'placing' supposes that an individual instance of Ψ is a particular; for what he in detail wishes to argue is that we may analyse the notion of a particular instance of Ψ, e.g. a cat, in terms of the notion of a corresponding feature together with the notion of 'placing'. It is to be noted here that there is deep embarrassment in speaking of a particular instance of *a* cat. One cannot have a particular instance of *a* cat any more than one can have a particular instance of *a* fall of snow. A cat *is* a particular instance; *a* fall of snow *is* a particular instance. One might say that one could have an instance of *cat*, but this would turn a sortal into that which I have claimed it is not and cannot be, viz., a predicate. This indicates, indeed manifests, the absurdity of talking of a particular instance of a sortal.

In that the above cases illustrate that Strawson is committed to holding that a particular instance of the sortal is necessarily introduced by reference to the sortal itself, then he is committed to an impossible account of what may constitute an instance of a sortal universal for the reasons advanced in Chapter IV. And in that he is committed to such an account he is further committed to what I have earlier[17] described as an 'unintelligible account' of what constitutes an instance of a sortal universal, namely, that sortals in general are of necessity required to specify what constitutes an instance of a sortal. In that these cases may only be held to illustrate the more general thesis that sortals are required in order to specify what instances them, then, alternatively, he is committed to an unintelligible account of sortal instantiation.

The only other accounts of 'sortal instantiation' open to Strawson are, I claim, *no* accounts. To hold that a sortal universal is only *mentioned* in the specification of its instances; to hold that in saying that what instances the sortal *man* or *apple*, is the *particular man* or the *particular apple;* that '*man*', '*apple*' in their second occurrence here only *mention* the respective sortals, they are not used, is to offer *no* account of what can instance a sortal universal. To merely mention something does not specify or describe anything, or offer us an account of anything. It will not do either, for reasons offered earlier,[18] to say that what constitutes an instance of a sortal may be specified, not by reference to that sortal itself, but by reference to another sortal in the sense that what may instance e.g. the sortal universal *cat* is a particular feline animal. This presents us with no account of what constitutes an instance of a sortal since we either indirectly have to introduce the sortal universal *cat* or we cannot introduce the instances of the sortal universal *cat* (e.g.) as instances of *that* sortal universal.

B.2. One final philosophical comment is in place here. It may well be pressed: 'What, after all, is wrong with holding with Strawson that "Every individual particular is necessarily an instance of something general"?', where 'something general' here is a sort or a kind (as is clear from his writing in 'Particular and General'). Granted the truth of this, it follows that the *only* way we can specify an instance of a sortal is by reference to the sortal itself; it is necessarily true concerning any instance of a sortal that it can only be introduced, specified, by reference to the sortal itself. It is thus, it will be said, necessarily not surprising that for Strawson, or indeed for anyone, an instance of a sortal, say *man* or *apple* is a particular man or a particular apple; or, again, that an instance of a sortal such as *lump of coal*, or *fall of snow* is a particular lump of coal or fall of snow.

What is wrong is precisely that granted the truth of the thesis then the *only* way we can specify an instance of a sortal, say what constitutes an instance of a sortal, is necessarily *no way* which shows that sortals cannot be said to have instances. *This* is reflected in what is the truth of the matter here:

> *Necessarily*: Every individual, particular, is an instance of something general (where 'something general' is to be construed as a sort or kind).

where *this* in turn is construed as:

> *Necessarily*: Every individual, particular, is of a certain sort or kind.

C. Conclusion

I thus rest my case that in spite of initial considerations to the contrary, we may cite Strawson as an example of one who is committed to an unintelligible account or to no account (in the ways earlier introduced in Chapter IV) of what can instance a sortal. This results from failing to recognise the true nature of sortals in that it treats sortals as predicates and thus permits one to raise the question: 'What does the instancing in the case of sortals?'. On the contrary, sortals cannot *have* instances; they are required in order for us to *speak of* instances.

Notes

1. pp. 215-216; for the general position *see* Introduction, p. 4, ch. IV, pp. 153-155.
2. ch. IV, secs. C.1. and C.2.
3. *Individuals* (London: Methuen, 1959), p. 169.
4. *Ibid.*
5. 'Particular and General', repr. in *Logico-Linguistic Papers* (London: Methuen, 1971), p. 35.
6. *See Individuals*, pp. 167-173.
7. 'Particular and General', *op. cit.*, p. 35.
8. *Ibid.*, p. 36.
9. *Individuals*, pp. 202ff., ch. 6[6].
10. 'Particular and General', *op. cit.*, pp. 36-42.
11. ch. I, sec. A.5.
12. *Individuals*, pp. 204-205.
13. My italics.
14. *See* ch. IV, sec. C.1., pp. 153-154.
15. 'Particular and General', *op. cit.*, p. 38.
16. *Ibid.*, pp. 39-42.
17. *See* ch. IV, sec. C.2., pp. 154-155.
18. *See* ch. IV, sec. C.1., pp. 153-154.

Chapter X

Strawson: Further Consequences of Failure to Recognise the Nature of the Sortal

A. That to treat Sortals as Predicates (indirectly) leads to the Search for a Basis for the Introduction of 'Particulars in General'

I claimed in my 'Introduction'[1] that failure to recognise the category of the sortal viz., to treat sortals as themselves predicates (indirectly) leads to the search for a basis for the introduction of 'particulars in general'. The reason for introducing 'indirectly' was illustrated in Chapter IV. I have claimed that such a search is unsuccessful in fact and misconceived in principle. In Chapter IV I later said[2] that I would illustrate this claim by reference to the work of Strawson. It is the aim of the first section of this chapter (Section A) to do just this. The attainment of this aim however will involve discussion of other pertinent issues formally to be considered in later sections.

In Chapter IV I further said[3] that Strawson's writing produced a particular illustration of a further consequence of failure to recognise the true nature of the sortal, namely a search for a basic notion of 'an instance' or explanation of the notion of an 'individual instance' (independently of the notion of a 'sortal'). I shall discuss this in the second section of this chapter (Section B) and press the claim that this enterprise too is misconceived in principle.

Finally, I claimed[4] that failure to recognise the category and nature of the sortal, to reduce sortals to predicates, indirectly leads to the search for an *underlying basis* for the subject-predicate distinction as manifested in a language in which the categories of name and predicable (predicate) are introduced or present, or in a mode of discourse in which particulars are introduced or spoken of, in the sense of a search for a *certain type* of empirical proposition which contains no reference to sortals and is formed independently of any reference to sortals. I claimed earlier (in my 'Introduction') that the search for some type of *empirical* proposition which ultimately underlies the subject-predicate distinction was mistaken in principle. I shall illustrate my thesis that the search for an *underlying basis* for the distinction, in the sense of a search for some one type of empirical proposition which contains no reference to sortals and is formed independently of any reference to sortals, is misconceived in principle by reference to Strawson's[5] search in section C of this chapter. I shall argue that Strawson's search is so misconceived in that a sortal has to be invoked in order to *articulate* the proposed basis.

A.1. Before engaging in detailed illustration and discussion of the first topic, it is necessary to explain how failure to recognise the category of the sortal in that sortals are treated as predicates *indirectly* leads one to a search for a basis for 'particular introduction in general'.

In that one treats sortals as predicates, one is committed to a doctrine of sortals having instances, of sortals being instantiable in particulars. In that one then holds (e.g. with Strawson) that every instance is *necessarily* an instance of something general, where 'something general' is to be construed as a sortal,[6] but may *not* hold *necessarily* every instance is an instance of something general viz. *necessarily* every particular is a particular under some sortal; *necessarily* to be a particular is to be a particular of some sort (or kind),[7] *then* one may have an inquiry into the basis of particular introduction in general precisely in that one may *not* subscribe to this later thesis, and as I have already argued, treating a sortal as a predicate is incompatible with subscription to such a thesis. Once one acknowledges the category of the sortal, does not treat the sortal as a predicate, but as rather necessary (as I have argued) for the introduction of names (individuals) and predicates and *in that* one (*pace* Strawson) treats principles of counting and distinction formed from 'matter' terms as sortals, then there can be no inquiry into the basis of particular introduction *in general*, but, at the most only an inquiry into the basis of particular introduction in the sense of the basis of (for) the introduction of particulars introduced by reference to *sortals*. Or again in that one *does not* (*contra* Strawson) treat principles of counting and distinction formed from 'matter' terms (expressions for 'feature' universals) *as* sortals then at the most there can only be an inquiry into:

(a) the basis of particular introduction in the sense of the basis of (for) the introduction of particulars introduced by reference to sortals;

(b) the basis of particular introduction in the sense of the basis of (for) the introduction of particulars introduced by reference to principles of counting and distinction formed from 'matter' terms (expressions for 'feature' universals).

In what follows I shall seek to show that Strawson's attempt to furnish a basis for particular introduction in general is unsuccessful and that the way it is unsuccessful, viz. it is of necessity unsuccessful, *shows* that it is misguided in principle to even seek for a basis for particular introduction in the senses introduced in (a) and (b) above, let alone seek for a basis for particular introduction in general.

A.2. To set out my immediate programme in some detail. I shall first consider the claim that we must distinguish between two types of 'sortal' expression in Strawson's extended sense of 'sortal' and I shall hold that in the light of discussion of this claim, he may *not* hold that *all* sortals, in this extended sense, rest for their introduction upon 'feature' universals (or expressions) [A.3.]. I shall then argue that he may not hold that sortals, in his (and my) original sense may be said to rest for their introduction upon 'feature' universals (or their expression): that feature universals supply a basis for their introduction [A.4.]. I shall then turn to substantiate my case for holding that Strawson cannot claim that feature universals, together with the notion of 'placing', may supply a basis for the introduction of types of particular introduced by sortals in his original sense and the sense which I have introduced [A.5.]. Finally, I shall attempt to substantiate my case for saying that Strawson cannot even claim that his 'feature' universals, together with the notion of 'placing', may supply a basis for particulars introduced by reference to 'functional sortals' i.e. principles of counting and distinction formed from 'matter' terms; 'feature' universals such as 'expanse of snow'; 'bit of gold'; etc. [A.6.]. To assuage the sceptic I shall then add a further 'indirect' argument to support the thesis advocated in A.5. and A.6. [A.7.].

A.3. *Functional and non-functional expressions of the form 'X of M'*

It might be held that we must clearly distinguish between two types of sortal in Strawson's extended sense of 'sortal', viz. between two types of expression of the general form '*X* of *M*', where '*M*' is a 'matter' term or expression for a 'feature' universal, and '*X*' is a 'parcel' term. That is, we must clearly distinguish between sortals (in this extended sense) which are of a *functional* form:

> '---- of *M*'

and those which are of a non-functional form:

> '(----) of *M*'

under the former we may include:

> 'expanse of snow';
> 'lump of gold';
> 'lump of coal';
> 'stretch of water';
> 'bit of wood';

under the latter:

> 'vein of gold';
> 'grain of coal';
> 'flake of snow';
> 'pool of water';

that is, we must distinguish between principles of counting and distinction *formed from* 'matter' terms or 'feature' universals and those formed by *reference to* 'matter' terms and 'feature' universals. In the case of the former there is a necessary reference to the 'matter' term or 'feature' universal: in the case of the latter the reference is only contingent. In the case of the former we have the case of a division *of matter*: in the case of the latter we have the case of matter being divided. In the case of the former:

> *necessarily* any expanse, lump, dump, stretch, bit is an expanse, lump, dump, stretch, bit of *some* matter;

in the latter:

> *contingently* any vein, ingot, grain is a vein, etc. of some matter.

In the case of the former we may only determine that we have *one* expanse, lump, dump and stretch in that we have one lump, dump, expanse or stretch *of* snow, etc.; the unity of the parcel is essentially determined by that of which it is a parcel and again we may only determine that we have the *same* expanse, lump, dump or stretch in that we have the *same* expanse, lump, dump, or stretch *of* snow etc. The continued identity of the parcel is essentially determined by that which it is the parcel.

In the case of the latter it is not the case that we may only determine that we have *one* vein, grain, flake or pool in that we have the vein, grain, flake or pool etc. of *M*: the unity of the parcel, what makes it the parcel, is not essentially determined by that of which it is a parcel. We may, for example, count the number of flakes, pools, grains or veins without being able to say what they are flakes, pools, grains or veins *of*. We may do so because flakes, grains, pools and veins are of relatively determinate *shape* or *form*. However, we seem to be in difficulties when we come to the question of sameness. For it does not seem that we can say that we may determine that we have the *same* flake, grain, vein, pool, etc. independently of what it is the flake, grain, vein, pool etc., *of*. Indeed, it will be pressed we only have the *same* grain or flake in that we have the same grain or flake of *M*; equally with vein, pool or ingot. Crudely, if the matter is different, then the parcel is different; difference in matter is a sufficient condition of difference in parcel, and indeed a necessary condition of difference in parcel. And again, sameness in matter is a necessary condition for sameness of parcel. It thus seems that there is an ambiguity in these latter cases between what appertains concerning *unity* of parcel and what appertains concerning *identity* of parcel. Various attempts may be made to eradicate this ambiguity.

1. By denying that in the case of unity we may be able to count the number of flakes, pools, grains, etc. independently of what they are flakes, etc. *of*, in the following way. It was commented earlier that we may count the number of flakes, grains, etc. independently of what they are flakes, grains, etc. *of* because such items have a determinate or

relatively determinate shape or form. But, it will be said, *necessarily* every shape or form is the shape or form of something or other; in such cases the shape or form of some matter: hence, 'in the last analysis' we may not count the number of grains, flakes, pools, etc. *per se*. But this clearly will not do as a general thesis concerning shape; for one may, for example, count the number of circular or triangular shapes independently of that which they are the shape of; and equally, we may note here, we may speak of having the *same* circular shape (the same circle) irrespective of what it is a circular or triangular shape of. Circles and triangles, e.g., form principles of counting and distinction in their own right. The examples which are at the centre of discussion here are clearly to be differentiated from these latter cases (Platonic mathematical Forms), for in our examples we have examples of terms (symbols) which supply principles of counting in their own right but not, thus far, examples of principles of distinction in their own right.

2. By holding that in the case of unity, whilst it is not the case either: (a) *necessarily* every flake, grain, etc. is a flake, grain etc. of *something or other*; (b) *necessarily* every shape is the shape of *something or other*, it is the case that every flake, grain, etc. is *necessarily* a flake, grain, etc. of something or other and equally that every shape is *necessarily* the shape of something or other. This equally will not do since:

A is *necessarily* p

entails:

Necessarily: A is p.

3. By claiming that in the case of unity, it can *still* be held that *necessarily* every flake, grain, etc. is a flake, grain, *of something or other*, in the following way. Whilst we cannot hold in these cases that *necessarily* every flake, grain, etc. is a flake, grain, etc. of some *particular* kind or stuff or matter, we may hold that *necessarily* every flake, grain, etc. is a flake, grain, etc. of something or other - some stuff or matter *in general*. But this claim is based on a mistake. There is no such X as stuff or matter 'in general' - indefinite matter, any more, than as Russell observed, there is such an X as an indefinite man.[8] To hold that *necessarily* every grain, flake, etc. is a grain, flake of something or other, is to hold that *necessarily* every flake, grain, etc. is a flake, grain, etc., of *either* p v q v r ...; but to hold that if we did not have the notion of *either* p v q v r then we could not have the notion of one parcel of matter in the sense or grain, flake, etc. is false; for, as we have seen, we may count flakes, grains, etc. independently of what they are grains, flakes, etc. *of*. Rather, what one may say here is that it is *empirically* necessary that every flake, grain, etc. is a flake, grain, etc. but not *logically* necessary; or more correctly: it is empirically necessary that

in that we have one flake, grain, etc. then we have one flake, grain, etc. *of* something or other (some matter), but not *logically* necessary.

The above attempts to eradicate the ambiguity have been in the form of, in effect, trying to reduce what appertains in the case of unity to what appertains in the case of identity. It may now be thought that we may reduce what appertains in the case of identity to what appertains in the case of unity: but I do no see how such a thought can be reasonably be held - for it is the thought that we may speak of the *same* flake, grain, etc. independently of what it is a flake, grain, etc. *of*, which does appear impossible to sustain.

In the examples of 'sortals' in the extended sense so far considered, I have instanced what may be said to be natural divisions of matter in the sense of natural parcels by the means of which matter is divided: the natural 'forms' in which matter comes. But we may produce other examples viz. examples which illustrate the artificial dividing up of matter into parcels and concerning which the same asymmetry holds, e.g. 'cup of milk', 'glass of water', 'ingot of gold', 'ladle of molten lead'. In these cases we have one unit of matter concerned independently of the stuff itself: for example, we have one glass of water in that we have one glass which is of, or contains water: similarly in the case of 'ladle'. Also, although we cannot say that we have one ingot in that we have one ingot which *contains* gold, even so we have one ingot in that we have one ingot which is of gold. We can clearly count the number of ingots without knowing what they are ingots of, as we can count the number of cups (i.e. cupfuls) or ladles (ladlefuls) without being able to say what they are cups (cupfuls), ladles (ladlefuls) of. On the other hand we only have the *same* cups (cupful), *same* glass (glassful), *same* ingot or ladleful in that we have the same cupful, glassful, ingot or ladleful of *M*. As with the earlier cases the matter concerned is necessary to the identity of the parcel and sufficient for difference of parcel.

In the light of such asymmetry we must distinguish between: (a) sortals (in the extended sense) which are *tout court* of a functional form; (b) sortals (in this sense) which are not *tout court* of a functional form viz. speaking at a 'linguistic level': expressions of the general form '*X* of *M*' which are to be construed as functional *vis-à-vis* furnishing principles of sameness and distinction for particulars (for parcels), but are *not* to be construed as functional *vis-à-vis* formulating principles of *counting* for particulars (for parcels). Such 'sortals' furnish principles of counting in their own right but of necessity, empirical necessity, have a reference to 'matter' (as explained earlier), but they do not furnish principles of sameness and distinction in their own right in that they are to be functionally construed.[9]

I thus conclude that Strawson may not claim that *all* sortals in the extended sense rest for their introduction upon 'feature' universals. The most he may claim is that sortals which are *tout court* of a functional form do; and that other 'sortals' of the form '*X* of *M*' do only in respect of sameness and difference but *not* in respect of unity. He may only claim that 'sortals' of such a form which do not furnish principles of *distinction* in their own right may be said to rest upon the introduction of 'feature' universals presumably in that *qua* furnishing such principles of distinction they are to be functionally construed.

A.4. Having claimed that Strawson may not hold that *all* 'sortals' (in the extended sense discussed above) may unrestrictedly be said to rest for their introduction upon feature universals, I now turn to the thesis that sortals in the sense originally introduced by him[10] and by me, rest for their introduction upon feature universals: that feature universals supply a basis for their introduction.

As a general thesis this may not be maintained. It cannot, in general, be held that sortals in this sense *indirectly* rest for their introduction upon 'feature' universals in that they rest upon functional universals and functional sortals in turn rest for their introduction on feature universals, since to so claim is to be open to the objections raised against such a thesis earlier,[11] and, in particular, it would deny that sortals in this sense (the ordinary sense) may supply principles of counting and distinction in their own right for particulars.

Furthermore, such a thesis is not one which Strawson himself, in the immediate context of his discussion in *Individuals*, may consistently hold. To illustrate this I now turn to that work.[12] There the thesis that 'feature' universals, whilst not providing a principle for distinguishing, enumerating, and reidentifying particulars of a sort (for example, particular veins of gold or lumps of coal), may be 'modified' in such a way as to yield several such principles and yet not presuppose the introduction of any sortal, (or a term which may be construed as a sortal), lands Strawson in some immediate difficulty. For on his thesis here[13] we have the view that such 'sortal universals' as 'veins of gold', 'flakes of snow', are formed from feature universals, by *attaching* 'vein', 'flake', etc., to the feature expression 'snow', 'gold', etc., by the means of the relational expression 'of' so that the complete expressions (e.g.) 'vein of gold', 'flake of snow' are thus analysable into their elements: 'vein', 'of', 'gold'; 'flake', 'of', 'snow'. The immediate difficulty which he faces on such an account, on such a thesis of modification, is that he has to allow the introduction of *apparently* sortal universals, such as 'flake', 'vein', 'lump' of coal etc. He might retaliate to such a charge by claiming that such expressions as 'vein', 'flake', 'lump', etc., do not introduce sortal universals; that such expressions *are* essentially incomplete being of a functional form - 'vein of ----', 'lump of ---' where the blanks takes as their value a 'material name'. Such a retaliation however is incompatible with the thesis that feature universals may be 'modified' by being prefaced by such items as 'vein', 'flake', 'lump', etc., which items, we should note, Strawson writes in *italics*. On the thesis that 'sortal universal' expressions such as 'vein of gold', 'flakes of snow', 'lump of coal' *are* functional expressions they are precisely *not* analysable into three distinct elements, each such element having an independent sense, nor are they analysable, as he seems to require[14] viz. as 'lumps *of coal*' or 'veins *of gold*', where 'lump' and 'vein' have independent sense.

Thus the 'modification' thesis, advocated in both *Individuals* and 'Particular and General', is incompatible with the functional analysis of such 'sortals' as 'vein of gold', 'lump of coal' etc. Yet it is only upon a functional analysis of such expressions that Strawson may claim that such sortals rest for their introduction upon 'feature' universals: that feature universals may supply a basis for the introduction of such sortals and hence indirectly a basis for the introduction of sortals in his original sense.

A.5. I now turn to substantiate the case for saying that Strawson cannot claim that 'feature' universals together with the notion of 'placing' may supply the basis for the introduction for types of particular introduced by sortals in his original sense and the sense introduced by me.

Such a case may, at the general level, be substantiated by reference to my case that the introduction of sortals in *this* sense may not be said to indirectly rest upon the introduction of feature universals; but we need substantiation of the case at the particular level and to produce substantiation it is necessary to examine and produce good grounds for rejecting Strawson's case as regards the introduction of particulars by reference to such sortals.

Strawson's way of establishing the thesis that feature universals, together with the notion of 'placing', may supply the basis for the introduction of types of particular introduced by sortals (in the original sense), is to establish that we may provide an explanation of the notion of 'an individual instance' in the case of 'substance names'[15] (introducing sortal universals), analogous to that provided for 'material names'[16] (introducing feature universals) although he admits that this seems 'much harder' to do.[17] Although it is much harder to do he claims that it is perhaps worthwhile: 'for if it succeeds, we may find that we have then an adequate basis for the explanation of the notion of an individual instance in other cases, and for the explanation of further kinds of general things'; i.e. a *basic* analysis of the notion of 'an individual instance'.

A.5.(i). His case proceeds as follows:[18]

> In order for the attempt to succeed, we must be able to envisage a situation in which instead of operating with the notion of an individual instance of a cat[19] or an apple we operate with the notions of a corresponding feature and of placing.

He mentions the objection that ordinary language does not seem to provide us with the appropriate feature-placing sentence, but since this is only mentioned, he presumably thinks the point unimportant, as it surely is. What *is* important and properly so, is the argument that the idea of such sentences is, in these cases, absurd. On account of its importance, I quote the argument Strawson considers in full:[20]

> (1) it might be pointed out that an all important difference between such things as snow and such things as cats lay in the fact that different instances of snow are, in a sense, indefinitely additive, can be united together as one instance of snow, whilst this is not true in the case of instances of cats .... (2) it might be added that we have no name for a general thing which could count as the required feature in the case of cats. It is true we speak of *the cat* in general, but

> 'the cat' ranks as a species-name, and the notion of species as surely presupposes the notion of individual members as the notion of a property involves that of individual things to which the property belongs or might belong. It is also true that we might speak of an instance (specimen) of *a* cat, as we may speak of an instance of gold; but here what follows the phrase 'an instance of', is not, as 'gold' is, a general thing name which could figure in a merely feature-placing sentence, but an expression which also serves as the indefinite designation of an individual.

And he comments:

> Does not all this strongly suggest that there *could be* no concept of the 'cat-feature' such as would be required for the analysis to work, that any general idea of cat must be the idea of *a* cat, i.e. must involve criteria of identity and distinctness for cats as individuals and hence the notion of an individual instance?

A.5.(ii). Strawson acknowledges that these objections have great force and importance, but holds that they are not decisive. It is my case that they *are* decisive, so I turn to an examination of (1). First I have to amend (1) to allow for the distinction between expressions of the general form '*X* of *M*' which I have discussed earlier, viz. the distinction between expressions of such a form which may be regarded as functional sortals as concerns furnishing a principle of counting and those that may not so, as concerns instances of '*X* of *M*'. Strawson's 'instance of snow' relates to both. I do not of course deny that e.g. patches of snow, expanses of snow, etc. (*X*'s of ----) are indefinitely additive - several drifts may be counted together in one large drift, several patches as one large patch; but *not all* 'instances of snow' are indefinitely additive; several flakes of snow do not make one large flake: several veins or nuggets of gold do not make one large vein or nugget any more than several cats added together make one large cat. The all important difference Strawson mentions holds *equally* in regard to some instances of snow (instances of *M*) as it does in the case of cats (substance names); the contrast which he wishes to draw therefore is not universally valid and hence any argument which attempts to play on this alleged difference - which the argument here does[21] - will itself not be generally valid, for it is assumed that the case has been made out for the analysis for the notion of an individual instance in the case of 'material names'.

The reason why we cannot introduce, e.g. 'there is a *P* (parcel) of cat here' or (in Strawson's terminology elsewhere)[22] - 'there is a slice of cat here'; 'there is *a* cat-slice here'* - for 'there is a cat here' such that we may have an alleged parallel between:

(a) 'There is a patch of snow here'
(b) 'There is a slice of cat here'

is *not* because, in the case of snow (or, in general, 'feature' universals), instances are indefinitely additive, for as we have seen this is not necessarily true; but because we cannot substitute a functional universal for a non-functional one either (a) in general or (b) in this type of case. We cannot in general substitute a functional sortal for a non-functional one since I have earlier produced a counter example. To reiterate: we cannot write for 'There is a flake (of snow) here', 'There is an *F* (of *S*) here'; 'There is a (flake of snow) here', 'There is an (*F* of *S*) here', for in the case of a functional sortal we may only have the notion of *an* instance in that we understand that of which it is an instance, whereas in the case of a non-functional sortal we have the notion of an instance independently of that of which it is an instance.[23]

It will be commented that this point assumes that in the case of 'cat-slice', 'slice of cat' we do indeed have a case of a functional sortal and this itself may be objected to. However, for Strawson's case to sense make 'slice of cat' *has* to be understood as a functional sortal.

We cannot, in this type of case, substitute a functional sortal 'slice of cat' for a non-functional sortal 'cat', let alone claim that the non-functional sortal rests on the introduction of the functional sortal, since, we may claim, in order to introduce the notion of 'slice of cat', we *already* have to be possessed of the concept of *a* cat. The notion of a (a single) 'cat-slice', gains it unity from that of *a* cat. We have one cat-slice only in that we have one slice of *a* cat since a cat-slice is the resultant of cat divided up: it is a division of cat, but in order to speak of 'cat matter' being divided up in some such way, we have first to introduce the notion of a *whole* divided and in order to do *this* we have to introduce the notion of *a* cat. A cat-slice is a division of a whole, but in order to understand what counts as one division we first have to understand what counts as one whole. We may thus only count the number of cat-slices in that we may first count the number of cats or at least one cat. Equally we only have the *same* slice, in that we have a slice of the *same* cat. What therefore constitutes the unity and identity of one such slice has a necessary back reference to the notion of *a* cat. Necessarily any slice of cat is a slice of some cat or other, though not necessarily some particular cat present to one's senses. Strawson would reject this, but as I shall contend in my discussion[24] of his negative case against (1) and (2) above, only upon a confusion. The thesis that we may only introduce the notion of 'cat-slice' in that we first introduce that of 'slice of *a* cat' may now be challenged at a general level. It might be claimed that we can produce criteria for distinctness and identity for cat-slices without reference to the notion of *a* cat, criteria of distinctness and identity for instances of the universal *cat* without reference to the notion of *a* cat. But in such a claim how do we determine that we have the *same thing*, i.e. cat-feature exhibited or present in all these instances? It is my case that we may only say that we have the same thing present, i.e. cat-feature present, in that we have the very same *sort of thing* present, namely a cat: we could not intelligibly claim that we had the same feature present but had no cat present. Equally, in that we are presented with a different sort of *A*, and *only* in that we are presented with a different sort of *A*, have we a different feature present. In other words, the identity of the feature universal is determined by reference to the sortal and thus may not intelligibly be said to replace it or be more fundamental than it.

Further, at the level of particular instances, we may validly contrast the idea of 'cat-feature' and, for example, 'snow feature'. Whereas in the case of cat-feature we only have *different* instances of the same *F* in that we have different instances of the *same sort* of thing, in the case of 'snow feature' it is distinctly not the case that we have *different* instances of snow *only* in that we have different instances of the *same sort* of thing, e.g. flake of snow, for we may indeed speak of different instances of snow by reference to items which invoke no sortal (in the original sense), e.g. 'patches of snow', 'expanse of snow'. Further, in the case of, for example, 'snow' we do not have the same stuff, or the same feature *only* in that we have the same single sort of thing present. It is not the case that we have the same stuff *only* in that we have, e.g. flakes (of snow). In the case of, for example, snow (or 'matter' in general), it is not the case that any snow ('matter' in general) is necessarily the snow (is matter) *of* something, i.e. some one thing and gains its identity from that which it is the snow or matter. Whereas, as I have argued above the notion of a cat-feature does gain its identity from that of a cat.

Having discussed Strawson's first argument against his own position, and attempted to show that the argument from 'addition' does not bring out clearly what I have advanced as the real difficulty for his thesis, I shall now turn to an examination of the second argument he presents against himself, *leaving for the moment further challenges to what I have advocated as the real difficulty for his thesis*.[25] I thus turn to (2) above.[26]

A.5.(iii). As we have seen, Strawson acknowledges that this argument, with its two subsections has great force but is not decisive. I hold that the argument is decisive but to see how it is so decisive involves rephrasing in such a way that the presentation is not prejudicial in Strawson's favour (as his own version is) and yet that such a presentation is one which may reasonably be advanced as the *correct* form of presentation and hence not prejudicial against him.

First, let it be said that I do not object to the implicit point that it is an unimportant matter that we have no name for a general thing which could count as the required feature in the case of cats. He argues that 'the cat' won't suffice since: '"the cat" ranks as a species-name and the notion of a species as surely presupposes the notion of individual members as the notion of a property involves that of individual things to which the property belongs ...'.[27] This form of presentation is prejudicial in Strawson's favour in that he wishes to separate the notion of 'being an individual member (of *A*)' from 'being an *A*'; in crude, to separate the notion of 'being an individual member' from 'being an *A*'. For he claims that the notion of a species presupposes the notion of an individual member. It is perfectly alright for him to be claiming that we may only have the notion of 'the species *cat*' in that we have the notion of an individual *cat*, or that we may only have the notion of 'the species cat' in that we have the notion of individual *cats*, or that we may only have the notion of the species *cat* in that we have the notion of an individual *A* (where '*A*' is some higher order species or genus), but this is *not* what he is claiming. He is, in effect, claiming that we may *only* have the notion of the species 'cat' in that we have the notion of individuals which are *cats* viz. the notion of individuals of which it is true that they are cats: thus once again turning a sortal into a description or predicate. What grounds have we for holding that the correct form of the objection is that we only have the notion of 'the species of *cat*' in that we have the notion of an 'individual *cat*', or more generally 'individual *A*' (where '*A*' is a species or sortal)? The ground I offer here is that the phrase 'being an individual member of *A*' must be construed as a functional expression 'being an individual member of ----', and *not* as Strawson construes it viz. either:

(i) 'being an individual member (of) ----'
(ii) 'being an individual member (of ----)'

Failure to so construe entails the absurdity that we may have individuals members of the species *A*, which are not *A*s. However once we admit that we may only have the notion of 'the species cat' in that we have the notion of individual cats, (*and* conversely), the argument against the introduction of 'cat-feature' from this standpoint is decisive.

Let us now consider the second sub-argument.[28] Strawson writes:

> It is also true that we may speak of an instance (specimen) of *a* cat, as we may speak of an instance of gold; but here what follows the phrase 'an instance of' is not, as 'gold' is, a general thing name which could figure in a merely feature-placing sentence, but an expression which also serves as an indefinite designation of an individual.

This form of the presentation is equally prejudicial in his own favour in that he regards the phrase 'an instance of *F*', *where F is a sortal (a 'substance name') or a 'matter' term (a 'material name')* as construable as 'an instance (of *F*)', thus treating sortals and 'matter' terms as ordinary predicates, and hence in *these* types of case divorcing the notion of an instance from what it is an instance of. The objection should not be in the form of what *follows* the phrase 'an instance of' but in the form of what *completes* the phrase 'and instance of ----'. Once again the question arises as to why this is the correct form of presentation. My case for this is that such a phrase as 'instance of' must be construed as 'instance of ----', in *these* types of case, otherwise we should be left with the notion of a 'pure instance', a position Strawson himself wants to deny.

As we have seen earlier,[29] Strawson confuses:

(1) Necessarily: every instance is an instance *of* something general;
(2) Every instance is necessarily an instance of something general.

where by 'something general' here is meant a sortal or a 'matter' term: and I have argued that (1) is the correct version of the matter, from which it follows that we must take 'an instance *of* something' as a functional expression in such cases.

To return to the text. The objection should *not* be that 'a cat' is not like 'gold', in that the latter can figure in merely feature-placing sentences but that the former cannot, since what follows the phrase 'an instance of' is *pertinently different*; but that *neither* 'gold' *nor* 'a cat' can be said to *follow* the phrase 'an instance of'. Once the objection is cast in this form, that in the first case what *completes* the functional expression 'an instance of ----' is an indefinite designation of an individual and that in the second case what *completes* the functional expression is a 'matter' term, 'a material name', the objection is in effect decisive. We may not, in such cases, divorce the concept of an instance from what it is an instance *of*, and hence advocate an analysis of the notion of an individual instance as regards 'substance names' in terms of feature universals and 'placing'. We may put the point more particularly and more crudely by commenting that as what constitutes an instance of 'gold' is not independent of the concept of 'gold', so too what constitutes an instance of 'cat' (to use Strawson's parlance) in not independent of our concept of *a* cat, i.e. some cat or other.

A.5.(iv). I now turn to Strawson's negative case against arguments (1) and (2) above with special reference to my case earlier,[30] namely, that in order to introduce the idea of a 'cat-slice' we already have to be possessed of the concept '*a cat*'. As I commented earlier,[31] he would reject this, but only, I claimed, on the basis of a confusion. I shall now develop this case.

Strawson explicitly writes as regard his negative case:[32]

> For they (i.e. the two arguments advocated)[33] do not show that it is logically absurd to suppose that we might recognize the presence of cat or signs of the past or future presence of cat, without ever having occasion to distinguish one cat from another as the cat on the left, or identify a cat as ours or as Felix.

The confusion here is that whilst it is not logically absurd to suggest that we might recognise the presence of cat without ever having occasion to distinguish one particular cat from another cat (i.e. one given cat from another given cat), nevertheless it *is* logically absurd to suggest that we might so recognise without having the notion of *a* cat i.e. some cat or other. To require the presence of a cat as sign of the past or future presence of cat, is to recognise the presence or sign of some cat or other, but of course not necessarily some particular one, Tibbles or Felix. As I have commented in relation to the parallel passage in *Individuals*,[34] Strawson wrongly infers from the consideration that one need not, in a given case, 'think identifyingly of particular cats', the thesis that in such a case one can introduce the notion of 'cat' (feature universal) as opposed to 'a cat' (sortal universal). This move would be valid only if 'thinking identifyingly of particular cats' were a necessary condition for introducing the item '(a) cat'; '(some) cats' into a proposition. But this is not so necessary. One can be said to recognise the presence of a cat or cats without necessarily being able to say *which* cat or cats it is of which one recognises the presence. I can well say 'Some cat has been here'; to use Strawson's parlance, I can genuinely introduce the item 'some cat' without being able to say which cat it is (without 'thinking identifyingly' of a particular cat); and further I can say 'some cat or other has been here', i.e. I can genuinely introduce the item 'some cat' and yet the question: 'Which one?' not clearly or even sensibly arise. In that, in order for Strawson's general negative case to hold here, he has to conflate the definite and indefinite sense of 'some cat'; his case hangs on taking the definite sense of 'some cat' as the *only* sense. In sum, I thus claim that the arguments Strawson advances against himself are in fact decisive and his general negative case plays on a confusion.

A.5.(v). I thus now turn to his positive case for the thesis that in regard to 'substance names' we may analyse the notion of 'an instance' in the way proposed. He allows[35] that the above arguments against him show something. The point about the species-name is acknowledged to be sound but he holds that the notion of a species, like that of a property, belongs to a level of complexity which he is trying to get below. However, as I have already argued, the notion of a species cannot be said to *presuppose* the notion of an individual member such that the notion of an 'individual member' of a species is logically independent of the notion of *a* species. '*A* is a member of a species *B*' cannot be written as: '*A* is an individual' and '*A* is a member of a species *B*' without, in the *last* analysis, subscribing either to a doctrine of 'pure individuals' (which Strawson himself, as argued, wants explicitly to deny) or to what I have described as an impossible or no doctrine of instantiation (which Strawson, as argued, is committed to accepting). To try to get *below* the level of complexity mentioned is only possible by subscribing to one or other of equally impossible doctrines.

More importantly here,[36] however, Strawson concedes that the first argument shows that if there is to be a general concept of the 'cat-feature' corresponding in the required way to the notion of an 'individual instance', it must already include within itself the basis for the criteria of *distinctness* which we apply to individual cats. Roughly, the idea of 'cat', and that of 'snow' would include the idea of a characteristic *shape*. To concede *this* however, he holds, is not to concede the impossibility of the analysis.

The problem which immediately arises is how the general concept of the 'cat-feature' can already include within itself the basis of such criteria of distinctness which we apply to individual cats. The argument *against* saying that it can so provide is that what counts as one instance of the characteristic shape is uniquely determined by what counts as *one cat*.[37] We have the instance of the cat-shape only in that we have the cat: *not*, I contend, as Strawson would want, that we only have one cat in that we have one instance of the cat-shape.

We may concede to him that cats do have a characteristic shape, but from this is does not follow that there is something, i.e. some one thing, namely a certain shape, which is independently identifiable and hence may be said to exist independently of a cat, which is what Strawson requires. Rather, for there to be a cat-shape is for there to be *A*'s *of that* shape, not for there to be *A*'s *exhibiting* that shape. I am *not* claiming here that: necessarily every shape gains its identity from that which it is the shape; indeed I have already acknowledged the contrary,[38] but I am claiming that: '*Necessarily* every *A* shape is the shape of some *A*', where '*A* shape' is introduced by reference to some sortal '*A*', and we cannot eliminate the sortal in favour of something else. Indeed, to claim that we *can*, runs into insurmountable difficulties. We could never give an account of 'cat-shape' (e.g.) in terms of *B*-ish shape, where '*B*-ish' attempts to spell out 'cat-shape' in terms of the qualities *it* exhibits, for the qualities (features) *it* exhibits has a necessary back reference to *it*. Hence any attempt at an analysis or equivalence would have a necessary back reference to 'it'. If someone, *per impossible*, were to press that we *could* present such an account or analysis, and that once the analysis were established we could then dispense with the original viz. with 'cat-shape', which I am claiming has a necessary reference to the notion of *a* cat, we must tackle the thesis that 'cat-shape' may, after all, be analysed out in terms of a set of qualities, features or predicates, that have no reference to the sortal 'cat' or indeed *any* sortal.

The attempt to explicate or offer an account of 'cat-shape' in terms which have no reference to the notion of *a* cat, or more generally an *A*, where '*A*' is a sortal, is a refined version of the thesis that we may reduce the idea of a cat to the idea of its predicates. Leaving aside the objection raised earlier concerning *its*, one may here advance a further but related objection, namely that a cat is something which *has* properties, features, and therefore cannot itself be reduced to or analysed out in terms of a set of properties or features.[39] As against this, however, it might be pressed, following Strawson, that we can reach a more sophisticated level of thought at which that which is said to have properties *itself* becomes a set of properties. But at such a supposed sophisticated level severe difficulties arise - at least in the first instance. The account of 'cat-shape' which does not mention 'cat', would have, it seems, to take the form (roughly) 'shape which is animalish or creaturish and which is ....', where the '....' here is filled out by some set of uniquely distinguishing features of cat (as opposed to animal). This account transforms 'animal' or 'creature' - *A*'s which *have* properties, into properties or (speaking linguistically) 'predicates', which again is open to the objection of absurdity. It transforms X's which furnish principles of counting and distinctiveness in their own right ('animal'; 'creature') into items which *depend* upon such principles for their countability and distinctness.

At this point it may well be protested that the case is not so absurd as I have tried to make out. It might be said that it is contingent that the items, symbols, which furnish principles of counting and distinction are the ones we have. It is possible to imagine *other* such items and thus render such items, symbols, as e.g. 'cat' or 'animal' (ordinary sortals) in terms of predicates or properties. It will be suggested that that which *has* properties may itself be translated into properties in another system; that those items which indicate that which has properties in language L may be translated into predicates in language L_1, subject to the provision that we may furnish an appropriate set of sortals to perform the role played by 'cat' or 'animal', etc., in L. In such a language envisaged, 'shape' e.g. might perform such a role such that we might have 'shape which is cattish' instead of 'cat', and 'shape which is cattish' or 'cattish-shape' could serve to distinguish one individual from another and form a principle of counting for individuals. I shall now maintain that we cannot introduce 'shape which is cattish' or 'cattish-shape' as logically basic, as such a principle able to play such a role. Now I have already acknowledged that it is not the case that every shape is such that it gains its identity from that which it is the shape - cases of geometrical shapes, Plato's mathematical 'Forms', are cases in point: but we cannot extend this line of thought to the present case, since 'cattish-shape', derived from 'cat-shape', is not the specification of a quite determinate shape. We simply have no parallel here with e.g. square and circle. We may count the number of squares, circles, triangles, rectangles without knowing what it is that is square, circular, etc., but we may not count the number of cattish-shapes without knowing it is cats which we are counting. To re-iterate the point I made earlier: we have one cat or cattish-shape in that we have one cat, rather than one cat in that we have one cattish-shape in that we can identify the cat or cattish shape *only* via the cat. A cat shape or cattish-shape is the *shape of a* cat: an *A*-ish shape (where '*A*' is a sortal) is the *shape of* an *A*: it is not a shape exhibited by a cat or an *A*; whereas e.g. being square or circular *is* a shape *exhibited* by something or other.

A.5.(vi). I thus hold that the general concept of the 'cat-feature' *cannot* include within itself the *basis* for such criteria of distinctiveness which we apply to particular cats. The idea of a characteristic cat-shape cannot, and *radically* cannot, form the basis for distinguishing between cats in general: generalising, the idea of a characteristic *A*-shape (where '*A*' is a sortal) cannot and radically cannot, form the basis for distinguishing between *A*s in general and thus Strawson's analysis of what he calls 'substance names' in terms of 'features' and the corresponding analysis of the notion of an individual instance furnished by a 'substance name' in terms of feature universals and placing, fails.

It is pertinent to add here that, in essence, Strawson raises the question: 'What makes two individuals two cats?', 'What makes two individuals two *A*s (where "*A*" is a sortal)?', and thinks that this is a possible question to raise. His answer, in particular, is that two individuals are two cats in that they have or exhibit the same characteristic shape. I contend that to raise this question is itself a misunderstanding. The question itself is only raisable upon a confusion which I have already explained earlier, namely the conflation of:

(1) *Necessarily*: any individual is an individual falling under something general (where 'something general' is a sortal in Strawson's original or extended sense);

(2) Any individual *necessarily* falls under something general.

For, as I have already pointed out, what Strawson works with is (2) when he gives some general and propaedeutic remarks about the notion of an individual or particular,[40] yet he fails to realise that (2) entails (1). (2) would appear to allow him to raise the question: 'What makes two individuals two cats?' ('What make two instances of *B* two instances of *B*'?) since (2) reads that it is a universally necessary truth *concerning* individuals that they fall under something general, but since (2) entails (1) this is appearance only. Strawson thus here treats 'being a cat' (in general being an '*A*') as a *property* which some individual necessarily has, whereas, as I have argued, the truth of the matter is rather that necessarily to be an individual is to be an *A*, where '*A*' is a sortal in either the original sense Strawson introduced, and I introduced, or in the 'extended' Strawsonian sense. We *may* it seems raise the questions: 'What makes any two *animals* two cats?' or 'What makes any two cats, two *cats*?' and get, say, as a rough and ready answer to the first: 'They are both *animals* having a certain shape, features *X*, *Y*, *Z*, ...'; that is, the *basis* on which we distinguish two cats as two, two *A*'s as two, is a more generic or higher order sortal - 'animal having ....'; '*A* having ....'. This *basis* therefore of our criteria of distinctiveness which one may apply to individual cats, or individual *A*'s, is *itself* some *higher order*, but still 'sortal universal' and such higher order 'sortal universals' themselves may not be reduced to or analysed out in terms of 'feature' universals, since the same problem in more general form will arise here as have arisen in the lower order cases.

A.5.(vii). Having thus, I hold, substantiated my case for saying that Strawson cannot claim that 'feature' universals together with the notion of 'placing' may supply a *basis* for the introduction of types of particular introduced by sortals in his original sense and the sortals introduced by me, it is necessary to emphasise the nature of my objection to Strawson. What my case against Strawson brings out is that not only does this case fail but that it *necessarily* fails: and this *shows* that it is a misunderstanding to seek for a basis for the introduction of types of particular introduced by sortals in the original sense. From another standpoint it emphasises the point that we may not speak of such sortals having instances, and seek to analyse such a notion of an instance in terms which will furnish a basis for the introduction of particulars by reference to such sortals.

A.6. I now turn to substantiate the case for saying that Strawson cannot even claim that feature universals, together with the notion of 'placing', may supply a basis for particulars introduced by reference to 'feature' universals, i.e. principles for counting and distinction formed from 'matter' terms, 'feature' universals such as 'expanse of snow', 'lump of coal', 'bit of gold' etc.

A.6.(i). In 'Particular and General'[41] and in an earlier parallel passage in *Individuals*,[42] where Strawson introduces the notion of 'feature-placing' sentences (or statements)[43] such as:

> 'Now it is raining'
> 'It is (has been) raining'
> 'Snow is falling'
> 'There is gold here'
> 'There is water here'

Such feature statements are ordinary empirical statements and whilst all these sentences contain either the material name of a general thing or a corresponding verb, 'none contain any expression which can be construed as serving to make a definite or indefinite mention of individual instances of those general things (i.e. falls or drops of rain, pieces of gold, pools of water and so on)'.[44] That is, such sentences do not bring particulars into discourse. But whilst such sentences do not introduce particulars into discourse, they provide the materials for this introduction.

He argues as follows:[45]

> Suppose we compare a feature-placing *sentence* ('There is snow here') with a *phrase* ('This (patch of) snow') in the use in which an individual instance of the feature is mentioned. It seems possible in this case to regard the notion of the individual instance as something logically complex in relation to the two simpler notions of feature and placing.

The argument for saying that the notion of an individual instance is 'logically more complex' is unconvincing. It is that that if we merely mention the individual without going on to say anything about it we fail to utter a complete sentence. But whilst this is true it does not follow that the phrase has no sense - viz. that it only has sense in the context of a completed sentence, which is what Strawson implies here; for he writes that 'yet what the feature-placing sentence does, explicitly is, in a sense, implicit in this mere mention'.[46] And we may challenge this too: how exactly is what the feature sentence does, in a sense, explicit in this mere mention? In what sense? Is he claiming that the 'This patch of snow' *entails* 'There is snow here'? If so, there is a sense in which this cannot be denied, but the entailment of such an empirically true proposition is not sufficient for a thesis of *logical* priority. As I have earlier argued,[47] what is empirically prior is not to be identified with what is logically prior and empirical priority or 'basicness' is not a sufficient condition of logical priority. Yet what Strawson is here advocating is a *logical* thesis, for he writes:[48] 'So, as the basic step in an explanatory schema, we may regard the notion of a particular instance of *certain sorts* of general thing as a kind of *logical compound*[49] of the simpler notions of a feature and of placing.'

So we now turn directly to the question: 'Why may we not regard the notion of a particular instance of, e.g. "patch of snow", "lump of gold", "expanse of water", as such a logical compound?' My answer to this is in order to determine that we have one case of snow *here*, one case of gold or water *here*, we have *already* have had to introduce the notion of 'one patch of snow *here*', 'one lump of gold *here*', 'one expanse of water *here*'. That is, we may only have one *item* of snow, gold or water, introduced by snow, gold or water *here*, in that we have one patch of snow, lump of gold, expanse of water *here*. In other words in order to determine the precise reference of *here* in regard to the 'matter', snow, gold or water, we have to introduce the notion (in general) of a *parcel of matter*; the prime reference of *here* cannot be given in any other way. Strawson implicitly recognises this problem when he writes:[50] 'But what about the criteria of distinctiveness and identity which were said in general to be necessary to the notion of an individual instance of a general thing?'. And he is of the view that he has the answer as follows:[51]

> The *basis* for the criteria of distinctness can already be introduced at the feature-placing level, without yet introducing particulars. For whilst we can say 'There is snow here' or 'There is gold here', we can ask, perhaps, more exactly, though not more correctly, say 'There is snow (gold) *here* - and *here* - and *here*.

But this, I claim, will not do since we may only fix the prime reference of 'here', in relation to snow, gold or whatever, in that we have the notion of a parcel, lump or expanse of the stuff in question. At this point it is pertinent to go back to the earlier case of 'substance names'. There Strawson acknowledges that if there is to be the general concept of the cat-feature, corresponding in the required way to the notion of an individual instance, it must already include within itself the *basis* for the criteria of distinction which we apply to individual cats - roughly the idea of a characteristic shape. Now one has to agree with Strawson that the idea of snow, gold, etc. does not include the idea of a characteristic shape or form, but from *this* it does *not* follow that the idea of snow, gold, etc. does *not* include within itself the *basis* for the criteria of distinctness which we apply to individuals falling under the general term. The idea of snow, etc. does so include such a basis, viz. the idea of *some* shape or form, as opposed to some characteristic shape or form.[52] To deny this would be to deny that matter comes in or is organisable under some shape or form and to deny this would render it impossible for us to individuate and count material particulars formed from such matter. It is contingent that matter comes in the *specific* form it does or is organised in the specific form it is, but it is not contingent that it comes in or is organised in *some form of other*. And here I am resorting to the Aristotelian principle that for matter to be (to exist) is for matter to be under some form.

Strawson would deny this. He would argue, and indeed in effect argues, that it is possible for us to individuate and count material particulars formed from such matter simply by reference to the notion of 'placing'. The problem for such an argument is that it in no wise enables us to individuate and count such material particulars since it does not tell us *what* we are supposed to be distinguishing or counting; and to press that feature universals, together with the notion of 'placing', can yield us with the notion of *what* we are supposed to be counting or individuating will not suffice, since I have argued above we may only have *one* (case of) snow *here*, etc. in that we have, e.g. one *patch of snow*, etc. here. In crude, we cannot form the notion of 'patch of *M*' (in general, 'parcel of *M*') from the feature universal together with the notion of 'placing' - 'Snow *here*' etc., since, as I have argued above, we may only fix the prime reference of *here*, by reference to the notion of *a parcel*. Strawson holds that[53] 'The considerations which determine multiplicity of placing become, when we introduce particulars, the criteria for distinguishing this *patch of* snow from that, or the first *fall of* snow from the second'. Whilst however it may be maintained that spatial or temporal positions may, and indeed do, serve as criteria for distinguishing this patch of snow from that patch of snow or the first patch of snow from the second, this does not entail, for reasons offered, that the notion of a patch or fall of snow may be derived from that of the notion of a feature plus placing.

A.6.(ii). I thus hold that in the case of what Strawson has designated 'material names', he may not carry out his programme. He cannot, that is, even claim that feature universals together with the notion of 'placing' may supply a basis for particulars introduced by reference to 'functional sortals'. Once again it is necessary to emphasise the nature of my rejection of his case. What this case against him brings out is that not only does his case fail but that it *necessarily* fails, and this *shows* that it is a misunderstanding to seek of a basis for the introduction of types of particular introduced by such 'functional sortals'. From another standpoint, as with the earlier type of case, it emphasises the point that we may not speak of such sortals themselves as having instances and seek to analyse such a notion of an instance in terms which will furnish a basis for the introduction of particulars by reference to such sortals.

A.7. *A further argument against Strawson's position*

Even if the above arguments in Sections A.5. and A.6. fail, one can produce a further, though indirect argument, for showing that Strawson's programme fails and necessarily fails.

His case for holding that, in relation to A.5., the notion of 'cat-feature' together with that of 'placing', may furnish a basis for the introduction of cat particulars, invokes the idea of *our* speaking of 'cat here' or 'cat *here* and *here*; and *here*': equally his case for holding that, in relation to A.6., the notion of 'snow-feature' (e.g.) together with that of 'placing' may furnish a basis for the introduction of snow particulars, invokes the idea of *our* speaking of 'snow *here*' or 'snow *here* and *here* and *here*'. That is, the idea of the individuation and identity of the placing of the feature concerned has a necessary reference to the notion of a *person* pointing; thus the individuation and identity of the placing itself, even on Strawson's own account, has a necessary implicit reference to the notion of a sortal in his own original sense. Thus we may not carry out Strawson's programme without reference to at least one sortal namely that of a person.

Strawson may reply to this in one or other of several ways, each of which I shall argue fails, and necessarily fails:

(1) that reference to a person or persons is contingent: the reference to 'here' may be determined by reference to place description, e.g. by snow here I mean the snow which is between the farm yard gate and the cow shed; or again: the snow which is bounded by that hedge, ('the hedge which is ....', where '....' is some uniquely individuating description), the stream and the road: or again the snow which has fallen between the dump of coal, the stretch of water and the

farmhouse. But in all such cases, whilst we *might* eradicate reference to context and person speaking by supplying a uniquely individuating 'description', in a manner favoured by Quine, we *still* determine the reference of 'here', the individuation and identity of the place, by reference to position relative to items introduced by reference to sortals, either in Strawson's original or extended sense;

(2) that reference to items introduced by reference to sortals in either sense is still contingent. The reference of 'here' may be determined by place description in the sense of locating the place concerned on, say, a 'grid' system or, more sophisticatedly, on an ordnance survey map using a method in addition to mere co-ordinate location. However, even on a grid system, we first have to introduce the notion of a *grid*, which *itself* introduces a sortal and further at the very minimum, we have to understand the notion of *a point* or *an intersection*, which again introduces sortals. Further, as regards the grid, we have to understand how to interpret it - e.g. a square in a grid represents *an area* of land of so many square kilometres. And note that we cannot get out of the problem here by arguing, e.g. that an intersection is an intersection of two lines, since *line* itself is a sortal. Likewise in the case of an ordnance survey map, we already have to understand and have introduced a sortal, namely *a map*: and further we require to understand 'an intersection', 'a point', 'a (grid) square'. In this case too we have to understand, i.e. be able to interpret, e.g. a grid square as representing *an area* of land; a point of intersection as representing a spatial *location*; '***NT***' as representing a *National Trust Property*; '**P**' as representing a *post office*, etc. In other words, even with such 'modes of representation', sortals are implicitly present;

(3) that the notion of '*a* person' itself may be analysed out in terms of a 'feature' universal together with the notion of 'placing'. But such a suggestion meets with all the objections produced earlier against the proposed analysis of 'a cat' in terms of 'cat-feature' together with the notion of 'placing'. And likewise so does any attempt to eradicate the sortals introduced in (1) and (2) in any such manner.

Further, since the notion of a person pointing is central to, indeed is a prerequisite of our understanding of, the analysis of the notion of 'an instance' by reference to feature universals together with the notion of 'placing' - for it is the foundation point of the whole system or language which Strawson is advocating as basic - he cannot consistently allow that it itself is to be analysed out in terms of the system or language itself; for if *it is* to be so analysed out then we cannot understand the very system or language that he is advocating as basic. It will of course be said that the reference to the person pointing is only foundational to one type of case, namely the type of case that explicates the notion of 'an instance' in terms of spatial location. But 'placing' covers also *temporal* location, as witness the following example:[54] 'And when we can say

"It snowed today", we can also, perhaps, more exactly, but not more correctly say "It snowed twice today"'.

However, even here: (a) reference to persons cannot clearly be eradicated; (b) reference to sortals cannot be eradicated. As concerns (a): by 'today' is meant the time at which person 'A' is speaking; (b): 'today' itself introduces the idea of *some* day viz. itself introduces a sortal - and we cannot get rid of the idea of some sortal even if we claim that we may analyse the notion of 'a day' in terms of 'a *period* of time'. Further, of course, Strawson could not coherently claim that the notion of 'day' or '*period* of time' could be analysed out in terms of a feature universal together with the notion of temporal placing, for a reason parallel to that offered above in connection with 'person'.

B. A Second Consequence of the Failure to recognise the Nature of the Sortal: the Search for a *Basic Notion of an 'Instance'* or *Explanation of the Notion of an 'Individual Instance'*

I now turn to a second consequence of the failure to recognise the nature of the sortal as present in the writings of Strawson: namely the search for a *basic notion of an instance* or *explanation of the notion of an individual instance.* Before proceeding to make out my case it is necessary for me to explain how failure to recognise the true nature of sortals indirectly leads one to such a search.

B.1. It indirectly leads one to such a search in that sortals are first treated as predicates[55] which may be said to *have* instances and not as terms or expressions which furnish us with principles of counting and distinction *for instances*. Once sortals are treated in such a way, the road is open for the search for a more fundamental concept of 'an instance' or an explanation of the notion of 'an instance' which has of necessity no reference to sortals. Indeed, as I explained[56] earlier, the search for the basic notion of 'an instance' or explanation of the notion of 'an individual instance' confuses the (true) thesis:

(1) *Necessarily:* every (individual) instance is an instance of something general, with the (false) thesis:

(2) Every (individual) instance is *necessarily* an instance of something general failing to realise that (2) entails (1) but not conversely.

Once one accepts (2) failing to realise that it entails (1), i.e. once one accepts that it is necessarily true concerning any individual instance that *it* is an instance of something general (in the prime case, of some sortal), then the way is open either to the positing of a basic notion of 'an instance' or to an explanation of the notion of 'a particular instance' furnished by a sortal. Such a way is *not* open to one who accepts (1) or that (2) entails (1).

B.2. That Strawson *is* committed to the search for a basic notion of 'an instance' or an explanation of the notion of an individual instance furnished by a sortal, in either his original or extended sense, is abundantly clear from his writings discussed in Section A above. As I there pointed out,[57] Strawson's way of establishing the thesis that a feature universal together with the notion of 'placing' may supply the basis for the introduction of types of particular introduced by sortals (in the original sense), is to establish that we may provide an explanation of the notion of an 'individual instance' in the case of 'substance names' (introducing sortal universals) analogous to that provided for 'material names' (introducing feature universals). He claims that if we can provide an explanation of the notion of 'an individual instance' in the case of 'substance names' then we may find that we have an adequate basis for the explanation of the notion of 'an individual instance' in other cases and for the explanation of further kinds of general things i.e. a *basic* analysis of the notion of 'an individual instance'. The basic analysis of the notion of 'an individual instance' is given in terms of the notions of a corresponding feature and placing.[58]

That his enterprise of searching for a basic notion of 'an instance' or an explanation of the notion of an individual instance furnished by a sortal is unsuccessful in fact and misconceived in principle, has, I contend, also been already shown earlier in that:

(i) I have shown that his case for holding that we may provide an explanation of the notion of 'an individual instance' in terms of 'feature' universals together with the notion of 'placing' in the case of 'substance names' fails, and necessarily fails (*see* sec. A.5 above). This *shows* that it is a misunderstanding to seek for a more basic notion of 'an instance' than that furnished by a sortal in Strawson's original sense, or to seek for an explanation of the notion of 'an individual instance' in terms other than that so furnished by a sortal;

(ii) I have shown that his case for holding that we may provide an explanation of the notion of 'an individual instance' in terms of a 'feature' universal, together with the notion of 'placing', in the case of 'material names' fails and necessarily fails (*see* sec. A.6 above). This too *shows* that it is a misunderstanding to search for a more basic notion of 'an instance' than that furnished by a sortal in Strawson's extended sense (both functional and non-functional; sortals having reference to 'matter' ['mass'] terms), or seek for an explanation of the notion of 'an individual instance' in terms other than so furnished by such sortals.

I thus conclude that the search for a basic notion of 'an instance' or an explanation of the notion of 'an individual instance' furnished by a sortal, either in its original or extended sense, in terms of a 'feature' universal together with the notion of 'placing' is unsatisfactory in fact and misconceived in principle.

C. A Third Consequence of the Failure to recognise the Nature of the Sortal: the Search for an *Underlying Basis* for the Subject-Predicate Distinction in the sense of the Search for a certain type of *Empirical* proposition

I now turn to a third consequence of the failure to recognise the nature of the sortal as present in the writings of Strawson, namely, the search for an *underlying basis* of the subject-predicate distinction as manifested in a language in which the categories of name and predicable (predicate) are introduced or present, or in a mode of discourse in which particulars are introduced or spoken of, in the sense of a search for a *certain type of empirical proposition* which contains no reference to sortals and is formed independently of any reference to sortals.

That the search for such an underlying basis in *Strawson's sense* is misguided in principle, in that it is a search for a certain type of *empirical* proposition, I have argued in my 'Introduction'.[59] That such a search is misconceived in principle and a consequence of his failure to recognise the true nature of the sortal I hold can be illustrated as follows.

Strawson contends that the type of empirical proposition which may be said to underlie the subject-predicate distinction is of a general type:

'There is a *P* (parcel) of *M* here'

where '*M*' is either (i) a 'matter' or 'mass' term ['material name'] or (ii) a 'substance' term ['substance-name']. In Strawson's terms, an existential empirical proposition formed from a 'feature' universal, together with the notion of 'placing'.[60] As illustrations of such a type of empirical proposition we have:[61]

'Now it is raining'
'There is gold here'
'There is water here'

and also:[62]

'There is *a* cat-slice here'

That such a type of empirical proposition cannot be held to be what ultimately underlies the subject-predicate distinction and that such a type of proposition may be thought to do so, is a result of the failure to recognise the category and nature of the sortal, has I claim, been indirectly shown in sections A and B above. For (i) in that I have shown in section A that the search for a basis for the introduction of 'particular introduction in general' is unsuccessful in fact and misconceived in principle, I have thereby indirectly shown that the search for an underlying basis for the subject-predicate distinction, in *Strawson's sense*, is *itself* so unsuccessful, and misconceived in principle. In seeking for a basis for the introduction of 'particular introduction in general' Strawson, at least indirectly, is seeking for an ultimate basis (in his sense) of the subject-predicate distinction. This is so in that in a proposition of a subject-predicate type, particulars feature as subjects, identifying reference being made to particulars in such propositions.[63] (ii) in that I have shown in section B that the search for a basic notion of 'an instance' or an explanation of the notion of an 'individual instance' is unsuccessful in fact and misconceived in principle, and that this too is a consequence of failure to recognise the nature of the sortal, to acknowledge the category of the sortal, then I have indirectly shown that the search for an underlying basis for the subject-predicate distinction in Strawson's sense is *itself* so unsuccessful and misconceived. For the search for an explanation of the notion of an 'individual instance' is indirectly a search for an underlying basis for the subject-predicate distinction in that an individual instance features as subject in subject-predicate propositions.

I am not claiming here that the search for an *underlying basis* for the subject-predicate distinction is *of itself* misguided or misconceived for it has been my claim that what underlies the distinction, in the sense of what makes the distinction *possible*, is the presence of the category of the sortal in language and the activity of identifying particulars and specifying particulars in speech, discourse. What I have been concerned to show here, in effect, is that to seek for some *further underlying basis*, an underlying basis for sortals, sortal identifying, and sortal specifying, is misconceived in principle in that in order to *articulate* such a proposed basis *sortals themselves* have to be invoked.

Notes

1. p. 4.
2. p. 157.
3. *Ibid.*, and *see also* Introduction, p. 4.
4. ch. IV, pp. 155-156; *see also* Introduction, p. 4.
5. as earlier promised in ch. IV, p. 157.
6. As in 'Particular and General', repr. in P.F. Strawson *Logico-Linguistic Papers* (London: Methuen, 1971), p. 34 and p. 36 and in *Individuals* (London: Methuen, 1959), *see* ch. VIII and IX above.
7. Strawson, as I shall observe, construes 'piece of *M*' as a sortal; *see also* ch. IX above.
8. 'On Denoting', *op. cit.*
9. Saying they do not furnish principles of sameness and distinction (difference) in their own right, does not, of course, reduce them to 'characterizing universals' or their expressions to expressions for characterising universals. It simply emphasises their functional nature.
10. *Individuals*, pp. 167-168.
11. In ch. I, sec. A.2.
12. *See*, p. 202 (as cited earlier in ch. IX, sec. A.3.).
13. Later modified in his paper 'Particular and General', *op. cit.*
14. *Ibid.*
15. 'Particular and General', *op. cit.*, sec. VII, pp. 39ff.
16. *Ibid.*, sec. VI, pp. 37-39.

17. *Ibid.*, p. 39. My criticism of the explanation of the notion of 'an individual instance' in the case of 'material names' in terms introducing feature universals, will be presented in the next section.
18. *Ibid.*, pp. 39-40.
19. I have already offered objection to this.
20. *Ibid.*, p. 40.
21. *Ibid.*
22. *Individuals*, p. 208.
23. I am not saying here that in the case of non-functional sortals we *only* have the notion of that of which it is an instance, in that we have the notion of an instance.
24. *See later*, sec. A.5.(iv).
25. *See* (e.g.) sec. A.5.(iv) in the form of Strawson's negative case against arguments (1) and (2) above, pp. 241-242.
26. *See* pp. 236-237 above.
27. 'Particular and General', *op. cit.*, p. 40.
28. *Ibid.*, *see also* p. 237 above.
29. *See* p. 230; *and also* ch. IX, pp. 222-223; ch. IV, pp. 154-155.
30. *See* p. 238 above.
31. *See* pp. 238-239.
32. 'Particular and General', *op. cit.*, p. 40.
33. My parentheses.
34. In my 'Feature Universals and Sortal Universals', in *Analysis*, Vol. XXXI, No. 2, December 1970, pp. 49-52.
35. 'Particular and General', *op. cit.*, p. 41.
36. *Ibid.*, pp. 40-41.
37. cf. my discussion of 'cat-slice', p. 238 above.
38. *See* sec. A.3 above, pp. 232-234.
39. cf. my paper 'Essence and Accident', in *Mind*, Vol. LXXXIV, N.S. No. 336, October 1975, pp. 595-600.
40. 'Particular and General', *op. cit.*, p. 35.
41. *Ibid.*, p. 37.
42. *Individuals*, pp. 202ff.
43. Strawson's alternatives.
44. 'Particular and General', *op.cit*, p. 38. Strawson does not distinguish the types of case here as I have done earlier (sec. A.3).
45. *Ibid.*
46. *Ibid.*
47. *See* sec. A.3, pp. 233-234; and *see also* Introduction, pp. 4-5, ch. III, sec. I.1., pp. 139-140.
48. 'Particular and General', *op. cit.*, p. 38.
49. My italics.
50. 'Particular and General', *op. cit.*, p. 39.
51. *Ibid.*
52. This point is not inconsistent with the point made earlier (sec. A.5.(ii), p. 239) since there I was making the point that we do not have the same *stuff* only in that we have the same (single) sort of thing present.
53. 'Particular and General', *op. cit.*, p. 39.
54. *Ibid.*
55. As I have earlier argued Strawson does (chs. VIII and IX).

56. *See* sec. A.5.(iii), p. 240 above. *See also* note 29 above.
57. In secs. A.5. and A.5.(i), pp. 236-237 above.
58. *See also* 'Particular and General', *op. cit.*, pp. 37-42; *Individuals*, pp. 202ff.; and my discussion in secs. A.5, A.5(i), A.5(ii), A.5(iii), and A.6, A.6(i), A.6(ii) above.
59. *See* pp. 4-5.
60. *Individuals*, pp. 202ff.; 'Particular and General', *op. cit.*, sec. VI, pp. 36-39.
61. *Individuals*, p. 202; 'Particular and General', *op. cit.*, p. 37.
62. *Individuals*, p. 208.
63. *See* (e.g.) *Individuals*, p. 185.

Chapter XI

Quine: Sortals and Canonical Notation

A. That Quine fails to recognise the Category of the Sortal and analyses Sortals out in favour of the Variables of Quantification and Predicates

This hardly needs establishing as such a position is essential to all Quine writes and in particular to his theory of canonical notation. I shall however illustrate this first by reference to his writing on the theory of canonical notation and then by reference to other extracts from his writings prior to his explicit advocacy of the theory of canonical notation in *Word and Object*. In relation to this second set of extracts or references I shall draw particular attention to a passage from his *Mathematical Logic*, which I hold to be the key to much of what comes in his later writings.

A.1. In his section 34 ('Quantifiers and Other Operators') of his chapter on 'Regimentation',[1] Quine writes that the essential forms of indefinite singular terms reduces to: 'Every *F*', 'Some *F*' (where '*F*' may be a sortal)[2] which in turn reduces to: 'Every *F* is an object *x* such that .... *x* ....'; 'Some *F* is an object *x* such that .... *x* ....'. We may make such a 'striking economy', since:

> these two classes of indefinite singular terms are in turn dispensable in favor of just the two indefinite singular terms 'everything' and 'something'. For, as noted ... 'every *F*' and 'some *F*' are needed only in the positions 'Every *F* is an object *x* such that .... *x* ....' and 'Some *F* is an object *x* such that .... *x* ....';

And further these in turn reduce to, may be paraphrased as:

(1) Everything is an object *x* such that (if *x* is an *F* then .... *x* ....)
(i.e. $(x)\ Fx$)[3]

(2) Something is an object *x* such that (*x* is an *F* and .... *x* ....)
(i.e. $(\exists x)Fx$)[4]

In parallel fashion, the canonical form for a singular 'description' (The *F*) becomes:

(3) The object *x* such that .... *x* ....[5]

and Quine does not radically distinguish between 'The *F*' where '*F*' is a sortal expression and where it is not.[6] He does mention[7] the class name of a further set of definite singular terms, but what he has in mind here is a definite singular term such as 'mankind', not 'the *F*'.

Further, commenting more generally later in the same chapter, he writes:[8]

> Where canonical notation is cut off, leaving unanalysed components, will usually vary with one's purposes (§33). But what typically remains unanalysed has the form of a term; indeed a general term, for we shall see how to eliminate singular terms. ... Moreover this residual general term regularly ends up in predicative position ... under regimentation of notation. Thus 'I now have a dog' and 'Every dog barks' show the general term 'dog' as part of an indefinite singular term: their paraphrases:
>
> ($\exists x$)(x is a dog and I now have x)
> (x)(if x is a dog then x barks)
>
> give it predicative position.

And again,[9] under the heading of 'A Framework for Theory', he summarises by saying:[10]

> Taking the canonical notation thus austerely, and holding also to the formal economies of Chapter V, we have just these basic constructions: predication, universal quantification ... and the truth functions (reducible to one).[11] The ultimate components are the variables and general terms; and these combine in predication to form the atomic open sentences.[12]

A.2. To cite some other extracts to illustrate my thesis. At *Methods of Logic*,[13] no distinction is drawn between: (i) 'the author of *Waverley*' and (ii) 'the prime number between 5 and 11'. *Both*[14] 'complex singular terms' are said to be descriptions and are writable in the form: (ιx)(x wrote *Waverley*); (ιx)(x is prime . $5 < x < 11$).[15] Thus the sortal 'number' becomes a predicate, (ιx)(x ....) being read as 'the object x such that it is a number . is prime'. Quine explicitly writes:[16] 'In general a singular term purports to name one and only one object and in particular a singular term of the form "(ιx)Fx" purports to name the one and only one object of which the predicate represented by "*F*" is true.'

Turning again to *Word and Object*, 'The man at Macy's' and 'The president of the USA' are both given as singular descriptions, writable as '(ιx)Fx'[17] and equally[18] 'the car which I bought from you' is given as a singular description thus writable. Further, in his programme for the elimination of descriptions[19] it is clear that singular 'descriptions' covers *both* phrases beginning with the definite article and which contain sortal expressions and those which do not: *both* are writable as (ιx)Fx.

Yet again from 'On What There Is'[20] we have for 'there is some prime number which is larger than a million', '*there is something* (bound variable) which is a prime number larger than a million'; and in 'Two Dogmas of Empiricism',[21] 'man' is explicitly said to be a 'one-place predicate' - he writes:

'Suppose now we consider a language containing just the following materials. There is a indefinitely large stock of one-place predicates (for example "*F*" where "*Fx*" means that "*x* is a man")'. Further in 'Reification of Universals',[22] '*x* is a whale' becomes *Fx*, when represented in the formula of the logic of quantification and he adds: 'the letters "*F*" and "*G*" are sometimes viewed as variables taking attributes or classes as values, for example whalehood ...', thus one might comment, on such a view sortals are eliminated or paraphrased in favour of abstractions such as whalehood, but I shall not pursue the matter here as Quine is not clearly committed to such a view. (To reduce sortals to *attributes* is objectionable for reasons already offered earlier in this book.)

Finally, turning to *Mathematical Logic*, we see the origin and proposed justification for the view that sortals are to be treated as predicates or more precisely can be written out, 'paraphrased', in terms of the variables of quantification and predicates. For here[23] Quine rejects the notation involving 'range indicators', (sortals) in favour of the notation involving the variables of quantification and predicates only. He contends that such a prefix as '(*x*) number', '(*x*) man' as previously introduced[24] can always be eliminated in favour of the 'simple kind' of prefix used in the representation of the statement 'Everything is identical with itself' viz. $(x)(x = x)$. thus the statement:

'Every man is mortal'

originally written as:

(18) (*x*) man (*x* is mortal)[25]

meaning:

(22) No matter what man *x* may be, *x* is mortal[26]

can be paraphrased as:

(23) No matter what *x* may be, if *x* is a man then *x* is mortal.[27]

And he crisply adds: 'Instead of limiting our assertion to men at the start, we admit all entities at the start and then employ the conditional form to spare the non-men'. Thus (18) gives way to:

(24) (*x*)(*x* is man . ⊃ . is mortal)

In general, the form '(*x*)***M***(------)' gives way to '(*x*)(*x* is an ***M*** . ⊃ . ------)'.

As this extract from *Mathematical Logic* is of particular importance and yields a quite particular insight into Quine's position, I shall take the liberty of offering some criticisms of such a position in advance of arguing for other consequences of Quine's failure to acknowledge sortals. My criticism of Quine here will lead to what I shall enunciate as the first, though not primarily, the most important consequence of his failure to acknowledge sortals.

A.3. My particular objections to Quine here are:

(1) that to speak of admitting 'all entities at the start' is apparently to assume that there is a class of things which simply *are*; viz. a class of entities as such: but as I, following Aristotle have argued, there can be no such class.[28] It may be claimed that it is *only* in *appearance* that Quine is thus committed; that when he speaks of admitting 'all entities at the start' this is simply a formal way of expressing 'all *Fs*, *Gs*, *Hs* at the start', where '*F*', '*G*', '*H*' are sortals. But such a thesis is incompatible with his (23) above, since, if to speak of 'admitting all entities at the start' is simply such a formal way of speaking, then for him (23) is writable as, 'No matter what *F x* may be, if *x* is a man then *x* is mortal', which it precisely is not. Further, it can hardly be claimed that the original here is simply a formal way of expressing what is expressed by the formulation in terms of sortals since Quine is explicitly eliminating sortals, yet a formal way of expressing something only has sense against the background of a non-formal mode, viz. 'formal concepts' only have intelligibility against the background of non-formal, i.e. proper concepts. Yet Quine removes this very background. I shall be returning to this point in my second objection below. I thus claim that Quine is committed to holding that there is a class of things which simply are; that he fails to realise that *necessarily* to be an individual is to fall under some sortal in either my original sense or the extended sense introduced by Strawson, which I have argued is the case.

(2) That he is committed to a class of things which simply *are* in his elimination of sortals in favour of the variables of quantification and predicates, in that he treats 'object' as a 'proper concept' as opposed to a 'formal concept', construing the 'pseudo concept', "object" as itself a 'proper concept'.[29]

In so far as sortals go out in favour of the variables of quantification, they go out in favour of variables which have 'no meaning beyond the pronominal meaning which is reflected in translations such as (20)'.[30] Now (20) reads: 'Whatever you may select, it = it' and this is given as a paraphrase of 'Everything is identical with itself'. My first complaint here is that as regards 'Everything is identical with itself', this is only intelligible in so far as it is writable as, is the formal expression of: 'Every *F* (where *F* is a sortal) is identical with itself', since (necessarily) to be an individual, an object, a thing, is to be an *F* (sortal). To hold that we may simply write 'Everything is identical with itself', where this is *not* merely a formal way of writing 'Every *F* is identical with itself', is to be committed to a class of things which simply are and Quine is committed to holding that 'Everything is identical with itself' is *not* merely a formal way of expressing 'Every *F* is identical with itself', precisely because he wants to eliminate to sortal '*F*' altogether in favour of 'thing' and predicates. But in so far as 'thing', 'object' is to be regarded as a 'formal' concept, and not as a pseudo 'proper concept', then sortals are *not* thus eliminable, since 'formal' concepts may not *replace* 'proper' concepts; they *are* formal and have no sense or content *independently* of the 'proper' concepts they present in a formal way. I now come to my second complaint which concerns the translation of (20). This translation is open to the objection that it itself cannot dispense with sortals - or may do so only on the pain of eliminating a sortal in favour of a 'formal' concept 'object' and thus treating a formal concept as a pseudo 'proper concept'. (20) should read *not* 'Whatever (i.e. whatever object) you may select, it = it', but 'Whatever *F* (where '*F*' is a sortal) you may select, it = it'. Quine cannot get out of the charge of construing a formal concept as a pseudo proper concept here by claiming that (20) is simply a purely formal way of writing: 'Whatever *F* you may select, it = it' precisely because sortals are to be eliminated in favour of 'object' and predicates, and thus the necessary conditions, background, for a 'formal' construction are removed. To pursue the original point that the variables of quantification 'have no meaning beyond the pronominal meaning which is reflected in translations such as (20)'; this is to say that they have no meaning in the sense of 'sense' but have reference, or 'back reference' only: 'they serve merely to indicate cross-references to various positions of quantification',[31] but what the variables refer to or 'can refer' to, by analogy with the case of 'it' in (20) is *not* some *F*, but some object, the formal concept of an object construed as a 'proper concept'. Once again we cannot free Quine from this charge here by saying that the reference of the variables of quantification are purely 'formal' objects, even though Quine speaks of cross reference of 'positions', since he rejects the very background against which the formal concept of object has its sense. With reference to Quine's talk of *positions*, the question arises: 'How is it possible for the variables of quantification to indicate such cross reference?' As in the grammatical case the condition of a pronoun indicating cross reference is that one is first able to specify *what* is referred to or cross referred to, which necessitates the introduction of *some* sortal, even though it is contingent what is referred to or cross referred to; so also, in parallel fashion, in the case of the variables of quantification we have to be able to say *what* makes it possible for such variables to indicate such positions - yet such a condition, in analogous fashion, is withdrawn. In short, as we require the notion of a sortal in order to introduce that of 'reference to an individual', so also we, in general, need the notion of a sortal, in order to introduce Quine's notion of 'referential position'[32] and subsequently of 'position', yet sortals are to be eliminated.

Neither can we present the following as Quine's position: as a pronoun may refer back to anything whatever, so also the variables of quantification may refer back or 'cross refer' to anything whatever. Whilst it is true that a pronoun may refer back to anything whatever, this is simply a formal way of expressing the truth that a pronoun may refer back to any *F* whatever. But once again Quine cannot allow that 'a pronoun may refer back to anything whatever' is simply a formal way of expressing the above truth, precisely because he wishes to eliminate sortals in favour of pronouns (grammatically speaking) and predicates and this once again destroys the very basis on which a 'formal' account is possible. In analogous fashion, he cannot allow that to say that the variables of quantification refer or cross refer to anything whatever is simply a formal way of expressing that the variables of quantification refer to, or cross refer to, any *F* whatsoever, for the very same reason.

Thus I see no alternative here to saying that for Quine, as what a pronoun refers back to is not some *F*, but some object, construing a 'formal' concept as a 'proper one': construing the pseudo concept 'object' as a 'proper' concept; so, in analogous fashion, what a variable of quantification refers to or cross refers to, is not some *F*, but some 'object'. Construing the formal concept of 'object' as a proper one: making the 'pseudo' concept 'object' into a non-pseudo, genuine one - or one might add, to invoke Quine's own locution here, construing a 'pseudo concept', 'formal concept' *position* as a non-pseudo, non-formal, i.e. proper concept.[33]

It might be said that such a criticism of Quine is outrageous: how can he rightly be accused of making such an error as to construe a formal concept as a proper one? That he may be rightly accused, I contend, is revealed in his famous doctrine that 'to be is to be the value of a variable'. 'To be assumed as an entity is, purely and simply, to be reckoned as the value of a variable'.[34] Geach[35] has criticised this doctrine, writing: 'Verbally at least, Quine's slogan involves what Frege would have called a confusion between concepts of different level and would have regarded as almost the grossest that could be committed'. He expands as follows:

> "There is a square root of 4" is true iff, for a suitable language **L**, "a square root of 4 is a value of a variable in **L**" is true. But although "a square root of 4" is the grammatical subject of both the sentences just quoted, its logical roles differ. Of the number 2, which is the square root of 4, we may truly say: "2 is a value of a variable in **L**". But we cannot say "There is 2"; the gap in "There is ------" used this way ... can be filled only by a predicable expression, not by a proper name. "Is a value of a variable in **L**" is predicable of objects, "there is" is not;

In the light of what I have said earlier concerning Frege's construction of sortals as predicables, I should have to amend Geach's point here to read: 'the gap in "There is -----", can only be filled by a *sortal* or *predicable*'. But this does not detract from the point Geach makes, namely that Quine confuses two different levels of concept, the concept of 'object' and the concept of 'concept', and in that he is committed to holding that the gap in 'there is ----' can be filled by a proper name, or more correctly by a *pronoun*, then he is guilty of the above charge since what the proper name (or pronoun) will name necessarily is an object of no sort or kind, which entails that *necessarily* what a proper name names (or a pronoun refers to) is an object of no sort or kind. It names or refers to, simply, *an object*.

It will be objected that what a proper name or a pronoun refers to in Quine's thought is not a 'pure' object, thus construing a 'formal concept' as a 'proper one', but simply to 'any object whatsoever', but this move is not open to Quine, since, as I have previously argued, 'any object whatsoever' is simply a formal expression of 'any *F* whatsoever'. Yet Quine eradicates sortals in favour of the variables of quantification and predicates and hence excludes himself from such a move.

My case may be further defended as follows. Wittgenstein says[36] that one cannot combine existence with a formal concept - 'There are objects'; but Quine is indirectly committed to this doctrine and thus of the confusion of which I accuse him, since he allows 'there is 2', 'there is Socrates' (or more correctly, 'there is *it*'); and "2" and "Socrates" are respectively the proper name of an object which *is* a man, the proper name of an object which *is* a number; so for 'there is 2', 'there is Socrates', Quine is committed to writing: 'there is an object designated (referred to) by "2" which is a number', 'there is an object referred to (designated by) "Socrates" which is a man', thus combining 'there is' with the concept 'object'. For Quine we may *not* write for 'there is 2', 'there is a *number* designated by the numeral "2"'; or again 'there is a *man* bearing the name or designated by the name "Socrates"', for this would be incompatible with his famous dictum. In Geach's terms it would be to allow 'variables of other styles',[37] which Quine explicitly denies. We may say here: 'of course it is contingent (in *one* sense) that "2" is the name of a number or again that "Socrates" is the name of a man', but it is *not* contingent that '2' is the name of *some F or other* or that 'Socrates' is the name of *some F or other*. Quine makes, it seems, two parallel moves, the first at a particular level, the second at a general level:

(A) from (1) *it is contingent that* '2' is the name of a number (a numeral designating a number), a particular *F*; that 'Socrates' is the name of a man, the name of a particular *F*;

to:

(2) it is contingent that '2', 'Socrates' is the name of *any F* whatever;

(B) from (1) it is contingent that a particular object is the *F* that it is;

to:

(2) it is contingent that *any* object is an *F*.

Whilst there is, or may be, a sense in which A(1) and B(1) are true, there is no sense, as I have argued, in which A(2) and B(2) are true. Failure to recognise this leads to the construction of the formal concept 'object' as a 'proper' concept and to the existence of objects which simply *are*.

B. Consequences of the Failure to recognise the Category of the Sortal

B.1. The first two consequences, which I claim have already been argued for above, are (1) that Quine is committed to the existence of a class of things (entities) which simply *are*; (2) that he construes the 'formal' concept 'object', 'thing', as a 'proper' concept which can be combined with the quantifier 'there is'. I shall now argue that there is a further, more dire consequence of such failure - namely that it results in the breakdown of his theory of canonical notation. I shall present my case for this by special reference to Strawson's criticism of Quine in the classic paper 'Singular Terms and Predication'.[38]

B.2. In his paper Strawson considers two attempts made by Quine to explain the idea of *singular term* and *general term in predicative position*, ideas which play a central role in the theory of canonical notation. He rejects both attempts, but my concern here is with the rejection of the first; I have already maintained that the rejection of the second, namely that predicative position is accessible to quantifiers, is not proven.[39] Strawson's rejects Quine's first attempt on the grounds that 'our theoretical grasp of the nature of canonical notation rests upon our theoretical grasp of the identificatory function of singular terms'[40] or, more specifically, that the distinction between *singular terms* and *general terms in predicative position* is to be explained in terms of the contrasting roles in predication of the definite singular term and the general term in predicative position, where the role of the definite singular term is to identify. Strawson claims that the identifying function of singular terms in predication is not fully acknowledged by Quine but that such acknowledgement is necessary for a satisfactory explanation of the distinction.

I too wish to claim that our theoretical grasp of the nature of canonical notation may indeed by shown to rest on the identificatory function of singular terms - more particularly on the identificatory function of the *definite* (sortal) singular term; but I differ from Strawson, in that I hold that our grasp does not rest upon such a function in the way Strawson claims. For it is *his* thesis that the sense in which our theoretical grasp of the nature of canonical notation rest upon the identificatory function of singular terms, is that Quine's notion of 'referential position'[41] - the position of singular terms under predication - is elucible, analysable, in terms of the *identificatory function* of singular terms. To so claim, I hold, amounts to redescribing or redefining the notion of 'referential position' in a way which: (a) is in general unacceptable; (b) is incompatible with Quine's doctrine. That such a redefining of the notion of 'referential position' is in general unacceptable I have already argued for earlier;[42] it fails to acknowledge, as I have argued, that referring *presupposes* identifying: it would allow that 'definite descriptions' or in my terms, 'the definite form of the sortal', 'the *F* ...', may *basically* occur in a referring role; that the *prime function* of such symbols is to refer, and concomitantly, that the *prime function* of names is to identify and hence that a name may be regarded as an abbreviated definite description. It leads, in terms of Strawson's early writing, to the (bastard) notion of 'identifying reference'. That such redefining is incompatible with

Quine's doctrine will be shown below. It is first necessary to establish that Strawson *does* implicitly claim that the notion of 'referential position' is so elucible or analysable; this may be seen as follows.

He writes in his summary:[43]

> how was the distinction between *singular terms* and *general terms in predicative position* explained? It was explained in terms of the contrasting roles in predication of the *definite* singular term and the general term in predicative position. This contrast in roles is our fundamental clue to all the theoretical notions employed.

That by 'explanation' here is meant 'elucidation' or 'analysis' can be seen from the fact that Strawson simply invokes the alleged contrasting roles in predication of the definite singular term and general term in predicative position. *All* that is invoked is the contrast in roles. For him, for a definite singular term to occur in the role of logical subject, is for a definite singular term to occur in an identifying role. Indeed he has written earlier in answer to the question 'What is the characteristic difference in the mode in which singular and general terms alike may fairly said to be applied to a single concrete spatio-temporally continuous object?':

> The characteristic difference, I suggest, is that the singular term is used for the purpose of *identifying* the object, of bringing it about that the hearer (or, generally, the audience) know *which* or *what* object is in question; while the general term is not. It is enough if the general term in fact applies to the object: it does not also have to identify it.[44]

Thus, Strawson is claiming that, as a matter of fact, the characteristic feature of the singular term in predication is that of identifying: in that he claims to be describing the *characteristic feature*, he is, by implication, claiming to give an elucidation or analysis of Quine's notion of 'referential position' in such terms; and this is what he means by 'explanation'.

The thesis that Quine's notion of 'referential position' and, more generally, the contrast between singular and general terms in predication is able to be explained, i.e. elucidated, analysed, in such a way rests on an assumption which is incompatible with his fundamental doctrine. That assumption is that Quine recognises a class of singular symbols (for him 'terms'), namely the definite singular symbol which is such that it may play the role of identifying an individual. But Quine may require *no* such class of singular symbols and hence Strawson may only claim to have 'explained' Quine's distinction at the expense of committing him to an inconsistency.

That Quine may require no such class of singular symbols is clear. To require such a class of singular symbols entails at least either (or both):

(1) that proper names, of their essence, nature, have a sense in their own right;

(2) that some 'definite descriptions' *present* a sense and *present* a sense in their own right;[45]

but:

(3) Quine's doctrine of the elimination of singular terms in favour of the variables of quantification precisely entails that singular terms may not be such.

Again, (a) Quine's treatment of names a 'pro-pronouns'[46] is incompatible with any doctrine that proper names of their nature or essence have a sense, or indeed any doctrine of names having a sense, since pronouns and hence 'pro-pronouns' have no sense. Yet a name having (i) a sense and (ii) a sense in its own right is a necessary condition of a name serving in an identifying role or performing an identifying function. That (i) is true is clear: a name which has no sense could never serve in an identifying role: the most such a 'name' could do would be to serve in a referring role; that (ii) is the case can be seen from the consideration that to hold that a name does *not* have a sense in its own right commits one to the thesis that a name only has sense in the context of a proposition - a thesis I have rejected earlier[47] - and, in that it only has sense in the context of a proposition, then it cannot serve to identify the subjects of propositions.

Further (b), Quine's treatment of 'definite descriptions' as eliminable in favour of the variables of quantification and predicates,[48] is incompatible with his recognition that some 'definite descriptions', some definite singular terms of the form 'the *F*...', where '*F*' is a sortal, present a sense and present it in their own right. To analyse out such definite singular terms in favour of the variables of quantification, or their equivalent, in ordinary grammar (pronouns) and predicates, is precisely to deny that such singular terms have sense in their own right, just in that they are so analysable out, for no predicate has a sense in its own right and no pronoun has a sense at all. Yet such a class of definite singular terms are needed for an identifying function to be performed. Thus I maintain that Strawson may not claim that Quine's notion of 'referential position' may be explained, in the sense of analysed, elucidated, in terms of identifying or the identificatory function of singular terms without landing Quine in a fundamental inconsistency; he is nevertheless, in an important sense, correct in holding that Quine's failure to acknowledge the identificatory function of singular terms presents a damaging criticism of the latter's theory of canonical notation. His mistake is to hold that *in predication* the identificatory function of singular terms is not fully acknowledged by Quine, for 'in predication', for *Quine*, singular terms may not play such a role. If Strawson replies that in our 'ordinary conception' of predication singular terms typically do play such a role, Quine would reply that he is not concerned to give an account of predication which merely reflects ordinary usage or practice, but an account of predication which underlies such an ordinary conception and is logically distinct and perspicacious. The sense in which Strawson's point may be seen to present a damaging criticism of Quine's theoretical apparatus is as follows.

Our theoretical grasp of the nature of Quine's canonical notation immediately rests upon our grasp of the notion of 'referential position',[49] which in turn rests upon Quine's concept of singular terms as eliminable in favour of the variables of quantification, in ordinary grammar, pronouns, i.e. a singular symbol, term, which of its essence has no sense and hence could not perform an identificatory function. The concept of 'pronoun', however, theoretically rests upon the concept of a 'singular term', which may perform an identifying role, since concerning any pronoun we must always be able to ask *what* it refers to, or more correctly, what it refers back to. Such a singular term may be an ordinary proper name, but: (i) I have argued that although an ordinary proper name *may* perform an identifying role this is not its basic function; the basic function of a proper name is to refer; (ii) it cannot be generally held that the introduction of pronouns rests upon the introduction of ordinary proper names - i.e. logically rests upon such an introduction; for a pronoun may refer to, or refer back to, an individual introduced simply by reference to a 'definite description', in the sense of a 'definite description' containing a sortal, e.g. 'the man who broke the bank at Monte Carlo first decided to take a holiday and then he could not resist going back to the scene of the crime'. We may not hold here that 'the man who broke the bank at Monte Carlo' is a 'many worded name', precisely because it *does* directly serve to identify, and there is no necessity to introduce any name here: indeed we may not know the name of the man concerned and still understand the use of the pronoun. However we do need *some* singular term which may perform an identifying (identificatory) role for the introduction of pronouns in a referential role for the reason already given, and the singular terms we basically need are those formed by the singular form of the sortal 'the *F*...'. To claim that such singular terms *themselves* are essentially pronominal has the result of generating an infinite regress, thus revealing by one plausible criterion (Ryle's), a category mistake. Singular terms which serve to specify the back reference (or cross reference) of pronouns, which serve to identify individuals and which are basic to this activity (as sortals are)[50] and upon which the intelligible introduction of pronouns rest, may therefore be held to *underlie* and form the *ultimate basis* for the introduction of Quinean 'real' singular terms, viz. pronouns or their equivalent in logical theory, the variables of quantification, which may solely serve to refer or indicate referential position. As the notion of a pronoun theoretically rests (*contra* Quine) upon that of the noun, in the sense of the sortal noun or singular definite sortal which presents a sense in its own right, so also the notion of 'pure reference' and hence 'referential position', *rests on*, *presupposes* identifying by reference to some sortal or other. If nothing has been previously identified or prior identification is not taken as understood, then, since such a pronoun typically refers back to something, a pronoun used or occurring in such a context can neither succeed nor fail in reference and the sentence containing it can express no proposition. The condition for its intelligible use or occurrence has not been met. Alternatively, if we were to hold with Strawson that there can be propositions or statements which are neither true nor false, then the propositions (statements) containing the pronoun are neither true or false.

In short, pronominal reference, and hence the notion of '(pronominal) referential position', *rests upon*, *presupposes* sortal identification in the sense of identification by reference to some sortal or other. It is in *this* sense that our theoretical grasp of the nature of canonical notation ultimately rests upon the identificatory function of the definite singular term, not in the sense Strawson maintains. Such identificatory function by the definite form of the sortal is what makes Quine's explanation *possible*: it is not entailed by his explanations as Strawson in effect holds, but is *presupposed* by such explanations and lies outside their ambit.

Strawson[51] criticises Quine for not elucidating more fully on those notions which he dismissed as vague and picturesque;[52] but one may readily account for such omissions. For him to do so would lead to an inconsistency in his own position. For he cannot maintain both that all singular terms are eliminable in favour of the variables of quantification, or in ordinary grammar, eliminable in favour of pronouns *and* that his thesis reflects the instinctive idea that a term occupying referential position is 'used purely to specify its object for the rest of the sentence to say something about'.[53] A term which can properly occupy referential position is precisely excluded from 'specifying an object' (or in Strawson's elucidations 'identifying an object'). A term properly occupying referential position could only be said to *reflect*[54] such and idea if such a term could be said to have some specificatory or identificatory force, i.e. if such a term were itself in some way an ordinary definite singular term or analogous to such a term, but this is precisely what is *not* the case in Quine's theory. Strawson, in giving Quine, as it were 'the benefit of the doubt', has assumed that what may occupy referential position in the theory of canonical notation is a term which reflects the identificatory, specificatory function of singular terms and is somehow implicit in the notion of 'definite singular term'. But the use of pronouns neither (a) reflects the specificatory function of a singular term nor (b) is implicit in the notion of 'definite singular term', in the sense that the use of pronouns is not entailed by the use of definite singular terms, but rather, as I have argued, *rests upon* such a use, is *presupposed by* such a use. What needs to be pointed out, as I have attempted to point out, is the precise sense in which our grasp of the nature of canonical notation rests upon our grasp of the identificatory function of the definite (sortal) singular term - viz. how the identifying function of such terms is the underlying foundation of the intelligibility of Quine's theoretical apparatus. It is in *this* sense that we may offer the identificatory function of the definite (sortal) singular term as an explanation of Quine's distinctions; it provides us with an account of how his distinction between a term in purely referential position and general term in predicative position is made *possible*. Failure to recognise the category of the sortal entails failure to recognise the identificatory function of the definite sortal 'term', yet recognition of *this* is the underlying foundation for Quine's distinctions and hence the theory of canonical notation; hence the failure to recognise the category of the sortal removes the very foundation upon which the theoretical apparatus of canonical notation is built.

C. A Suggested Implication of my Thesis Rejected

Having claimed that Quine's thesis is radically undermined by such failure, it may be suggested that my position implies that I am rejecting any theory of canonical notation, indeed any theory at all. This does not follow. I agree that we must be able to formalize, but that we may not formalize at the expense of rejecting or eliminating the very foundation upon which any formalization of language rests.

It is the implication of my discussion that any theory of canonical notation must take full and due account of sortals which in effect implies:

(a) that we may not admit proper name variables as such or *per se*, not simply as Geach suggests,[55] that we must admit bound variables of kinds other than proper name variables;

(b) that fundamentally, we may only admit proper name variables in that we admit *sortals*, (where 'sortal' is used in Strawson's extended sense as described earlier) or *sortals* and *kinds*, using 'sortal' in the original sense.

It is to be noted here that I am *not* holding that one way of putting the implication of my thesis is simply to capitalise on Quine's slogan and say that 'to be' is to be a value of a *sortal variable*, in the first sense of sortal just mentioned. For to so hold would be land myself in inextricable difficulties. To hold that to be is to be the value of a sortal variable, would, in effect, deprive one of being able to speak of an individual existing at all; for to hold such a thesis entails that what constitutes a value for such a sortal variable is an individual sortal, i.e. a specific sortal, not an individual: a specific general name, not an individual name. And I intend no such implication; and no such implication is entailed, for the intended implication of my thesis is not that to be is to be the value of a sortal variable, but that to be is to be an *F* (where '*F*' is a sortal in the first sense given above). My claim is that sortals *are not variables at all* but are terms, symbols, necessary for the introduction of variables, whether these be individual variables (proper name variables) or other types of variable (e.g. general name variables). My capitalising on Quine's famous dictum could therefore perhaps be expressed as follows.

'To be is to be a value of a proper name variable', provided: (a) it is realised that it is necessary that proper name variables are determined[56] - viz. that to be a proper name variable is to be a variable such that *necessarily* what is substitutable for it is the proper name of some *F* (sortal); concomitantly (b) it is realised that *necessarily* to be a value of a proper name variable is to be an *F*.

It is now appropriate to look at Quine's aims and claims for the theory of canonical notation in order to develop my case. Such aims and claims are conveniently set out in the first part of Chapter V of *Word and Object*.

D. Quine's Aims and Claims for the Theory of Canonical Notation

D.1. Quine begins by commenting[57] that practical temporary departures from ordinary language do sometimes recommend themselves in the interests of clarity and communication - a point with which I should not wish to take serious issue - but he then goes on to say that in relation to the concerns of *Word and Object*: 'those departures have interested us less as general aids to communication than as present aids to understanding the referential work of language and of clarifying our conceptual scheme.'[58] My problem here is how such departures may enable us to understand the referential work of language and clarify our 'conceptual scheme', when: (a) the very conditions of reference - the conditions under which we may intelligibly speak of referring - namely the presence of sortals and sortal identification have been removed; and (b) likewise what is essential to our conceptual scheme, i.e. our speaking of objects and making true or false predications of objects, has also been removed. Quine adds joyfully: 'Now certain such departures have yet a further purpose that is decidedly worth noting: simplification of theory'.[59] Whilst one may argue that simplification of theory is a desirable end and a *sine qua non* of scientific procedure, simplification of theory may not be bought at the cost of failing (and necessarily failing) to produce, in this case, an aid to understanding the referential work of language and a clarification of our conceptual scheme. Such simplification is epitomised, Quine holds, in the use of parentheses: the use of parentheses has far reaching importance and its importance cannot be over emphasised:

> They enable us to iterate a few selfsame constructions as much as we please instead of having to continually vary our idioms in order to keep the grouping straight. They enable us thus to minimise our stock of basic functions or constructions and the techniques needed in handling them.[60]

Whilst these virtues need not be denied, Quine's use of parentheses, in so far as he has the specific concerns mentioned in mind is open, I claim, to the following important objections. Paraphrases eliminate the need for that of which they are paraphrases. They are not synonymous with their originals as Quine rightly presses.[61] Yet they serve the same purpose as the original. Quine himself writes:[62]

> In neither case is synonymy to be claimed for the paraphrase. Synonymy, for sentences generally, is not a notion we can readily make adequate sense of (cf. §§ 12, 14); and even if it were it would be out of place in these cases. If we paraphrase a sentence to resolve ambiguity, what we seek is not a synonymous

> sentence, but one that is more *informative*[63] by dint of resisting some alternative interpretations.

And he later writes that in the paraphrase S^1, S^1's relation to S (original) is: 'just that the particular business that the speaker was on that occasion trying to get on with, with the help of S among other things, can be managed well enough to suit him by using S^1 instead of S' and concludes his immediate discussion by writing:[64] 'The real point is simply that the speaker is the one to jüdge whether the substitution of S^1 for S in the present context will forward his present or evolving program of activity to his satisfaction.'

Now agreed (1) a paraphrase is in one sense more informative than the original in that it makes clear the sense of the original, or the sense in which the original is to be taken, but in Quine's attempt to paraphrase 'ordinary language', can it be held that 'the particular business that the speaker, on some occasion was trying to get on with, with the help of S ... be managed well enough using S^1 instead of S?' My answer is firmly *not so.* Let me now put my argument in detail.

(A) I have said that a paraphrase eliminates the need for the original. Quine in effect holds this and further holds that a paraphrase is more *informative* than the original;

(B) I have said that a paraphrase serves the same purpose as the original: Quine too holds this.[65]

As regards (A), one major problem for Quine's position is that in that paraphrases eliminate the need for that of which they are the paraphrases, such paraphrases as *Quine* offers eliminate the very foundation on which the intelligibility of the paraphrases depends. This can be seen, for example, in the paraphrasing of:

> 'Every *F* is an object *x* such that .... *x* .... '

as

> 'Everything is an object *x* such that (if *x* is an *F* then .... *x* ....)[66]

for the formal concept (quantifier) 'Everything' rests for its intelligibility upon 'Every *F*'.[67] Or again in the paraphrasing of 'The *F*' as 'The object *x* such that .... *x* .... '.[68]

In that a paraphrase is 'more informative' than the original, it can hardly be said, e.g. that 'the object *x* such that .... *x* .... ' is more informative than 'The *F*', since as I have argued our being able to write 'the object *x* such that .... *x* .... ' depends for its intelligibility upon our being able to write 'The *F*'.

As regards (B), it cannot be held that, e.g. 'The object *x* such that .... *x* .... ' may serve the same purpose as the original, for the typical, (standard), purpose of phrases of the form 'The *F*' is to identify an object, but the paraphrase could never serve to identify an object - or at least could only do so in that phrases of the form 'The *F*' may serve to do so,[69] but this condition is necessarily withdrawn here. In relation to Quine's real point, that the speaker is the one to judge whether the substitution for S^1 for S in the present context will forward his present or evolving 'program of activity' to his satisfaction, my comment is that, for reasons given, it is necessarily not the case that the substitution of S^1 for S in *any* present context will forward his present or evolving 'program of activity' to his satisfaction in so far as he is, by his use of 'The *F*' in S concerned to identify or indeed in so far as he is by his use of 'The *F*' in S concerned even to refer.

D.2. Quine admits that[70] 'on the whole the canonical systems of logical notation are best seen not as complete notations for discourse on special subjects, but as partial notation for discourse on all subjects'; but a system of canonical notation which is even a partial notation for discourse on all subjects, cannot ignore implicit reference to sortals in such notation, as I have argued. Simplification of theory is indeed a central motive of 'the sweeping artificialities of notation in modern logic'[71] and indeed 'it would be folly to burden logical theory with quirks of usage that we can straighten', but the presence and use of sortals is no 'quirk of usage'. The simplification of theory as present in the theory of canonical notation presented by Quine is distinctly *over simple* and *over sweepingly* artificial, introducing as it does just three basic constructions: predication, universal quantification and truth functions, holding that the ultimate components are the variables and general terms and that these combine in predication to form the atomic open sentences.[72] We must admit *identification* to the list of constituents if my thesis throughout this book has been cogent and correspondingly amend:

(1) *Universal Quantification*, instead of: '(x)', to read:

$[(x)F]$[73]

(2) *Particular Quantification*, instead of: '$(\exists x)$', to read:

$[(\exists x)F]$[74]

and, in corresponding fashion, hold that the ultimate component of any system of canonical notation which is adequate to and an understanding of the 'referential work of language' and 'clarifying our conceptual scheme' or indeed be a 'partial notation for discourse on all subjects', must contain an implicit reference to sortals[75] as well as an explicit reference to general terms. Such are needed for an adequate 'canonical grammar' and 'vocabulary'.[76]

Quine writes of his own system:[77] 'What thus confronts us as a scheme for systems of the world is that structure understood so well by present day logicians, the logic of quantification or the calculus of predicates'. This structure, I claim, has to be radically amended in a way which permits implicit reference to sortals if it *is* to be a scheme for 'systems of the world', for no 'system of the world', i.e. system in terms of which we organise the world, can coherently be explicated without reference to sortals (in my original sense) or to principles of counting and distinction formed from 'matter' terms ('sortals' in Strawson's extended sense).[78] For in every such system we need to be able to introduce either (i) the notion of 'an object', count such objects and to identify and re-identify such objects or (ii) the notion of 'a particular' and for this *too* we need to introduce sortal in an extended sense, namely principles of counting and distinction in their own right formed from 'matter' terms.[79]

Notes

1. *Word and Object*, paperback edn. (Cambridge, Mass.: The M.I.T. Press, 1964), p. 162.
2. My parentheses: Quine writes '(in the sense of a "certain *F*") where "*F*" stands for any general term in substantival form'.
3. My insertion in parentheses.
4. My insertion in parentheses.
5. *Word and Object*, p. 164.
6. cf. also below (A.2.) extract from *Methods of Logic*.
7. *Word and Object*, p. 163.
8. *Ibid.*, p. 174.
9. *Ibid.*, ch. 6, sec. 47.
10. *Ibid.*, p. 228.
11. As *per* Sheffer's reduction: cf. Quine's n. 2, p. 228.
12. cf. also *ibid.*, p. 231.
13. *Methods of Logic* (New York: Henry Holt and Co., 1950), p. 216.
14. cf. also Russell who likewise made no such distinction; *see* ch. VI above.
15. cf. note 6 above.
16. *Method of Logic*, p. 216.
17. *Word and Object*, p. 107.
18. *Ibid.*, p. 111.
19. *Ibid.*, pp. 183ff.
20. In *From a Logical Point of View*, 2nd rev. edn. (New York and Evanston: Harper and Row, 1963), p. 12.
21. *Ibid.*, p. 30.
22. *Ibid.*, p. 107.
23. *Mathematical Logic*, rev. edn. (Cambridge, Mass.: Harvard University Press, 1961), p. 69.
24. *Ibid.*, p. 67.
25. *Ibid.*
26. *Ibid.*, p. 69.
27. *Ibid.*
28. *See* ch. I, sec. B.3, pp. 28-31; ch. II, sec. C, pp. 46-47 and Aristotle *Posterior Analytics*, 92b 14.
29. For 'formal' and 'proper' concept cf. Wittgenstein earlier, ch. I, *see* B.3, pp. 28-29.
30. *Mathematical Logic*, p. 70; for (20) *see* p. 68.
31. *Ibid.*

32. For 'Referential Position' cf. *Word and Object*, pp. 142ff. and later on in this chapter.
33. As, for example, in *Methods of Logic*, pp. 205-206.
34. 'On What There Is', in *From a Logical Point of View*, p. 13; *see also*: *Methods of Logic*, pp. 205-206.
35. *Reference and Generality*, pp. 161-162.
36. cf. *Tractatus*, *passim* (*see also* e.g. ch. I, sec. B.3, pp. 28-31).
37. *Reference and Generality*, p, 161.
38. repr. in *Logico-Linguistic Papers*, pp. 53-74.
39. *See* ch. III above, sec. F.2, pp. 113-119.
40. *Logico-Linguistic Papers*, p. 64.
41. cf. *Word and Object*, pp. 142ff.
42. ch. II, pt. I, sec. A.2, pp. 42-43; sec. B, pp. 43-44; sec. C, pp. 44-48.
43. *Logico-Linguistic Papers*, p. 64.
44. *Ibid.*, p. 59.
45. Proper names may be said to *have* a sense and have it in their own right, but it can hardly be said that other definite singular 'terms' such as 'definite descriptions' containing a sortal, e.g. 'the man who wrote *Waverley*' *have* a sense: rather they may be said to *present* a sense. They present, for example, the sense of proper names. (cf. Frege's doctrine of names which represent proper names).
46. *Word and Object*, p. 186, following Peirce's 'mot'.
47. ch. I, pp. 25-26.
48. cf. Quine's programme for the elimination of singular terms, *Methods of Logic*, p. 222.
49. cf. *Word and Object*, pp. 142ff., 146ff., 152-155, 167ff.
50. It cannot be held that 'definite descriptions' not containing sortals are thus basic, for reasons already offered (cf. ch. II, p. 41).
51. 'Singular Terms and Predication', *op. cit.*, p. 64.
52. cf. *Word and Object*, p. 96.
53. 'Singular Terms and Predication', *op. cit.*, p. 60.
54. *Ibid.*, p. 61.
55. *Reference and Generality*, pp. 161-162.
56. In ch. I, sec. C.2, p. 32; sec. C.3, pp. 32-35; sec. C.4, pp. 35-36, esp. p. 36.
57. *Word and Object*, p. 157.
58. *Ibid.*, p. 158.
59. *Ibid.*
60. *Ibid.*
61. *Ibid.* p. 159.
62. *Ibid.*
63. My italics.
64. *Ibid.*, p. 160.
65. cf. the extract from p. 160 above.
66. *Ibid.*, p. 162.
67. And similarly Quine's treatment of 'Some *F*' (*ibid.*).
68. *Ibid.*, p. 164.
69. cf. my ch. I, p. 27; and sec. C of this chapter.
70. *op. cit.*, p. 160.
71. *Ibid.*, p. 158.
72. *Ibid.*, p. 228.

73. Where '*F*' is a sortal; *see* ch. III, sec. I.2, pp. 140-142; sec. I.4, p. 145.
74. Where '*F*' is a sortal; *see* ch. III, sec. B.5, p. 101; sec. C, p. 102; sec. H.3, pp. 131-132; sec.I.2, p. 141.
75. Sortals here will include sortals in my original sense and also in the sense of sortals formed from 'matter' terms. For universal quantification in regard to 'matter' terms, *see* ch. III, sec. I.2., p. 142.
76. I refer directly to *Word and Object*, p. 231.
77. *Ibid.*, p. 228.
78. *See* ch. IX, p. 225; ch. X, sec. A.3.
79. *See* ch. I, sec. A.4 and note 78 above.

Conclusion

This work has discussed sortals and their relationship to the subject-predicate distinction. It has been a principal contention that the sortal, as an independent and distinctive category of symbol in language and as performing distinctive and special functions in speech, has not been recognised in much modern classical writing in Philosophical Logic and the Philosophy of Language.[1] It has been claimed that the sortal must be recognised as an independent and distinctive category of symbol in that sortals furnish us with principles for distinguishing and counting particulars in the sense of 'objects' and do so in their own right, relying on no antecedent principle or method of so distinguishing and counting. It has been argued that it is not possible to claim that sortals in the sense defined in Chapter I and in Strawson's original sense are reducible to 'matter' terms (Aristotle) or 'mass' terms (Quine). Correspondingly, it has been claimed that particulars (objects) introduced by reference to sortals in my original sense may not be said to rest for their introduction upon such 'matter' terms. It has not been denied that 'matter' or 'mass' terms precursed in a certain way (or at least a sub-class of these terms) may form principles of counting and distinction in their own right for particulars, but the stronger thesis that such terms are logically prior to sortals in the sense I have originally introduced has been denied and defended (ch. I). *Both* sortals *and* terms formed from 'matter' terms precursed in a certain way have to be acknowledged as fundamental principles of counting and distinction: (i) for 'objects'; (ii) for particulars, in the sense (crudely) of items of matter (*see* ch. I). Furthermore, it has been argued that *neither* (i) individuals ('objects') introduced by reference to sortals in my original sense *nor* (ii) individuals ('particulars') introduced by reference to 'matter' terms precursed in a certain way, may be said to rest for their introduction upon the (allegedly) *more basic* notion of a 'feature' universal together with the notion of 'placing', as interestingly suggested and developed by Strawson (*see especially* ch. X).

Even more centrally, it has been claimed that sortals are not reducible to either (a) names or (b) predicables/predicates, whilst acknowledging the strong tradition, following Frege (but apparently denied by Strawson) that, to put the matter in Geachian terms, names and predicables are 'absolutely distinct'. It has been argued (in the 'Introduction' and ch. I) that whilst the categories of 'name' and 'predicable' are indeed *exclusive* they are not *exhaustive* and that logical theory has failed to take due cognisance of this. This work has endeavoured to press home the point that *both* the category of the name (particular [proper] and general) *and* the categories of the predicable and predicate *presuppose* the category of the sortal. Names (either proper or general) so do in that a name, in the sense of a proper name, at least (to gloss Quine), 'purports to stand for just one object'; and it has been argued that we only have the concept of 'one object' in that we have the concept of 'one *F*', where '*F*' is a sortal. Equally a general name purports to name several objects, yet it has been argued that we only have the concept of 'several objects' in that we have the concept of 'several *F*'s', where '*F*' is a sortal (see especially ch. VII). The category of the predicable also presupposes the category of the sortal in that a predicable is a fundamentally incomplete symbol of the form: '---- is *F*'; a 'propositional fragment' which becomes a proposition (a 'predication' in Geach's sense) when completed by a name. In that a predicable *is* thus fundamentally incomplete, awaiting completion in the form of a name and names themselves presuppose the introduction of sortals, then, at least indirectly, predicables themselves presuppose the introduction or prior presence of sortals. The category of 'predicate' in an analogous fashion also indirectly presupposes that of the sortal in that it is held that a predicate 'describes a subject', 'ascribes something to a subject', 'tells us something about a subject'; for a predicate, in order to 'describe', 'tell us something about' a subject, supposes that the subject of the predication is a distinct object (or objects). Yet it has been argued that we need the notion of a 'sortal' to introduce the notion of 'an object' and hence of a distinct object (or objects) which the predicate 'describes' or 'tells us something about'. Equally, if we hold that predicates are 'true or false of objects', then we need the notion of a 'sortal' to specify and count the objects.

It has been further contended that at the level of the use of language in speech, the role of the sortal has not been duly acknowledged. Sortal identification has been reduced to definite describing, referring or predicating. It has been claimed here that whilst sortals may occur in a referring role or again in a descriptive role, this is *not* their *primary* or *distinctive* role. Rather, it has been argued, the distinctive role of the sortal in its definite form, is to *identify* objects in the sense of 'picking out' or 'discriminating'; in its indefinite form to *specify* an object; to say what an object *is;* to answer Aristotle's 'τί ἐστι;' question. It has been held that describing, predicating, and indeed referring, *presuppose* identifying and hence the use and understanding of the nature of sortals in such a role.

Failure to recognise the true nature of the sortal and its role in discourse, it has been claimed, has led to many modem writers failing to recognise or appreciate what *underlies* the subject-predicate distinction in the sense of what makes the distinction *possible.* Whilst some writers (following Frege) have rightly pressed that the subject-predicate distinction is an *ultimate* distinction, they have further supposed that the distinction is *fundamental* or *basic* in the sense that there is at least no need and at the most no possibility of seeking a *basis* or *foundation* for the distinction. I suggest that the underlying reason for this is that they have made or assumed the categories of *name* and *predicable/predicate* to be not only *exclusive* but also *exhaustive*. If the categories of name and predicable (predicate) are contingently exhaustive then there is no *need* to seek for a basis or foundation for their introduction; if such categories are necessarily exhaustive, then there is no *possibility*. On the contrary it has been argued here that such categories are *not* exhaustive and, at the level of speech, discourse, the activities of referring on the one hand and describing, predicating, on the other are equally not exhaustive. As names and predicables (predicates) presuppose sortals, so referring, predicating etc., presuppose (in standard cases) identifying, specifying, by reference to sortals. It has thus been claimed that we may seek for an *explanation*, (*basis, foundation*) for the subject-predicate distinction in the sense of what makes the distinction *possible*[2] and this explanation, basis, foundation, lies in the presence of sortals in language; in the presence of identification and specification by the means of sortals in speech, discourse.[3] I have also held[4] that we may seek for an *explanation*, (*basis, foundation*) for the subject-predicate distinction in the sense of an answer to the question: 'Is there any *kind* of proposition in a language which employs proper names and predicables (predicates), in a mode of discourse which introduces individuals, in the sense of "objects", the truth of which is necessary for the introduction of propositions of a subject-predicate kind in that language, mode of discourse?' In this case *too*, I have argued that we need to invoke sortals to express the appropriate *kind* of (existential) proposition.[5]

It has been acknowledged that of the writers I have considered, Strawson *does* allow that we may seek for a *basis* or *foundation* for the subject-predicate distinction but it has been argued that to seek for such a basis in *his* sense is misguided in principle and misconceived in principle. It is misguided in that it seeks for some one kind of *empirical* proposition which will ultimately underlie the distinction; whereas I have pressed that this is misleading in that a *logical distinction* which is a transcendental distinction, cannot be said to rest on a *matter of fact* (see 'Introduction'): it is misconceived in principle in that Strawson claims that we may seek for a basis for the introduction of particulars introduced by sortals and for sortals themselves; whereas it has been argued (ch. X) that particulars introduced by sortals and sortals themselves are basic or foundational and that Strawson's attempt to introduce such a basis as *he* seeks has *itself* implicitly to invoke sortals in order to *articulate* the proposed basis. His account, it has been pressed, also presents a misunderstanding in that it treats sortals *themselves* as predicates,[6] whereas it has been argued that the notion of 'a predicate' *itself* indirectly presupposes the notion of a sortal.

My thesis concerning the primacy of sortals, I have suggested, has implications for logical theory. Due cognisance of sortals must be taken into account in any theory of canonical notation. In our formulations of e.g. universal and existential quantification there must always be *implicit* reference to sortals as illustrated (e.g.) in Chapter III and in my discussion of Quine in Chapter XI.

In this book I have sought to make clear the true nature of the sortal symbol and its role in discourse and have attempted to offer:

(i) an *explication* of the subject-predicate distinction (ch. III, sec. A; sec. B, p. 94);

(ii) an *explanation* of the distinction in the sense of an answer to the question: 'What differentiates an expression occurring in subjective role from an expression occurring in predicative role?' (ch. III, secs. C, D, E);

(iii) a *criterion* for the distinction (ch. III, sec. G);

(iv) *explanations* of the distinction in the sense of the *bases, foundations* for the distinction (ch. III, sec. I.1).

I have sought to draw out the *general consequences* of failure to acknowledge the category of the sortal and its place in language and discourse (ch. IV), and tried to work out the *specific* consequences of such a failure in such writers as Gareth Evans (ch. II, Part II); Frege (ch. V); Russell (ch. VI); Geach (ch. VII); Strawson (chs. VIII, IX, X) and Quine (ch. XI) whilst, at the same time, acknowledging their lasting, ever exciting, and challenging contributions to the Philosophy of Logic and more generally the Philosophy of Language.

It has to be admitted, and many will recognise, that the initial stimulus for this work lies in my understanding of Aristotle's doctrine of 'second substance', together with Professor Anscombe's original and constantly rewarding paper on 'Aristotle - The Search for Substance' published in *Three Philosophers* (*op. cit.*); but further stimulation and rewards have indeed been gleaned, and will continually be gleaned, from the writers whose works I have had the privilege of discussing.

Notes

1. It may be said that contrary to what has been claimed in this book, modern writers *have* recognized the true nature of the sortal in that they have recognized that sortals serve to stipulate a *domain of objects* for first level functions and subsequent generalizations invoking quantifiers. However, in that such writers (Frege *et al*) have also construed sortal expressions as first order functional expressions and sortals as predicates (as I have sought to demonstrate) and that this is a crucial feature of their writing, then it is impossible that sortals may play the role of stipulating or specifying a domain of objects since predicates *presuppose* specification of objects for their introduction. And in that sortal expressions are *genuinely predicative*, they could never perform this role owing to their *essential* incompleteness.

   Further, as concerns the notion of '*a domain of objects*', those who introduce such a notion are faced with the following possibilities, either:

   (A) It is *contingent* that any object falls under some domain;
   (B) It is *necessary* that any object fall under some domain;
   (C) Concerning any object, it is necessary that *it* fall under some domain.

   As regards (A), for reasons offered in the course of this book, we should never be able to refer to objects. As regards (B), this may be said to be a version of the thesis

I have advocated, but if (B) is maintained then it is impossible (for reasons offered) that expressions which stipulate or specify domains *should also* be predicative or construed as functional expressions. As regards (C), we are faced with the problems I encountered in discussing Strawson and centrally with the problem of the reference of '*it*'.

2. Introduction, p. 3; ch. III, sec. I.1, p. 136.
3. The sense in which we may seek for a *basis* or *foundation* for the subject-predicate distinction in the sense of a search for what makes the distinction *possible*, is the sense in which we may seek for a basis as the distinction is *manifested* in a language, area of discourse, way of speaking, in which we speak of 'particulars' in the sense of 'objects' or 'parcels of matter' (*see* Introduction, p. 5; ch. III, sec. I.1, p. 136).
4. ch. III, sec. I.1, pp. 136-137.
5. It is to be noted that from the fact that I have claimed that an existential proposition invoking a sortal provides an *explanation* of, a *foundation* for, the subject-predicate distinction in *this* sense in such a language, mode of discourse, it does not follow that such an existential proposition is an *empirical* one. Specification of a logical kind or type of proposition does not confer any particular *status* on that proposition.
6. *see* chs. VIII and IX.

Postscript

Stephen Horton

A. Introduction

My purpose in this postscript is to give some indication of the significance, importance and relevance of Michael Durrant's work as presented in this volume to current deliberations in philosophy. Unfortunately, due to restrictions of space, I will be unable to do full justice to my case in support of this assertion. Nevertheless, as I will show, this work has implications for many areas of philosophical debate. These include not only (and obviously) such disciplines in philosophy as the philosophy of language, philosophical logic and formal, symbolic logic but also in other areas of philosophy such as, for example, the comparative study of analytic and continental philosophy.

I begin with an examination of the relevance of Durrant's work to philosophical logic and the philosophy of language. Since at least the work of Frege, philosophers have held that descriptive or fact-stating language is comprised of two basic categories of logical symbols: names and predicates (predicables). One of the most novel aspects of Durrant's work herein is his challenge to this dogma. The central contention of *Sortals and the Subject-Predicate Distinction* is that there are, in fact, *three* categories of logical symbol involved here, i.e. in addition to names and predicates (predicables), there are sortals which, it is rightly held, cannot be reduced to either of the aforementioned categories. The knock-on effect of this conclusion is that any theory which rests for its foundation on the assumption that names and predicates (predicables) is an exhaustive and basic logical distinction must either be subject to greater or lesser modification to take into account Durrant's position, or stand recognised as: (a) at best, inadequate; (b) at worst, fundamentally flawed.

B. Theories of Reference

Perhaps two of the theories or general positions in philosophical logic and the philosophy of language most obviously affected by Durrant's findings are the 'description theory of reference' and the 'causal theory of reference'. Both these theories accept, at least implicitly, the distinction between names and predicates as exhaustive and basic.[1] It is to a consideration of the effects of Durrant's work on these two theories I now turn. I hold that these effects are felt far more dramatically by the causal theory of reference than by its rival, the description theory. For this reason I start with a discussion of the causal theory of reference, returning to the description theory later. In the ensuing discussion I concern myself only with the introduction of *proper names* for individual, material objects. A parallel case may be made here for the introduction of kind names, however, such a case would require careful consideration of Durrant's findings regarding principles of counting and distinction furnished from 'matter' terms.[2] I do not undertake such an investigation here because my prime concern in this postscript is to focus upon what I consider the most important and exciting aspects of Durrant's findings in this present work which, in my opinion, are related to his examination of sortals. However, this is not to underrate the importance of these other findings, it is merely to concentrate upon Durrant's work on sortals as primary which, as can be seen from his own text, he considers to be likewise primary. Thus, as a result, although I speak of the 'causal theory of reference' and the 'description theory of reference', to borrow from and adapt Gareth Evans,[3] my considerations may more accurately be directed at what could be labelled the 'causal theory of proper names' and, by reflection, the 'description theory of proper names'.

The causal theory was popularised by Saul Kripke in his seminal work *Naming and Necessity*.[4] I begin with a brief exposition of the theory as originally given by Kripke.

The causal theory of reference, as described by Kripke, falls into two parts: the 'reference fixing' part and the 'reference borrowing' part. With regard to the second part of this theory, I will have little to say about this here. Indeed, I hold that if it can be shown that the reference fixing part of the theory fails, then any account of reference borrowing is irrelevant to the success of the theory.[5] According to Kripke, it is at an original dubbing ceremony or 'baptism' that a name is first introduced into a public language as the name of a particular person or object. It is at this moment that a name is referentially grounded in its object. Subsequently, the referential capacity of any name is preserved when other speakers learn the referential use of the name from someone at this original dubbing ceremony:

> Someone, let's say, a baby, is born; his parents call him by a certain name. They talk about him to their friends. ... Through various sorts of talk the (referential capacity of the)[6] name spreads from link to link as if by a chain.[7]

Thus, it is at the initial 'dubbing' ceremony that the reference of a name is fixed. At this ceremony Kripke informs us: 'the object may be named by ostention, or the reference of the name may be fixed by a description'.[8] Once the referential capacity of the name is grounded in its object, this capacity can then be passed on to other language users via the process of 'reference borrowing'.[9] However, we must be careful here as to the weight Kripke places upon the function of descriptions in the process of reference fixing. In a footnote on the same page as the previous quotation he tells us:

> Two things should be emphasised concerning the case of introducing a name by description in an initial baptism (the reference fixing part of the theory).[10] First, the description used is not synonymous with the name it introduces but rather fixes its reference. Here we differ from the usual description theorists. Second, most cases of initial baptism are far from those which originally inspired the description theory.[11]

As is well known, for a supporter of the 'description theory of reference', a speaker cannot refer to an object unless they first possess some identifying knowledge of the object in question in the form of a definite description.[12] However, and this is the point Kripke is making here, this is *not* so for supporters of the causal theory.

What, then, is the situation for a supporter of the causal theory of reference? In a footnote on the following page Kripke elucidates further:

> Once we realise that the description used to fix the reference of a name is not synonymous with it, *then the description can be regarded as presupposing the notion of naming or reference.*[13]

The central point here is that although a description *may* be used to introduce a name as the name of a particular object into a language by an initial baptiser, it is not, *in principle*, required to do so. Indeed, for Kripke, the fundamental reason for the failure of the description theory of reference is 'that some notion of ostension, or primitive reference, is required to back it up'.[14] In other words, referentially speaking, on this account the notion of 'naming' is basic.

However, Durrant's case, as stated herein, is that sortals: (i) are not reducible to names or predicables (predicates); (ii) are to be recognised as a distinct category of symbol; (iii) and, most importantly for my purposes here, are logically prior to the introduction of names and predicables (predicates).[15] Correspondingly, he also argues that the distinctive role which sortals play, namely that of identifying particulars in the sense of: (a) serving or being used to pick out or discriminate particulars; (b) serving or being used to specify what they are, is not reducible to naming/referring on the one hand or predicating on the other. Rather, this role must be recognised as a function, a mode of speech activity, which is *prior to both operations.*[16] Indeed, on Durrant's convincing account, it is *sortal identification* which is basic and *not*, as Kripke stipulates, *naming* when it comes to allocating a name to a particular object.[17]

This damaging implication of Durrant's position for the causal theory of reference can be brought out particularly well with reference to his discussion of the work of Gareth Evans in the second part of Chapter II of this present work. In the first section of the second part of the aforementioned chapter, Durrant is discussing Evans' notion of 'demonstrative identification'. 'Demonstrative identification', for Evans, consists of identifying, picking out or discriminating an object simply by the use of a demonstrative together with a pointing gesture. It is some such process as this that I take to be central to what Kripke means by the naming of an object by ostension and which is crucial to the reference fixing part of his causal theory. Again, it is such acts of ostension which Kripke seems to take as basic. This being so, then the case for the causal theory must surely fail given that Durrant has shown, to my mind conclusively, that there is no substance to the claim that such acts of ostension have any referential power without the support of, and presupposition of, *sortal identification*:

> Cases where we may claim that the use of a demonstrative on its own together with an 'appropriate' pointing gesture, may serve to identify, to discriminate an individual do exist - as in '*That's* the one', '*That's* it', but on investigation we are able to see that all such cases are writable as 'That *F* (where "*F*" is a sortal) is the *F* I mean': 'that *F* is the *F*'. In short, in order for 'that' together with a pointing gesture to identify, i.e. to discriminate, pick out an individual, we logically have to be able to specify *what* it is we are pointing at, without being able to do this no discrimination can be intelligibly claimed.[18]

Granted the above, then Kripke's case for the causal theory of reference must surely fall at the first hurdle. That is, we cannot claim that naming/referring is basic if such an 'act' presupposes the presence of *sortals* for its very success. For the causal theory to survive such scathing criticism would require some very major reworking of its basic tenets on the part of its supporters. However, I'm at a loss to see how any such reworking could avoid the invocation, on the part of the causal theorist, of the presupposition of a mode of identification not unlike that suggested by Durrant. This being so, then it becomes unclear, precisely, what it is that marks off the causal theory as being at all remarkable, i.e. any such modification to the theory as the one suggested above would seem to show that this theory shares much more in common with its rival theory, the description theory of reference, than it supporters would have us believe.

Having examined the implications of Durrant's finding for the causal theory of reference, next I want to say a few brief words about the effects of Durrant's thesis *vis-à-vis* sortals has upon the description theory of reference. The common thread held by supporters of this theory is that there is no straightforward referential access to the world, i.e. to borrow a metaphor from Anscombe,[19] we cannot simply nail names to reality *à la* the causal theory of reference. Rather, at least according to the classical theory, as noted by Devitt and Sterelny:

> A description theory of names explains the referential properties of one category of term, names, by appeal to those of another, definite descriptions. ... The referential properties of descriptions are explained, in turn, by appeal to those of general terms ...[20]

However, as they go on to note:

> This process cannot ... go on forever: there must be some terms whose referential properties are not parasitic on others. Otherwise, language as a whole is cut loose from the world. Description theories, which explain one part of language in terms of another, can give no clues as to how, ultimately,

> language is referentially linked to reality. These theories pass the referential buck. But the buck must stop somewhere.[21]

Now, whilst it cannot be said that Durrant's new insights into our analysis of the subject-predicate distinction, at least as they are reflected in our understanding of names and predicates (predicables) and the relationship which exists between these two logical categories, completely answers the worries expressed by Devitt and Sterelny here regarding the referential links between language and reality (see below), his insights do at least go quite way to ameliorating these worries. What Durrant in effects shows is that, yes indeed, the referential buck does stop somewhere, i.e. with the intervention of sortals. It is the intervention of the category of the sortal as first proposed by Strawson and so fruitfully developed by Durrant which yields us a third and supportive logical category which grounds the referential power of, at least, proper names[22] in that:

> sortals [in his, Durrant's, sense] present us with a category of symbol which furnishes a principle of counting and distinction for particulars in the sense of 'objects'... in their own right presupposing no other ... category of symbol. ... Sortals permit us to form the notion of 'a particular' in the sense of 'an object' and likewise 'one', 'many', 'several' particulars (objects). In this sense I claim sortals present us with a quite distinct category of symbol which needs to be recognised as such.[23]

Thus, with the intervention of sortals, we are presented with a category of terms which have the capacity to endow speakers with the ability to identify or pick out objects. This capacity of sortals is *not*, contra Devitt and Sterelny, *parasitic upon those of other terms*. That is, sortals, as Durrant so forcefully puts it, 'present us with a category of symbol which furnishes a principle of counting and distinction for particulars in the sense of "objects" ... in their own right presupposing no other ... category of symbol ...'. Again, it is sortals which serve, in their own right, to identify objects, thereby securing the referential power of, at least, proper names.

Granted Durrant's account of the sortal, the amendments necessary to the description theory to incorporate his findings are, I hold, relatively minor compared to those which would be necessary to the causal theory. I have already indicated above the probability that for the causal theory to accommodate Durrant's position on the true nature of the sortal would require modifications which would, in effect, eradicate any fundamental difference between the two theories. However, with regard to the description theory, what we would have to recognise is the crucial function of the sortal in constructing the type of descriptions necessary to generate a referential connection between a name and an object. Such adjustments require some modifications to the theory, but my instinct is that such modifications would require little by way of fundamental alteration in that they, *prima facie*, seem to fit quite comfortably with overall structure of the theory pretty much as it stands. I do not intend to develop this point here, but rather note, by way of abbreviation, that the inclusion of the sortal in identifying definite descriptions such as 'the *man* who broke the bank at Monte Carlo' requires little by way of modification to the theory.[24]

Having indicated the comparative effects of Durrant's work on the sortal upon both the causal theory and the description theories of reference,[25] and having indicated why I hold that the latter theory fares much better regarding its encounter with Durrant's findings, there remains one question raised implicitly by Devitt and Sterelny which is left as yet unaddressed. This question echoes a problem which I will return to later in this postscript, i.e. 'What is the exact nature of the relationship between language and the world?'[26] It is at this point, I believe, that we must glean a lesson from followers of Kripke in that we must give at least some account of a causal element in any, what could be claimed plausible, theory of reference. I will be elaborating upon this point a little later in this essay. Nevertheless, it is important to note here that whilst Durrant's insights give us a number of clues as to what a fully-fledged theory of reference might 'look like' from a speaker's point of view, it leaves unvoiced worries relating to the 'external' relationship between language and reality.[27] Again, whilst Durrant's work allows us to furnish supporters of the description theory with a potential 'buck-stop' from and internalist perspective on the question of reference, we are still left with the puzzle of how, exactly, language 'meets' the world.[28] I will return to consider this question in a little more detail shortly. It is my contention that Durrant, again, offers some valuable insights here – however, to understand the full potential of these insights I must speculate somewhat upon the direction in which the themes found within Durrant's work might lead us.

C. Formal Logic

Another vitally important implication of Durrant's work is the effect of his findings upon the discourse of formal, symbolic logic.[29] One of the key assumptions behind current systems of predicate logic is that propositions comprise of only subject-terms and predicates, the former represented by lower case letters such as '*x*', '*y*' and '*z*', and the latter by upper case letters such as, for example, '*G*' and '*H*'. However, given that it follows from Durrant's explanation of the subject-predicate distinction that sortals fall neither into the categories of subject nor predicate, then current systems of symbolic, predicate logic seem left wanting. This point does not escape the author of this present work.

Taking just one of his examples:

'Man is rational'

he points out that 'Man' here can be used as either the kind name for humanity or a sortal. If it is used in the former sense then we have a genuine subject-predicate proposition. However, if 'Man' is used in the second sense, i.e. as a sortal, then this is not the case. It is not the case since here 'Man' is writable as

'Every man', 'Some man' and in 'Every man is rational', 'Some man is rational', as Durrant notes, we are not expressing potential subject-predicate propositions, rather we are expressing quantified propositions.[30] In the normal parlance of predicate logic 'Every man is rational', for example, comes out as the universally quantified proposition:

'For all x, if x is a man, x is rational'.

Translating this proposition into symbolic, predicate logic, this gives us:

$(x)(Mx \supset Rx)$.

However, on Durrant's account, allowing for the inclusion of the sortal, this proposition must be informally expressed as:

'For every F, if F is a man, then F is rational'.

Reflecting the inclusion here of the sortal, Durrant gives the following as the formal translation of this proposition into symbolic, predicate logic:

$[(x)F](Mx \supset Rx)$

where, and I quote, '"F" is a "higher" order sortal, "M" is "man" and "R" is rational.'[31] Durrant is quick to note that the fact that 'M' occurs in predicative position here does *not* commit him to the thesis that a sortal is a predicate. It is only the limits of the language of current predicate calculus which forces this sortal to be so construed.[32]

With regard to the discourse of symbolic, predicate logic at least two important implications arise from the aforementioned example. Firstly, granted Durrant's own modest suggestions, we have only a partial allowance for sortals given these modifications to the current language of symbolic, predicate logic. That is, we have the inclusion of the higher order sortal 'F', but the lower order sortal 'M' still appears in the *apparent* position of a predicate. Nevertheless, given even this moderate re-casting, it is obvious, granted the correctness of Durrant's findings, that the language of symbolic, predicate logic needs to be modified to allow for the presence of sortals. As it stands, the knock-on effects of such an allowance upon the process and formulation of logical proofs in the current language of predicate calculus would seem relatively small. That is, although we may require some modification to the rules for the elimination and re-introduction of quantifiers (incorporating an allowance for the presence of higher order sortals), the rules for the calculation of logical proofs *after* the elimination of such quantifiers would seem to require little or no modification.

However, a second and, perhaps, more radical implication seems to follow from Durrant's position. To truly reflect the presence of a third logical category of symbol, i.e. sortals, within the language of symbolic, predicate logic would seem to require the accommodation of not simply the higher order sortal

'*F*', as given in the above example, but also the lower order sortal '*M*'. Indeed, as already mentioned in passing, Durrant himself seems to hold that this failure to so reflect results from the limitations of the language itself: 'It is a case of it being necessary to write a sortal as a predicate granted the language of predicate calculus.'[33] I do not intend to attempt to reformulate the language currently under discussion in a way that would be necessary to fully account for Durrant's findings on sortals and their function. My purpose here is merely to signpost the way to the potential for further important research which arises from the author of this present work's endeavours. Having said this, however, it would seem to me that any such modification or re-casting of the language of symbolic, predicate logic as those currently under consideration would entail much greater modifications than those I have suggested which result from Durrant's own relatively minor alterations to predicate calculus. For example, if the presence of what Durrant calls (by implication) 'lower order' sortals is to be accounted for in the language of symbolic, predicate logic, then this would seem to suggest the need for an accompanying rule for their elimination prior to the application of the rules for generating logical proofs and, upon completion of such proofs, the need for an accompanying rule for the re-introduction of such 'lower order' sortals when formulating the resulting universally or existentially quantified proposition. In any event, although the aforementioned is clearly mere speculation pending any fully-fledged research into the amendments necessary to the language of predicate logic to take complete account of Durrant's work on the true nature of the sortal, what remains abundantly clear is that given the correctness of the findings of this present work, the current language of symbolic, predicate logic will certainly require some modification - either radical or less radical.

Before moving on from my considerations of the implications of Durrant's work for the discourse of formal logic, there is one further supporting point I would like to make. Although most, if not all, modern philosophers of logic and language have moved away from the view that formal, logical languages represent what an ideal language might look like, there are still those, reflecting Ayer's viewpoint, that value such formal languages as heuristic devices, i.e. '[that] [a]part from their intrinsic interest, their examination throws light ... on the workings of our [natural] language'.[34] If this is the case, and such languages do perform such a heuristic function, then it seems to me that they can only do so if they reflect as truly as possible the constituents and workings of our natural languages. And if, as Durrant's work would seem to suggest, fact-stating language requires or, indeed, necessitates the presence of sortals in that without such an additional category of term, predication, naming and reference would be impossible, then any formal language which does not reflect their inclusion would prove, at best, inadequate as a heuristic device, at worst, useless. Thus, if one is a supporter of what I will label the 'heuristic school of thought' *vis-à-vis* formal, logical languages, then one must also admit of the changes to such languages which must arise from the necessity of taking account of Durrant's findings in relation to the sortal.[35]

D. Comparative Philosophy: The Analytic/Continental Divide

This, the penultimate section, is the most speculative part of this postscript. What is offered here is justified on the grounds that it highlights, at least in part, the relevance of Durrant's work to current debates in philosophy.[36] Historically there exists a divide between Anglo-American analytic philosophy and what has become labelled 'continental philosophy'. Whilst what determines the division between these two approaches to philosophy is not always clear it would, nevertheless, seem fair to say that one of the things that does distinguish analytic from continental philosophy is, at least in terms of linguistic philosophy, that the former wishes to maintain that language has some contact with reality whilst the latter is, on the whole, happy to relinquish this contention. However, although I hold that this stands as a reasonable, if partial, account of the distinction between analytic and modern continental philosophy, upon closer examination there seems ample *prima facie* evidence to suggest that there are at least some practitioners within the former tradition who have more than a little sympathy with our continental counterparts. In turn, others within the analytic tradition fear that such sympathy, if taken to its logical extreme, would damagingly undermine the difference between the two approaches to philosophy under discussion.[37] One method of surveying the evidence in support of, and thereby assessing the real weight of this contention is to return briefly to the subject of theories of reference.

It is held by some that one of the implications of description theories of reference is that they lead to a form of anti-realism. 'Anti-realism', in this context, is taken to stand for any theory about, or approach to language which tends towards the view that language, itself, plays a vital role in how we interact with and come to understand the world. The argument used to support this view runs, roughly, as follows:[38]

Any flirtation with the description theory of reference, or the acceptance of any theory which stresses too heavily our reliance upon language in our dealings with the world, entails:

(a) an acceptance of what Devitt calls 'weak realism':

> External objects exist objectively... only as things-in-themselves. As we know them - that is as familiar objects - they exist partly in virtue of our imposition of concepts ...[39];

(b) that once we accept weak realism, the likes of Devitt hold that it is thence only a short and, more importantly, irresistible step to our acceptance of full-blown *anti*-realism, i.e. the view that:

- the only independent reality is beyond the reach of our knowledge and language; it is the noumenal world of things-in-themselves;

- the known world is partly our construction; it is the phenomenal world of appearances, created by the imposition of our concepts upon things-in-themselves.[40]

Thus, in summary, Devitt seems to hold that if we even *partially* accept the insertion of what one could term a 'linguistic wedge' between language users and the world, then we are driven to accept that we cannot know the world other than it is presented to us by the concepts we acquire when we learn our language. In crude terms, what we might refer to as our 'mind-independent reality' ceases to be mind-independent. Examples of this 'tradition' within the sphere of analytic philosophy are not difficult to find. Going back as far as 1958 we find Peter Winch expressing the following sentiments:

> Our idea of what belongs to the realm of reality is given for us in the language that we use. The concepts we have settle for us the form of the experience we have of the world. (It may be worth reminding ourselves of the truism that when we speak of the world we are speaking of what we in fact mean by the expression 'the world': there is no way of getting outside the concepts in terms of which we think of the world. ... The world *is* for us what is presented through those concepts.)[41]

Refining his definition of anti-realism, Devitt ultimately labels any philosopher who holds, even partially, a similar view to that expressed in the previous quotation, i.e. that language significantly influences our understanding of the world, as a 'constructivist'. He defines this position as follows:

> The only independent reality is beyond the reach of our knowledge and language. A known world is partly constructed by the imposition of concepts. These concepts differ from (linguistic, social, scientific, etc.) group to group, and hence the worlds of groups differ. Each such world exists only relative to an imposition of concepts.[42]

In effect, what Devitt is claiming here is that the noumenal world of the weak realist is a mere 'fig-leaf' (his phrase) to what he calls 'the naked idealism of Constructivism'.[43] Once this fig-leaf is dropped, we are left, he informs us, with a 'modified constructivism' which seems to deny that there are any external constraints at all upon particular languages or sign systems, the net result being that '*we can think anything we like*'.[44] I will be disputing this, what I will call the 'anything goes' hypothesis, below.

In contrast, the 'realist', of which Devitt himself is one, takes himself or herself to be very much bound up with the 'hard rock' of reality and with how we can make linguistic 'contact' with it. Following Devitt, the basic tenets of realism can be summarised as follows:

(i) that realists are committed to the existence of such common-sense entities as stones, trees and cats and also what Devitt calls 'scientific types';[45]

(ii) that these entities exist independently;
(iii) that we can come to know these entities, and therefore the world, as it is, independent of our modes of perception of that world.

In other words, the realist holds that there is a mind-independent world 'out there', and that we can come to know that world - in effect, for the realist, the world is pretty much as we perceive it. As a result, it should come as no surprise that realists support the revival of Russellian-like approaches to theorising the relationship between language and reality,[46] which emphasise the causal effect of the world upon language, rather than our language's effect upon our relationship with the world.

Having briefly, if somewhat crudely, outlined the realist and anti-realist positions *vis-à-vis* the philosophy of language, I will return to the point I raised earlier. My claim was that what, at least in part, distinguishes analytic philosophy from continental philosophy is that the former still wishes to maintain that language has some contact with reality whilst the latter is, on the whole, happy to relinquish this contention. Unsurprisingly, the realist is keen to maintain this distinction between these two styles of philosophy. This being so, then the realist is apt to have, to say the least, very strong reservations about any trend within analytic philosophy which might lead to anything like a continental position regarding the relationship between language and the world. And, for the realist, just such a fear presents itself in what the likes of Devitt label the 'anti-realist' or 'constructivist' tendency within analytic philosophy noted above. Again, holds the realist, if a linguistic wedge is inserted between language users and the world then, to borrow and slightly modify a phrase from Harris, 'languages ... have nothing in the world "outside" to secure and shore up their internal organisation'.[47] The net result being, to repeat part of a quotation I used earlier from Devitt and Sterelny, that 'language as a whole is cut loose from the world'.[48]

The most influential purveyors of this trend in analytic philosophy, holds the realist, are writers such as (the later) Wittgenstein,[49] Quine[50] and Kuhn.[51] For the realist, it is the work of these philosophers which puts the distinction between what I have labelled analytic philosophy and continental philosophy at utmost risk. And, indeed, to be fair to the likes of Devitt, it is not uncommon for just such 'analytic' philosophers as, for example, (the later) Wittgenstein to be claimed by those of that 'other' tradition, thereby emphasising this fear. A prime example is Lyotard, an arch-postmodernist of his day, who is more than happy to claim Wittgenstein as one of his own. In his, Lyotard's, exposition of a postmodernist view of language he has no hesitation in calling upon Wittgenstein to support his position:

> The social bond is linguistic, but is not woven with a single thread. It is a fabric formed by the intersection of at least two (and in reality an indeterminate number) of language games, obeying different rules. Wittgenstein writes: 'Our language can be seen as an ancient city: a maze of little streets and squares, and of houses with additions from various periods; and this surrounded by a multitude of new boroughs with straight regular streets and

> uniformed houses.' And to drive home that the principle of unitotality ... is inapplicable, he subjects the 'town' of language to the old sorites paradox by asking: 'how many houses or streets does it take before a town begins to be a town?'[52]

Thus, to summarise, as far as the realist is concerned any theory of reference or language which tends towards the admission that language plays a part in the construction of our reality must be rejected because it threatens to undermine what, perhaps, they would see as the most vital distinction between analytic and continental philosophy, i.e. that for the former, language maintains its contact with reality, whilst for the latter, language is simply 'cut-loose from the world'.

It is just such a fear as that outlined in the last paragraph, I contend, which has led some analytic philosophers to support causal theories of reference, overly atomistic theories of meaning and to attempt to reject what I regard, along with many others, as an obvious truism - that language and its conceptual content plays a fundamental role in our understanding of the world. The essence of this fear, as felt by the realist, is summed up nicely in the following quotation already alluded to earlier:

> unless names are somewhere nailed to reality without the mediation of senses[53] which hold true of objects, the relation between sentences and reality which constitutes their truth will in no way have been explained.[54]

Perhaps it is this fear which has lead philosophers, such as Kripke, to posit a theory of reference which seems to demand of us that we can pick-out objects, whilst possessing no identifying knowledge of them, and to philosophers as eminent as Russell to positing linguistic items which have meaning in their own right independently of the meaning of all other words, i.e. his 'logically proper names'.[55] As a result, these philosophers are left positing some very strange 'nails' indeed.

However, given Durrant's work as presented herein, I hold that the necessity of such extremes is avoidable. To briefly reiterate, the core of my contention thus far is that part of the reason why philosophers such as Kripke and Russell have been driven to the positions they have adopted, is because they fear that any introduction of a linguistic wedge between language users and the world will result in language being cast adrift from the world, i.e. in Devitt's terms, that we will be forced to adopt some form of anti-realism/constructivism. All this, despite the fact that it seems all to obvious that our language plays a fundamental role in our relationship with the world. What I intend to do in the rest of this section of this postscript is to illustrate how Durrant's work opens up the possibility of a potentially productive area of research into the question of how we can allow that language, to some extent, shapes our perceptions of the world, whilst simultaneously granting to that world its effect upon language and the shaping of the concepts therein. However, as will become clear, although the 'solution' hinted at here may not go as far as the realist would like in maintaining the distinction between analytic philosophy and continental philosophy, what it does provide is the basis for building some bridges between what Devitt refers to as the realists and the constructivists. This being so, then this 'solution' does at least demonstrate that there exists the potential of preserving a vital distinction between the analytic and continental approaches to philosophy, whilst allowing analytic philosophers to admit of the influence of language and concepts upon our dealings with the world without, concomitantly, giving up the influence of that very material reality upon our understanding.

I think that the key to understanding how Durrant's work can help here lies in unpacking what I have called above the 'anything goes' hypothesis. The central contention of this hypothesis is that the net result of allowing language any influence upon our dealings with the world leads to the cutting-off of language from our mind-independent reality. This, in turn, so the realist's story goes, results in the absence of any external constraints upon language and thus, presumably, upon concept formation. It is this move which, at least in part, yields the accusation that 'we can think anything we like'.[56] In other words, if there is no extra-linguistic constraint upon language, then how we construct our reality depends upon how we construct the content of our language and crucially *not* upon the content or shape of the world itself. This being so, then the fear is that language would become little more than, to coin a phrase from another realist, Christopher Norris, 'a realm of open-ended textual "freeplay" or floating signifiers devoid of referential content'.[57]

Implicitly, however, there seems to be at least one other element to the 'anything goes' hypothesis. That is, if we are truly able to think anything we like, then language itself must be open to, to say the least, quite radical manipulation by speakers and groups of speakers. Again, it seems that, if this hypothesis is correct, speakers must be able to change the conceptual content of a language more or less at will. Having said this, we must be careful here as to the weight we place upon the phrase 'more or less at will'. Even realists such as Devitt would allow to the constructivist some conventionalist aspect to their theory. 'Conventionalism' here is the idea that there exists within a particular language or language-game certain rules or mechanisms which constrain thought and speech. These 'rules' or mechanisms are acquired when one learns a language. One of the most sophisticated conventionalist theories on offer is that proposed by Nelson Goodman in his *Ways of Worldmaking*.[58] Here Goodman describes how it is possible to manufacture new 'worlds' or, perhaps more accurately, new world-views from our old ones. Ultimately, the only constraint upon such worldmaking is what has gone before, i.e. the world-views we already inhabit. Nevertheless, the potential new 'worlds' we can come to inhabit are, by implication, only limited by what we can imagine given the conceptual content of our old worlds so, at least to a degree, Devitt seems correct in his assertion that according to the constructivist we can, pretty much, think anything we like. This being so, then even on a conventionalist account, language and its conceptual content are open to quite considerable manipulation.[59]

Given the above, it is my contention that the 'anything goes' hypothesis rests on the satisfaction of two conditions:

1. that there is no contact between language and reality;
2. that speakers can change the conceptual content of a language more or less at will.

Condition '1' can only be fulfilled providing language *is* actually cut-off from the world completely. It seems to me, as I will endeavour to show below, that this is so does not automatically follow from the realist definition of either anti-realism or constructivism.[60] Thus, I hold, it is possible to adopt something like a constructivist position and yet still allow the world to impinge upon language and concept formation. I will take this point up again below.

In terms of condition '2', if we can simply think anything we like, then this seems to imply that language and the concepts therein are, to say the least, somewhat plastic. The thesis this hypothesis seems to suggest is that we are able to warp language and concept to mirror that which we choose to think. Thus, to misquote Wittgenstein, 'the limits of our imaginings are the limits of our world'. This point will also be taken up again in what follows.

In order to undermine the weight of Devitt's criticism here as directed towards constructivism, we would need to demonstrate that these two conditions do not automatically follow from the adoption of a constructivist-like position, i.e. any position, for Devitt, which admits of the influence of language upon our understanding of the world. It is not my intention to attempt a full such demonstration here, but rather to hint at what such a demonstration would need to accomplish if it were to be strong enough to resist, at least partially, Devitt's criticisms. Any demonstration to this effect would have to show that neither of the conditions as given above are satisfied by what Devitt calls 'constructivism'. Thus, firstly it would have to be shown that there exists at least the possibility of some form of causal contact between language and the world such that our language could mirror, to some extent, our mind-independent reality. Secondly, it must be shown that language is not, after all, for the constructivist, 'plastic' in the sense suggested above.

Turning, then, to address the first of these issues, how is it possible to hold that, from a constructivist perspective, language has some contact with our mind-independent reality? Crucial to understanding the nature of this question is the point that if one adheres to the view that one cannot think of, or refer to a thing unless one can think of it in a particular way, then it follows that it is impossible to compare reality as it 'really' is, to how we think about that reality. In other words, there is no position we, as language users, can assume to check the world against how we think or speak of that world - in Putnam's terminology, we have no access to the 'God's Eye point of view'.[61] This being so, then the question of how 'well' language maps on to our mind-independent reality cannot, at least directly, be asked without invoking the accusation, *à la* Wittgenstein, that we are, strictly, talking *nonsense*. However, all this having been said, we have still not yet denied language any contact with the world and thereby any representational power,[62] we seem to have implied, rather, that this 'contact' is somewhat more indirect than perhaps the realist would like. I will return to this point anon. Before doing this, however, I want to consider the second of the issues given above, namely the implied supposed 'plasticity' of language.

The point at issue here is the claim that one of the contributory factors to language collapsing into 'open-ended textual "freeplay"' is that it has no significant internal stability.[63] It is at this point that I want to return directly to the consideration of Durrant's work on the sortal. It will be remembered from the discussion in section 'B' of this postscript,[64] that the writer of this present work describes sortals in the following terms:

> sortals ... present us with a category of symbol which furnishes a principle of counting and distinction for particulars in the sense of 'objects'... in their own right presupposing no other ... category of symbol. ... Sortals permit us to form the notion of 'a particular' in the sense of 'an object' and likewise 'one'; 'many', 'several' particulars (objects). *In this sense I claim sortals present us with a quite distinct category of symbol which needs to be recognised as such.*[65]

Clearly, the central point here *vis-à-vis* the sortal (and the point which grants the sortal its distinctive characteristics as part of a language), is that the sortal is that linguistic item which has the capacity, *in its own right*, to serve as that which provides us with a principle for counting and distinguishing particular material objects. In that the sortal so serves *in its own right*, suggests that the sortal, itself, possesses a capacity *not* shared by the other constituents of a language, which enables it to perform its peculiar function. If this were not the case, then it is difficult to see how it could serve in this capacity. One possible basis for this capacity of the sortal is that sortals share a relative conceptual stability within a particular language. As to how this relative stability is generated or maintained is an open question. However, for my present purposes, the mere presence of the possibility that sortals possess such an internal, conceptual coherence (relative to the other constituents of a language) is enough for me to offer a potential answer to the worry about constructivism under discussion. In the following I shall refer to this possible relative conceptual stability of sortals as their 'conceptual inertia'.[66]

Returning directly to the problem of the plasticity of language as outlined above, i.e. that, in terms of condition '2',[67] we can simply think anything we like, then this seems to suggest that we, as speakers, are able to warp language and concept to mirror that which we choose to think. However, if there is, as I have suggested, at least an implication in Durrant's work that the sortal could provide a language with terms which possess what I have called a *conceptual inertia*, then the plasticity under discussion may have certain well definite limits *vis-à-vis* how we can think about the world. Again, if the sortal possesses sufficient conceptual inertia, then it is simply not true that we can think what we like regarding what we conceive of as our world, since the sortal provides a constraining mechanism to the excesses of our imaginative thought.

Having posited the notion of 'conceptual inertia', before moving on to make full use of it, it is important for me to make a clear distinction between this notion and that of mere 'conventionalism' as spoken of above.[68] It will be remembered that the sense of 'convention' outlined above amounted to the positing of certain rules or mechanisms which constrain thought and speech. By implication of the above discussion, these conventions are open to quite dramatic manipulation by speakers, thereby suggesting a high degree of plasticity on the part of these conventions. In comparison, what I want to suggest here *vis-à-vis* the notion of 'conceptual inertia', is that although the force of the conceptual content of sortals to shape our thinking, too, can be overcome by speakers, this content is able to resist the attempts of speakers and groups of speakers to change it more readily than that of mere conventions. This contention gleans its support from Durrant's work herein, i.e. from:

(a) the already mentioned fact that sortals, *in their own right*, serve as that which provides us with principles for counting and distinguishing particular material objects. Comparatively speaking, conventions, as will be seen below in my discussion of Quine, have no such *independent*, internal consistency, relying for their coherence, as they do, upon a relatively amorphous supporting network of truth-values;

(b) the vital role sortals play in naming material objects, referring to material objects and describing material objects.[69] This suggests, as I will attempt show, to borrow a phrase from Quine, a certain 'germaneness' on the part of sortals which whilst, on the one hand, may justify changing the conceptual content of a *particular* sortal, lends resistance to such changes on the part of others.

I will develop these points a little further below when I turn directly to discuss the work of Quine.

Having at least partially laid my foundations, what I must do now is sketch a constructivist model of language and its relationship with reality which avoids falling into the trap of the 'anything goes' hypothesis by avoiding the satisfaction, on the part of the constructivist, of the two conditions which support this hypothesis outlined above.[70] I must do this by showing that constructivism is capable of:

1. allowing some representational contact between language and reality;
2. allowing that although speakers can change the conceptual content of a language, they *cannot* do this more or less at will.

Due to restrictions of space, I can only hint at the sort of model that would be required to avoid meeting the two conditions which support the 'anything goes' hypothesis. To help me here I will draw upon certain aspects of the work of both Quine and the later Putnam. Quine's work will provide the basis for generating what I consider to be a more robust form of constructivism in that it avoids the problems outlined above, and Putnam's work will provide the basis for some refinements to this model.

Famously, many years ago, Quine wrote his 'Two Dogmas of Empiricism'.[71] One way of interpreting Quine's work in this essay is that he is positing the notion that language is a holistic system which is a 'man-made fabric' which impinges upon experience only 'along the edges':[72]

> The totality of our so-called knowledge or beliefs, from the most causal matters in geography and history to the profoundest laws of atomic physics or even of pure mathematics and logic is a man-made fabric which impinges on experience only along the edges. ... A conflict with experience at the periphery occasions readjustments in the interior of the field. Truth values have to be redistributed over some of our statements. Reëvaluation of some statements entails reëvaluation of others, because of their logical interconnections - the logical laws being in turn simply further statements of the system, certain further elements of the field. Having reëvaluated one statement we must reëvaluate some others, which may be statements logically connected with the first statement or may be the statements of logical connections themselves. But the total field is so underdetermined by its boundary conditions, experience, *that there is much latitude of choice as to what statements to reëvaluate in the light of any single contrary experience. No particular experiences are linked with any particular statements in the interior of the field, except indirectly through considerations of equilibrium affecting the field as a whole.*[73]

The model of the relationship between language, speakers and the world created here by Quine was, to say the least, quite revolutionary for its day. To unpack the above quotation a little, the model we have here is constituted by the following important elements:

- that the totality of our knowledge and beliefs which, for my purposes, I take to be embedded in our shared language,[74] butts up against, meets or, in Quine's terms, 'impinges' upon experience, i.e. our experience of realty/the world, 'only along the edges';
- that this 'totality' is:

  (a) man-made;
  (b) and includes not only our mundane statements about the world, but also the statements of history, geography, the laws of science, mathematics and even the laws of logic;

- that this totality can come into conflict with our actual experiences of the world. In other words, although we are conceptually conditioned to expect the world to behave in a particular way, it, the world, can confound these expectations;
- that if the unexpected does happen, then we have to adjust our totality in order to accommodate the unexpected anomaly;
- that such adjustments, according to Quine, can take place *anywhere* within the totality, i.e. it is open to us to adjust, for example, the laws of science, mathematics and, if we choose, even the laws of logic;
- that the only limits upon *what* we may alter within the 'total field' are:

  (a) considerations of equilibrium affecting the field as a whole;

(b) and, correspondingly, what he refers to elsewhere in the 'Two Dogmas' as 'germaneness' - ultimately, this notion is couched out in terms of the likelihood of our altering certain statements rather than others because such alterations will disturb the 'total system' of the field as little as possible.

Staying with the twin notions of 'equilibrium' and 'germaneness' for the moment, Quine clearly holds that we, as speakers, are naturally conservative in terms of the alterations we are likely to make to our language, belief and knowledge system. That is, he holds, we are less likely to alter the highly theoretical statements of, say, logic - which he says are very close to the centre of the network of the field - as this would require a quite massive alteration to the field as a whole. This 'conservativeness' is re-enforced by his notion of 'germaneness' which he supports by positing certain statements which are nearer to the periphery of the totality than others and therefore, he holds, have a greater empirical content. However, even granted such statement as these, and this is the key point, for Quine *speakers still have almost complete sovereignty over what is and what isn't altered within the totality*. Indeed, when speaking of his notion of 'germaneness' he talks in terms of a 'loose association' of such statements with our sense experiences and the 'relative likelihood' that we will choose to alter these statements rather than others, such as statements relating to the profoundest laws of atomic physics or even those of logic. It is here wherein lies the seeds of the illicit danger which disturbs the likes of Devitt, represented by what I have called the 'anything goes' hypothesis. Again, it is here that the likes of Devitt gain the necessary ammunition to depict any constructivist-like theory as undermining the possibility of there being any sort of a representational relationship between language and the world. Indeed, if, ultimately, the scope of readjustment to the interior of the field is, to say the least, somewhat elastic then the plasticity of language spoken of above becomes very much an actuality.

However, I hold that with the assistance of Durrant's work and a little help from Putnam, Quine's model as given in the 'Two Dogmas' can be re-worked so as to provide a basis for a constructivism which is far less worrying than the one outlined by Devitt above. I begin by returning to Quine's notion of 'germaneness'. Quine, as noted above, holds that some statements are more germane to change than others. Nevertheless, this germaneness, as it stands, does not unduly prevent any major alterations to the internal content of our totality. This is because, ultimately, Quine relies solely upon conventionalism to support his position, i.e. in the form of the resistance to change provided by the necessity of redistributing truth-values over some statements, and the resulting knock-on effect upon other statements within the totality. But, he tells us, such constraints still allow of '*much latitude of choice as to what statements to reëvaluate in the light of any single contrary experience*'. To avoid such elasticity we need to provide more by way of resistance to such alterations whilst, simultaneously, introducing greater discretion in relation to where such alterations should or could be made, thereby introducing somewhat more stability into the field of the totality than is allowed for by Quine. One way of doing this is to supplement his notion of 'germaneness'.

What I propose here is the supplementation of Quine's notion of 'germaneness' with some of the material provided by Durrant herein. This would yield, as a consequence, a much more precise notion of 'germaneness'. As a result, what would emerge would be a much tighter account of what happens *vis-à-vis* possible statement alterations, when a conflict occurs at the periphery of the totality between our conceptual expectations and our actual experiences. This would offer, in turn, a much less destructive account, for the constructivist, of why some statement alterations (and their accompanying necessary conceptual changes) are more permissible than others. A sketch of such a supplementation might go something like as follows:

1. granted a distinction between the relative systemic instability associated with Quine's reliance upon what I have referred to above as 'mere conventions', and the comparatively greater conceptual stability supplied by sortals, which results from what I have called their 'conceptual inertia', then when our experience of reality comes into conflict with our conceptual expectations about the world, it is much clearer *which* range of statements are the most germane with regard to the alterations necessary in order to take account of such conflicts. To elaborate further: the statements delineable as most germane in such circumstances are those statements about the world which are formed from material-object sortals or rely upon material-object sortals for their construction, i.e. precisely because these are the statements *most directly apposite in our references to, descriptions of, and understanding of our material world and its constituents* compared to, for example, the statements of logic and pure mathematics;[75]
2. in addition, the notion of 'conceptual inertia' could be employed to develop ever greater theoretical distinctions between statements resting for their construction on *particular* material-object sortals. Thus, with sufficient conceptual clarification, it may become possible to make distinctions *within* the class of statements derived from material-object sortals. The net result would be even greater discretion with regard to *which* statements are the most germane to *particular* experiential conflicts;
3. once such distinctions are arrived at, then we will have given Quine's notion of 'germaneness' much more concreteness. In doing so, we will also have made it clear that, after all, the changes necessary and possible within the field of a particular language, or part of a language, to reflect conflicts with our expected experiences of reality are far more constrained than those made possible by Quine's sole reliance upon mere conventions, or those suggested by what I have called the 'anything goes' hypothesis.

Thus, the potential benefit of the above supplementations to Quine's notion of 'germaneness' is rather less plasticity, in relation to the internal structure and content of a particular language on the part of the constructivist, than that suggested by Devitt. Again, it is not up to particular speakers or groups of speakers to alter language in whatever way they please, rather they are restricted by the limitations imposed upon them by the revitalised account of 'germaneness' suggested as possible here and grounded, as it is, in the notion of 'conceptual inertia'. The net result, for the constructivist, is that it appears that speakers do not possess, after all, the freedom necessary to change the conceptual content of a language in quite the way the likes of Devitt would have us believe. Thus, is does seem at least possible that the constructivist can avoid the second of the two conditions stated above, and upon which the 'anything goes' hypothesis rests, i.e. that speakers can change the conceptual content of language more less at will. Without the satisfaction of this condition, constructivism does not seem quite as much of a spectre as Devitt suggests.

However, to fully avoid the negative implications of Devitt's criticisms of constructivism, we must also avoid satisfaction of the first of the conditions suggested above, i.e. that of there being no contact between reality and language. If this condition *is* met, then the suggestion is that, on a constructivist model, language is altogether deprived of any representational potential. Here I want to pick up on Quine's notion of 'conflict with experience at the periphery'. What this seems to suggest is that it is possible, as I have already noted above, that our mind-independent reality will confound our conceptual expectations. These expectations, clearly, are conditioned for us by our language and the concepts therein. Thus, we 'approach' the world in anticipation of meeting or finding a particular phenomenon, but we are confounded as to that expectation. Again, we are conceptually conditioned to expect to find the world in a particular way, but discover that it is not as we anticipated - in modern parlance, our world-view does not meet with the world as we interact with it. Another way of thinking of such conflicts with experience, is that our present conceptual expectations of reality do not 'fit' with the world as we experience it. This being so, then some adjustment is required within the field of the totality of our beliefs and knowledge in order to accommodate this unexpected experience. As I have already indicated, not *any* adjustment will do, i.e. it must be 'germane', but not in Quine's somewhat elastic sense. Rather, this germaneness must be given a new, more precise form as in the way suggested above.

Granted the above revamped notion of 'germaneness', then it becomes possible, I hold, to develop the constructivist model under discussion still further to encompass the following idea: that given the limitations imposed upon us by what *is* and *is not* germane, then the alterations we make to a particular 'totality' due to conflicts with experience at the periphery would enable us to generate, within certain limitations (e.g. granted a particular conceptual starting point or world-view), an increasingly more accurate 'picture' of our mind-independent reality. Again, what I'm suggesting here is that because statement changes within the totality will have a greater relevancy to a particular conflict with experience, then we are more likely to end up changing our world-view in such a way as to more accurately reflect our mind-independent reality. Admittedly, a great deal more work needs to be done to substantiate this claim and, unfortunately, I do not have the space to do this here. However, the following quotation from Putnam does at least provide a hint at what I have in mind here:

> Internalism (a form of what Devitt would ultimately consider constructivism)[76] is not a facile relativism that says, 'Anything goes'. Denying that it makes no sense to ask whether our concepts 'match' something totally uncontaminated by conceptualisation is one thing; but to hold that every conceptual system is therefore just as good as every other would be something else. If anyone really believed that, and if they were foolish enough to pick a conceptual system that told them they could fly and to act upon it by jumping out of a window, they would, if they were lucky enough to survive, see the weakness of the latter view at once. Internalism does not deny that there are experiential *inputs* to knowledge; knowledge is not a story with no constraints except *internal* coherence; but it does deny that there are any inputs *which are not themselves to some extent shaped by our concepts*, by the vocabulary we use to report and describe them, or (that there are)[77] any inputs *which admit of only one description, independent of all conceptual choices.*[78]

Granted the suggestion that in this sense, indeed, *not* anything goes, the constructivist can allow of the notion that some world-views are better than others, given the fact that language does, after all, retain at least something of its representational capacity. Further, in that language does retain such a capacity, then this offers strong *prima facie* evidence for the existence of an underlying causal feedback mechanism which allows language to represent to us, at least to some extent accurately, our extra-linguistic reality.

Clearly, the model that Putnam has in mind, and one I hope is reflected in my discussion above, is a long way from the model of constructivism offered by Devitt. Indeed, granted the developments to the notion of the sortal made possible by Durrant's work herein, we may have ended up with a model that would meet with the approval of even the likes of Devitt. That this is at least a possibility can be seen from the following quotation which comes from Devitt's own *Realism and Truth*:

> Realism is 'modest', as Curtis Brown nicely puts it: '*when we develop a language we are not imposing an organization on the world, but selecting one of the world's organizations for our own use.*[79] On this view the world "in itself" has *more*[80] objects than we usually talk about, not fewer'. The Realist agrees with Putnam's metaphysical realist that there is 'a *ready made* world'; but this is not a commitment to the world having *just one*[81] 'built-in structure'.[82]

I hold that the sort of modesty proclaimed here by Devitt, if it was professed by the sort of constructivist I've been defending, would not be at all out of place.

Granted the discussion in this, the penultimate section of this postscript, I think that enough has been done to claim that, at the very least, contra Devitt, it does not automatically follow from the admission that language significantly influences our relationship with the world, that we damagingly undermine the distinction between analytic and continental approaches to the philosophy of language.[83] Indeed, given the sketch offered above of the sort of constructivism which can be generated using Durrant's notion of the 'sortal', we seem to have ended up quite a long way from the model of language much feared by the realist, i.e. that of language as 'a realm of open-ended textual "freeplay" or floating signifiers devoid of referential content'. Rather, we seem to have opened up the possibility for a constructivist model which supplies a language with not only its own internal conceptual coherence, but also allows of the sort of representational contact between language and reality which is necessary to claim back the world as that which provides language with its (language's) horizon of descriptive possibility. Thus, it appears that, after all, even on a constructivist model of the relationship between language and reality, we cannot think what we like.

E. Conclusion

The purpose of this postscript has been to give some indication of the significance, importance and relevance of Durrant's work as presented in this volume to current deliberations in philosophy. Granted even the scant offerings presented in this postscript this contention is by now, I hold, beyond doubt. Indeed, it is quite clear that Durrant's work on the subject of the sortal has the potential to contribute not only to some of the ongoing debates within the sphere of analytic philosophy e.g. with regard to theories of reference and formal logic, but serves also to open up areas of discussion which, to date, have been left relatively or wholly unexplored. A prime example of this latter category is the potential benefits to be gleaned from Durrant's novel re-examination of our historical roots for developments in modern, comparative philosophy of language. Indeed, this current work not only represents a critical examination and novel re-evaluation of our philosophical heritage, it also serves to point the way to a wealth of as yet undiscovered gold present within the work of the analytic tradition's founding thinkers. Thus, to my mind, this proves beyond doubt not only the ongoing validity of works such as this present one, in addition it also serves to emphasise the continuing vitality of the analytic tradition as a whole.

Notes

1. This is certainly true of the classical version of the description theory as it dates from Frege through Russell, and there is no reason to suppose that Kripke, in the construction of his 'causal theory', thought otherwise.
2. *See* present text, pp. 14ff.

3. Gareth Evans, 'The Causal Theory of Names', in *Aristotelian Supplementary Volume*, Vol. XLVII, 1973, pp. 187-208.
4. Saul Kripke, *Naming and Necessity*, rev. edn. (Oxford: Basil Blackwell, 1990).
5. *See also* n. 9 below.
6. My parentheses.
7. Kripke, *op. cit.*, p. 91.
8. *Ibid.*, p. 96.
9. Clearly, to confirm the point I alluded to earlier, if the 'reference fixing' part of the theory fails, then there is no reference to be passed on via the process of 'reference borrowing', thus if the first part of theory fails then, surely, so must the second part.
10. My parentheses.
11. Kripke, *op. cit.*, p. 96, n. 42.
12. As noted above, I will be returning to discuss the implications of Durrant's position for supporters of the 'description theory' below, whereupon this assertion will require some modification.
13. Kripke, *op. cit.*, p. 97, n. 44; my emphasis.
14. *Ibid.*
15. present work, ch. I.
16. *Ibid.*, pp. 40ff.
17. *Ibid.*, ch. II, pt. I, sec. A.1.
18. *Ibid.*, p. 53.
19. G. E. M. Anscombe, *An Introduction to Wittgenstein's Tractàtus*, 4th edn. (London: Hutchinson University Library, 1971), p. 50.
20. Michael Devitt and Kim Sterelny, *Language and Reality: An Introduction to the Philosophy of Language* (Oxford: Basil Blackwell, 1987), pp. 51-52.
21. *Ibid.*
22. I say 'at least' here as it *may* be possible to develop a case for sortals as securing the use of a certain type or types of kind name.
23. Present work, p. 37.
24. *See ibid.*, p. 142. Here Durrant notes the use of the sortal as a '*quasi* name' in that although the sortal may *appear* as being used to function as a name by filling the subject place in a proposition it, in fact, can only do so in that it presupposes the specificatory use of the definite form of the sortal. It seems to me to be only a short step from here to substituting a proper name (or possibly even a certain type or types of kind name) for such *quasi* names in one or a number of such definite descriptions, formed from the use of sortals, to securing the referential use of a particular proper name (or possibly a certain type or types of kind name). Indeed, Durrant is happy to admit of such substitution: 'in that [a *quasi* name] so occurs or is used a name may be coherently substituted for it', *ibid.*
25. At least with regard to versions of these theories as they apply to the use of proper names for individual, material objects.
26. Perhaps the phrase 'mind-independent reality' would be a better substitute for term 'world' here.
27. Mind-independent reality.
28. Mind-independent reality.
29. I take no account here of Durrant's discussion of 'matter' terms but, rather, choose again to concentrate upon his work on sortals. However, as can be seen from his text (pp. 142-143), it may also follow from this particular discussion that our system

of formal, symbolic logic requires some further re-working in a vein similar to that suggested in the following to take account of his findings regarding 'matter' terms.

30. present work, pp. 140-141.
31. *Ibid.*, p. 141.
32. *See* present text, *ibid.*
33. *Ibid.*
34. A. J. Ayer, *Central Questions of Philosophy* (London: Penguin Books, 1976), p. 49.
35. And perhaps, by parallel case, his findings *vis-à-vis* 'matter' terms.
36. As a result, I hope that this section of this postscript reflects what could be seen as an emerging trend, certainly in relation to the analytic/continental divide in philosophy, i.e. of those who admit that the new-wave of continentally influenced Anglo-American philosophy, which has emerged only relatively recently, still has a great deal to learn from 'old-style' analytic philosophy. What this 'new' trend represents is a middle-position between those analytic philosophers who claim not to have noticed a crisis within analytic philosophy at all, and those post-analytic philosophers who are currently proclaiming the death knell for traditional analytic philosophy. For example, regarding the latter school of thought, I recently read the following comment on the web site of the 'Centre for Post-Analytic Studies', based in the Philosophy Department at the University of Southampton (England, UK): 'Among the founder members of the Centre there is unanimous agreement that the later work of Wittgenstein is represented incorrectly when described as "analytic". Wittgenstein's work presents, rather, in our view, a systematic dismantling of many of the central tenets of analytic philosophy: it destroys, for example, the notion of "logical form" that plays a central role in the thinking of the Frege/Russell tradition; it undermines the search for a "theory of meaning", a search that the analytic tradition regards as the chief task of philosophy; and it repudiates the ahistorical perception of philosophical problems that is characteristic of the analytic method.' Prior to reading and editing Durrant's book, I may have even been tempted by this latter school of philosophy myself (*see* my 'Reading, Resistance and Disempowerment', in Ruth Evans (ed.), *Simone de Beauvoir's The Second Sex* (Manchester and New York, Manchester University Press, 1998), pp. 159-179). However, to my mind, what Durrant shows us herein is that it may well be a mistake to dismiss, out of hand, those tenets within analytic philosophy which are so maligned in the above quotation, i.e. those very tenents which allow of Durrant's development of Strawson's notion of the 'sortal'.
37. *See* below.
38. I say 'roughly' here as I have neither the time nor space to do full justice to this argument. The most clearly stated version of this argument is given by Michael Devitt in his *Realism and Truth*, 2nd edn. (Oxford: Basil Blackwell, 1991) and it is therefore his work I shall draw heavily upon here.
39. *Ibid.*, p. 72.
40. *Ibid.*, p. 157.
41. Peter Winch, *The Idea of a Social Science* (London: Routledge and Kegan Paul, 1958), p. 15; Winch's parentheses.
42. Devitt, *op. cit.*
43. *Ibid.*, p. 238.
44. *Ibid.*; my emphasis. Devitt does, briefly, consider two other views which he says could be ascribed to the constructivist. The first of these amounts to the constructivist merely asserting that external constraints to language exist, without

being able to substantiate such a claim. The second of these alternative views is, to my mind, a modification of the 'we can think anything we like' accusation. Given that Devitt, himself, seems to take the first accusation, i.e. the 'we can think anything we like' accusation, as the most damning for constructivism, it is this one I shall concentrate upon here.

45. 'Scientific types', he informs us (*ibid.*, p. 18.), can be 'observables', e.g. rare molluscs and the moons of Jupiter, or 'unobservables', e.g. electrons or muons.
46. *See*, for example, my discussion of Kripke above.
47. Roy Harris, *Language, Saussure and Wittgenstein: How to Play Games with Words* (London: Routledge, 1990), p. 60.
48. Devitt and Sterelny, *op. cit.*, pp. 51-52.
49. *See*, for example, Ludwig Wittgenstein, *Philosophical Investigations*. Note: Peter Winch was himself very influenced by the work of Wittgenstein (*see*, for example, his *The Idea of a Social Science*, *op. cit.*).
50. W. V. O. Quine, 'Two Dogmas of Empiricism', in his *From a Logical Point of View*, 2nd rev. edn. (Cambridge, Mass. and London: Harvard University Press, 1980), pp. 20-46.
51. Thomas S. Kuhn, *The Structure of Scientific Revolutions*, 2nd enlarged edn. (Chicago and London: University of Chicago Press, 1970). Kuhn, himself, was heavily influenced by the work of Quine.
52. Jean-François Lyotard, *The Postmodern Condition: A Report on Knowledge*, trans. Geoff Bennington and Brian Massumi (Manchester: Manchester University Press, 1989), pp. 40-41.
53. What I have labelled above a 'linguistic wedge'.
54. Anscombe, *op. cit.*, p. 50.
55. For a discussion of Russell's notion of 'logically proper names' *see* present text, ch. VI, sec. A.1.
56. Devitt, *op. cit.*, p. 238.
57. Christopher Norris, *Uncritical Theory: Postmodernism, Intellectuals and the Gulf War* (London: Lawrence and Wishart, 1992), p. 17. As already mentioned in the text, Norris is, himself, a realist and it is just such fears as those under discussion that has led him to adopt a realist approach to philosophy. For a discussion of Norris's views on realism *see* his excellent *Resources of Realism* (London: Macmillan, 1996).
58. Nelson Goodman, *Ways of Worldmaking* (Indianapolis: Hackett Publishing Company, 1978).
59. Incidentally, this allowance for conventionalist forms of anti-realism and constructivism also heads off a potential paradox for the realist since, if language is fully open to the manipulations of speakers and groups of speakers, then there arises the question of how language can, as is required by Devitt's definitions of both anti-realism and constructivism, constrain and condition how we think about the world. Again, it becomes at least problematic to suggest that the conceptual content of our language can condition our thoughts about the world, if this very content is subject to our every imaginative whim. However, granted a form of conventionalism, then language still has at least a limited power to conceptually constrain our thoughts about the world.
60. I will direct my comments below towards what Devitt calls 'constructivism' as I consider him to take this to be a more common position within analytic philosophy.

However, most of what is said could equally well be directed towards the position he labels 'anti-realism'.

61. Hilary Putnam, *Reason, Truth and History* (Cambridge: Cambridge University, 1981), p. 49.
62. By 'representational power' I mean the potential of language in some way to accurately reflect the true nature of the world and the objects therein.
63. I use the term 'significant' here to indicate that mere conventionalism (*see* above) cannot supply the stability necessary to avoid the weight of Devitt's criticisms. I develop this point in what follows.
64. p. 284.
65. Present work, p. 37 - my emphasis.
66. Clearly, to develop this idea further would require the tying-up of Durrant's work on the sortal with a theory of meaning. Again, unfortunately, I do not have the space to attempt this here. Nevertheless, it is worth noting that, contra one of the points made in quotation referred to in n. 36, consideration of theories of meaning are not quite so easily disposed of as one might imagine.
67. *See* p. 293.
68. *See* p. 292.
69. *See*, particularly, present work ch. II.
70. Present work, p. 292f.
71. Quine, 'Two Dogmas of Empiricism', in his *From a Logical Point of View*, *op. cit.* Although Quine attempted, many years later, to limit the 'damage', to some extent, of the content of this essay (*see* his 'Two Dogmas in Retrospect', in *Canadian Journal of Philosophy*, Vol. XXI, No.3, September 1991, pp. 265-274), nevertheless, his work in the 'Two Dogmas' has had a tremendous influence upon writers such as Kuhn and others who have adopted constructivist-like positions. In fact, along with the work of the later Wittgenstein, Quine's 'Two Dogmas of Empiricism' could be regarded as seminal in this respect.
72. I say 'one way' of interpreting Quine here since a distinction can be made between what some have labelled 'semantic holism' and 'epistemic holism' – the former relating to what we mean and the latter to what we know. However, for my purposes I will take Quine's position in the 'Two Dogmas' as relating to language since, as will have become clear from the discussion thus far, the constructivist would hold that what we know is, to say the least, heavily influenced by the conceptual equipment supplied to us by a language.
73. 'Two Dogmas of Empiricism', pp. 42-43 (my emphasis).
74. Please note: that although, in this essay, Quine seems to be talking of language as a sort of monolithic, holistic structure, it is not difficult to imagine Quine's points here applying equally well to a 'language-game', 'paradigm' or, in modern parlance, 'discourse'.
75. To clarify this point: the problem for Quine is, that although he seems to want to hold to the position that certain statements are more open to change than others in circumstances of conflict as a result of their 'location' within the totality, because of his reliance upon a form of conventionalism, ultimately he ends up undermining this position. What I try to develop here, using Durrant's work, is some support for Quine's initial inclination. I do this by offering a more concrete theoretical mechanism for distinguishing between statements nearer the periphery, and those located more towards the centre of the totality.
76. My parentheses.

77. My parentheses.
78. Putnam, *op. cit.*, p. 54.
79. My emphasis.
80. Original emphasis.
81. Original emphasis.
82. Devitt, *op. cit.*, p. 245, including a quotation from Curtis Brown, 'Internal Realism: Transcendental Idealism?', in Peter A. French, Theodore E. Uehling, Jr., and Howard K. Wettstein (eds.), *Midwest Studies in Philosophy, Vol. XII: Realism and Antirealism* (Minneapolis: University of Minnesota Press, 1988), pp. 145-155; p. 148.
83. *See* p. 288 above.

Bibliography

Anscombe, G.E.M., *An Introduction to Wittgenstein's Tractatus* (London: Hutchinson University Library, 1959).

Anscombe, G.E.M., *An Introduction to Wittgenstein's Tractatus*, 4th edn. (London: Hutchinson University Library, 1971).

Anscombe, G.E.M. and P.T. Geach, *Three Philosophers* (Oxford: Basil Blackwell, 1961).

Aristotle, *Categories* in *Aristotle: Categories, On Interpretation, Prior Analytics* [Löeb Classical Library Edition], trans. H.P. Cooke and H. Tredennick (London and Cambridge, Mass.: William Heinemann and Harvard University Press, 1955).

Aristotle, *Metaphysics*, Books I-IX [Löeb Classical Library Edition], trans. H. Tredennick (London and Cambridge, Mass.: William Heinemann and Harvard University Press, 1956).

Aristotle, *Posterior Analytics* in *Aristotle: Posterior Analytics* and *Topica* [Löeb Classical Library Edition], trans. H. Tredennick and E. S. Forster (London and Cambridge, Mass.: William Heinemann and Harvard University Press, 1960).

Aristotle, *Topica* in *Aristotle: Posterior Analytics* and *Topica, ibid.*

Aristotle, *Categories and De Interpretatione*, trans. with notes by J.L. Ackrill (Oxford: Clarendon Press, 1963).

Aristotle, *Physics I and II*, trans. with notes by W. Chalton (Oxford: Clarendon, 1970).

Aristotle, *Metaphysics*, Books Γ, Δ and E, trans. with notes by Christopher Kirwan (Oxford: Clarendon Press, 1971).

Attfield, Robin, 'How Things Exist: A Difficulty', in *Analysis*, Vol. XXXIII, No. 4, March 1973, pp. 141-144.

Attfield, Robin and Michael Durrant, 'The Irreducibility of "Meaning"', in *Nous*, Vol. VII, No. 3, 1973, pp. 282-298.

Ayer, A. J., *Central Questions of Philosophy* (London: Penguin Books, 1976).

Brown, Curtis, 'Internal Realism: Transcendental Idealism?', in Peter A. French, Theodore E. Uehling, Jr., and Howard K. Wettstein (eds.), *Midwest Studies in Philosophy, Vol. XII: Realism and Antirealism* (Minneapolis: University of Minnesota Press, 1988), pp. 145-155.

Copi, Irving, 'Essence and Accident', repr. in J.M.E. Moravcsik (ed.), *Aristotle - A Collection of Critical Essays* (London: Macmillan, 1968), pp. 149-166.

Devitt, Michael and Kim Sterelny, *Language and Reality: An Introduction to the Philosophy of Language* (Oxford: Basil Blackwell, 1987).

Devitt, Michael, *Realism and Truth*, 2nd edn. (Oxford: Basil Blackwell, 1991).

Dummett, Michael, *Frege: Philosophy of Language* (London: Duckworth, 1973).

Durrant, Michael, 'Feature Universals and Sortal Universals', in *Analysis*, Vol. XXXI, No. 2, December 1970, pp. 49-52.

Durrant, Michael, *Theology and Intelligibility* (London: Routledge and Kegan Paul, 1973).

Durrant, Michael, 'General Terms and Logical Subjects', in *Canadian Journal of Philosophy*, Vol. III, No. 4, June 1974, pp. 525-538.

Durrant, Michael, 'Essence and Accident', in *Mind*, Vol. LXXXIV, N.S. No. 336, October 1975, pp. 595-600.

Durrant, Michael, 'Plato, the "Third Man" and the Nature of the Forms', in *Southern Journal of Philosophy*, Vol. XVII, No. 3, 1979, pp. 287-304.

Durrant, Michael, 'Demonstrative Identification and Descriptive Identification', in *The Philosophical Quarterly*, Vol. XXXVIII, No. 151, April 1988, pp. 213-221.

Durrant, Michael, 'On the Thesis that a Name has Sense only in the Context of a Proposition; A Problem', in *Cogito*, Vol. XVII, No. 1, 1993, pp. 47-49.

Evans, Gareth, 'The Causal Theory of Names', in *Aristotelian Supplementary Volume*, Vol. XLVII, 1973, pp. 187-208.

Evans, Gareth, *The Varieties of Reference*, ed. John McDowell (Oxford: Clarendon Press, 1982).

Frege, Gottlob, *The Foundations of Arithmetic*, trans. J.L. Austin (Oxford: Basil Blackwell, 1950).

Frege, Gottlob, 'On Concept and Object', repr. in Peter Geach and Max Black (eds.), *Translations from the Philosophical Writings of Gottlob Frege* (Oxford: Basil Blackwell, 1952), pp. 42-55.

Frege, Gottlob, 'On Sense and Reference', repr. in *Translations from the Philosophical Writings of Gottlob Frege, ibid.*, pp. 56-78.

Geach, Peter, *Reference and Generality* (Ithaca, New York: Cornell University Press, 1962).

Geach, Peter, *Logic Matters* (Oxford: Basil Blackwell, 1972).

Geach, Peter, 'A History of the Corruptions of Logic', repr. in *Logic Matters, ibid.*, pp. 44-61.

Geach, Peter, review of P.F. Strawson, *Subject and Predicate in Logic and Grammar*, in TLS (1974); *see also* Zak van Streaton (ed.), *Philosophical Subjects* (Oxford: Oxford University Press, 1980).

Goodman, Nelson, *Ways of Worldmaking* (Indianapolis: Hackett Publishing Company, 1978).

Harris, Roy, *Language, Saussure and Wittgenstein: How to Play Games with Words* (London: Routledge, 1990).

Horton, Stephen, 'Reading, Resistance and Disempowerment', in Ruth Evans (ed.), *Simone de Beauvoir's The Second Sex* (Manchester and New York, Manchester University Press, 1998), pp. 159-179.

Kenny, Anthony, 'God and Necessity', in Bernard Williams and Alan Montefiore (eds.), *British Analytical Philosophy* (London: Routledge and Kegan Paul, 1966), pp. 131-151.

Kripke, Saul, 'Identity and Necessity', in Stephen P. Schwartz (ed.), *Naming, Necessity and Natural Kinds* (Ithaca, New York and London: Cornell University Press, 1977), pp. 66-101.

Kripke, Saul, *Naming and Necessity*, rev. edn. (Oxford: Basil Blackwell, 1990).

Kuhn, Thomas S., *The Structure of Scientific Revolutions*, 2nd enlarged edn. (Chicago and London: University of Chicago Press, 1970).

Lyotard, Jean-François, *The Postmodern Condition: A Report on Knowledge*, trans. Geoff Bennington and Brian Massumi (Manchester: Manchester University Press, 1989).

Norris, Christopher, *Uncritical Theory: Postmodernism, Intellectuals and the Gulf War* (London: Lawrence and Wishart, 1992).

Norris, Christopher, *Resources of Realism* (London: Macmillan, 1996).

Owen, G.E.L., 'Aristotle on the Snares of Ontology', in Renford Bambrough (ed.), *New Essays on Plato and Aristotle* (London: Routledge and Kegan Paul, 1965), pp. 69-95; repr. in G.E.L. Owen, *Logic, Science and Dialectic* (Ithaca, New York: Cornell University Press, 1986), pp. 259-278.

Owen, G.E.L., 'Plato on Not-Being', in Gregory Vlastos (ed.) *Plato: I Metaphysics and Epistemology* (Garden City, New York: Doubleday and Company, Inc., 1971), pp. 223-267; repr. in G.E.L. Owen, *Logic, Science and Dialetic*, *ibid.*, pp. 104-137.

Plantinga, Alvin, *God and Other Minds* (Ithaca, New York and London: Cornell University Press, 1967).

Putnam, Hilary, *Reason, Truth and History* (Cambridge: Cambridge University, 1981).

Quine, W.V.O., *Methods of Logic* (New York: Henry Holt and Co., 1950).

Quine, W.V.O., *Mathematical Logic*, rev. edn. (Cambridge, Mass.: Harvard University Press, 1961).

Quine, W.V.O., *From a Logical Point of View*, 2nd rev. edn. (New York and Evanston: Harper and Row, 1963).

Quine, W.V.O., 'On What There Is', in *From a Logical Point of View*, *ibid.*, pp. 1-19.

Quine, W.V.O., *Word and Object*, paperback edn. (Cambridge, Mass.: The M.I.T. Press, 1964).

Quine, W.V.O., *Ontological Relativity and Other Essays* (New York: Columbia University Press, 1969).

Quine, W.V.O., *Philosophy of Logic* (Englewood Cliffs, N.J.: Prentice-Hall, 1970).

Quine, W.V.O., 'Two Dogmas of Empiricism', in his *From a Logical Point of View*, 2nd rev. edn. (Cambridge, Mass. and London: Harvard University Press, 1980).

Quine, W.V.O., 'Two Dogmas in Retrospect', in *Canadian Journal of Philosophy*, Vol. XXI, No. 3, September 1991, pp. 265-274.

Russell, Bertrand, *Introduction to Mathematical Philosophy* (London: Allen and Unwin, 1930).

Russell, Bertrand, 'Lectures on the Philosophy of Logical Atomism', repr. in R.C. Marsh (ed.), *Logic and Knowledge* (London: Allen and Unwin, 1956), pp. 177-281.

Russell, Bertrand, 'On Denoting', repr. in *Logic and Knowledge, ibid.*, pp. 41-56.

Schiffer, Stephen, 'The Basis of Reference', in *Erkenntis*, Vol. XIII, 1978, pp. 195-196.

Searle, John, *The Philosophy of Language* (London: Oxford University Press, 1971).

Strawson, P.F., *Individuals* (London: Methuen, 1959).

Strawson, P.F., 'Singular Terms and Predication', repr. in P.F. Strawson *Logico-Linguistic Papers* (London: Methuen, 1971), pp. 53-74.

Strawson, P.F., 'Particular and General', repr. in *Logico-Linguistic Papers, ibid.*, pp. 28-52.

Strawson, P.F., 'On Referring', repr. in *Logico-Linguistic Papers, ibid.*, pp. 1-27.

Strawson, P.F., *Subject and Predicate in Logic and Grammar* (London: Methuen, 1974).

Wiggins, David, 'Identity, Designation, Essentialism and Physicalism', in *Philosophia*, Vol. V, 1975, pp. 1-30.

Winch, Peter, *The Idea of a Social Science* (London: Routledge and Kegan Paul, 1958).

Winch, Peter, 'The Unity of Wittgenstein's Philosophy', in Peter Winch (ed.), *Studies in the Philosophy of Wittgenstein* (London: Routledge and Kegan Paul, 1969), pp. 1-19.

Wittgenstein, Ludwig, *Tractatus Logico-Philosophicus* (London: Routledge and Kegan Paul, 1922).

Wittgenstein, Ludwig, *Philosophical Investigations*, trans. G.E.M. Anscombe (Oxford: Basil Blackwell, 1953).

Bibliography of Other Writings

Articles/Discussions

'Religion as the "Inexpressible" - Some Logical Difficulties' in *Sophia*, Vol. IV, No. 1, April 1965, pp. 14-21.

'Religion as the "Inexpressible"- Some Logical Difficulties' in *Sophia*, Vol. IV, No. 2, July 1965, pp. 3-10.

'Mr. Strawson on the notion of "Predicate"' in *Philosophy*, January 1966, pp. 79-84.

'Professor Austin on Whether every Proposition has a Contradictory' (joint paper with Charles Sayward) in *Analysis*, Vol. XXVII, No. 5, April 1967, pp. 167-170.

'Sources of Induction in Aristotle's Writings' (Abstract of paper) in *Proceedings of the Nebraska Academy of Sciences*, 1966, pp. 23-24.

'St. Thomas's "Third Way"' in *Religious Studies*, Vol. IV, April 1969, pp. 229-243.

'Professor Zink on "The Meaning of Proper Names"' in *Mind*, Vol. LXXVIII, N.S., no. 312, October 1969, pp. 571-575.

'God and Analogy' in *Sophia*, Vol. VIII, No. 3, October 1969, pp. 11-24.

'Feature Universals and Sortal Universals' in *Analysis*, Vol. XXXI, No. 2. December 1970, pp. 49-52.

'Professor Geach and the Gods of the Heathen' in *Religious Studies*, Vol. VII, 1971, pp. 227-231.

'The Use of "Pictures" in Religious Belief' in *Sophia*, Vol. X, No. 2, July 1971, pp. 16-21.

'Numerical Identity' in *Mind*, Vol. LXXXII, N.S., No. 325, January 1973, pp. 95-103.

'Aristotle's "Second Substance" and its Significance' in *Second Order*, Vol. XI, No. 1, January 1973, pp. 40-54.

'Is the Study of Philosophy irrelevant to the Study of Theology?' in *Education for Development*, Vol. II, No. 4, October 1973, pp. 29-38.

'The Irreducibility of Meaning' (joint paper with Robin Attfield) in *Nous*, Vol. VII, No. 3, September 1973, pp. 282-298.

'Is the Justification of Religious Belief a possible Enterprise?' in *Religious Studies*, Vol. IX, January 1974, pp. 449-455.

'General Terms and Logical Subjects' in *Canadian Journal of Philosophy*, Vol. III, No. 4, June 1974, pp. 525-538.

'Professor Penelhum on Persons as "Gap-Inclusive" Entities' in *Southern Journal of Philosophy*, Vol. XII, No. 4, 1974, pp. 513-518.

'Essence and Accident' in *Mind*, Vol. LXXXIV, N.S., No. 336, October 1975, pp. 595-600.

'*Parmenides* 127E-130E' in *Philosophical Papers*, Vol. IV, No. 2, October 1975, pp. 105-116.

'Some Comments on Professor Winch on "Meaning and Religious Language"' in Stuart Brown (ed.), *Reason and Religion* (Ithaca, New York and London: Cornell University Press, 1975), pp. 223-233.

'Cumulative Arguments in Theology' in *Sophia*, Vol. XV, No. 3, October 1976, pp. 1-6.

'Landesman and Austin on Performatives' in *Second Order*, 1978, pp. 85-92.

'Plato, the "Third Man" and the Nature of the Forms' in *Southern Journal of Philosophy*, Vol. XVII, No. 3, 1979, pp. 287-304.

'Particulars, Sortal Forms and Knowledge' in *Second Order*, Vol. VIII, 1979, pp. 47-54.

'Recent Developments in the Philosophy of Religion' in *Jubilee Lectures of the Faculty of Theology* (Cardiff: UCC Press, 1981), pp. 42-57.

'Prescriptivity and Justification' (joint paper with Robin Attfield) in *Philosophical Papers*, Vol. X, May 1981, pp. 16-23.

'Demonstrative Identification and Descriptive Identification' in *Philosophical Quarterly*, Vol. XXXVIII, No. 151, April 1988, pp. 213-221.

'Reference and Critical Realism' in *Modern Theology*, Vol. V, No. 2, January 1989, pp. 133-143.

'The Logic of God Incarnate - Two Recent Metaphysical Principles Examined' in *Religious Studies*, Vol. XXIV, January 1989, pp. 121-127.

'Scepticism: Three Recently Presented Arguments Examined' in *Philosophical Investigations*, Vol. XIV, No. 3, July 1991, pp. 252-266.

'The Meaning of "God"' in Martin Warner (ed.), *Religion and Philosophy* (Cambridge: Cambridge University Press, 1992) pp. 71-84.

'On the thesis that a Name has Sense only in the Context of a Proposition; A Problem' in *Cogito*, Vol. XVII, No. 1, 1993, pp. 47-49.

'Transcendence, Instantiation and Incarnation - An Exploration' in *Religious Studies*, Vol. XXIX, 1993, pp. 337-352.

'Meinwald's *Pros Heauto* Analysis of Plato's Apparently Self-Predicational Sentences' in *Australasian Journal of Philosophy,* Vol. LXXV, No. 3, September 1997, pp. 383-395.

'Plato's Quinean Beard: Did Plato ever grow it?' in *Philosophy,* Vol. LXXIII, No. 283, January 1998, pp.113-121.

'Modes of Existence' in *Cogito*, Vol. XIII, No. 3, November 1999, pp. 177-179.

'On Moore's Refutation of *Esse* is *Percipi*' in *Philosophical Investigations*, Vol. XXIII, No. 1, January 2000, pp. 26-47.

Books

The Logical Status of 'God' (London: Macmillan, 1973).

Theology and Intelligibility (London: Routledge & Kegan Paul, 1973).

Aristotle's De Anima in Focus, (ed.), (London and New York: Routledge, 1993) of which the author, as editor, contributed the 'Introduction' (pp.1-13), a revised translation of Books II and III (pp. 15-74) and a fully comprehensive 'Index' (pp. 217-225).

Book Reviews and Review Articles

D.Z. Phillips (ed.), *Religion and Understanding* in *Y Traethoddyd,* Cyfrol CXXIV, 1969, pp. 159-178. (Review article.)

Anthony Kenny, *The Five Ways: St. Thomas Aquinas's Proofs of God's Existence* in *Religious Studies*, Vol. XVII, 1971, pp. 187-189.

P.F. Strawson, *Logico-Linguistic Papers* in *Philosophical Books*, Vol. XIII, No. 1, January 1972, pp. 32-35.

John Hick, *Arguments for the Existence of God* in *Religious Studies*, Vol. VIII, 1972, pp. 183-185.

Terence Penelhum, *Problems of Religious Knowledge* in *Religious Studies*, Vol. VIII, 1972, pp. 270-272.

Kai Nielsen, *Contemporary Critiques of Religion* in *Religious Studies*, Vol. VIII, 1972, pp. 272-274.

Charles Landesman, *Discourse and its Presuppositions* in *Philosophical Books*, Vol. XIV, No. 2, May 1973, pp. 12-15.

Basil Mitchell, *The Justification of Religious Belief* in *Religious Studies*, July 1974, pp. 233-236. (Review Article.)

P.F. Strawson, *Freedom and Resentment* in *Philosophical Books*, Vol. XVI, No. 1, January 1975, pp. 28-32.

Keith Ward, *The Concept of God* in *Religious Studies*, Vol. XI, No. 4, 1975, pp. 490-494.

Walter Leszl, *Aristotle's Conception of Ontology* in *Mind*, Vol. LXXXVII, N.S., No. 345, 1978, pp. 130-133.

Patrick Sherry, *Religion, Truth and Language Games* in *Philosophical Books*, Vol. XIX, No. 2, May 1978, pp. 72-75.

William Craig, The *Cosmological Argument from Plato to Leibniz* in *Religious Studies*, Vol. XVII, 1981, pp. 289-291.

Richard Taylor, *Ethics, Faith and Reason* in *Philosophical Books*, Vol. XXVII, January 1986, pp. 59-60.

Peter Vardy, *God of our Fathers - Do we know what we believe?* in *New Blackfriars*, December 1987, pp. 577-578.

Thomas V. Morris, *Anselmian Explorations* in *New Blackfriars*, Vol. LXIX, No. 818, January 1988, pp. 48-49.

Janet Martin Soskice, *Metaphor and Religious Language* in *Religious Studies*, Vol. XXIII, 1988, pp. 297-300.

Daniel W. Graham, *Aristotle's Two Systems* in *Philosophical Books*, Vol. XXX, No. 1, January 1989, pp. 16-19.

William Charlton, *Philosophy and Christian Belief* in *New Blackfriars*, Vol. LXX, No. 831, October 1989, pp. 456-459.

Richard Swinburne, *Responsibility and Atonement* in *Philosophical Books*, Vol. XXXI, No. 3, July 1990, pp. 190-192.

Scott Dowell, *Atheist Priest? Don Cupitt and Christianity* in *New Blackfriars*, May 1990, pp. 258-260.

David Palin, *The Anthropological Character of Theology* in *New Blackfriars*, Vol. LXXII, No. 353, October 1991, pp. 452-454.

Paul Johnston, *Wittgenstein: Rethinking the Inner* in *Philosophy Today*, No.16, 1993, p. 6.

Kenneth J. Lucey (ed.), *What is God? The Selected Essays of Richard La Croix* in *Philosophical Books*, Vol. XXXV, No. 4. October 1994, pp. 281-283.

William H. Brenner, *Logic and Philosophy: An Integrated Introduction* in *Philosophical Investigations*, Vol. XVIII, No. 4, October 1995, pp. 373-376.

Joseph Houston, *Reported Miracles* in *Philosophical Books*, Vol. XXXVIII, No. 1, January 1997, pp. 66-67.

Index of Names

Anscombe, G.E.M., 11, 26, 38, 187, 188, 193, 278, 283, 302, 304
Aquinas, St. Thomas, 139, 158, 194, 207
Aritstotle, 3, 7, 13, 20, 38, 45, 85, 129, 130, 132, 143, 144, 145, 147, 158, 188, 215, 272, 275, 276, 278
Attfield, R., xiii, 27, 38, 85, 86, 147, 193
Augustine, St., 116
Ayer, A.J., 287, 303
Bennington, G., 304
Brown, Curtis, 300, 306
Charlton, W., 38
Copi, I., 133, 146
de Beauvior, S., 303
Devitt, Michael, 283-284, 285 288-290, 291-292, 293, 297, 299, 300, 301, 302, 303, 304, 305, 306
Dummett, M., 38, 157, 158, 169, 174, 176, 177, 178, 180
Durrant, M., xiii, xv, xvi, 27, 38, 39, 86, 133, 146, 193, 207, 254, 280, 280-281, 282, 283, 284-285, 285-287, 288, 291-292, 294, 295, 297, 298, 300, 301, 302, 303, 305
Evans, G., 9, 10, 12, 49, 52-75, 76-85, 85, 86, 87, 174, 175, 176, 180, 193, 278, 281, 282, 302
Evans, R., 303
Frege, G., 1, 3, 10, 11, 14, 21, 38, 39, 54, 84, 92, 157, 159-180, 194, 195, 201, 210, 211, 218, 261, 273, 275, 276, 278, 280, 301, 303
French, P.A., 306
Geach, P.T., 1, 3, 7, 8, 9, 10, 11, 12, 13, 14, 20, 21, 25, 28, 37, 38, 39, 45, 47, 74, 84, 85, 87, 88-94, 94, 103, 104, 105, 107, 108, 111, 137, 145, 146, 157, 162, 193, 194-208, 210, 211, 212, 217, 218, 219, 261, 262, 268, 273, 275, 276, 278
Goodman, N., 292, 304
Harris, R., 290, 304
Horton, S., xi, xiii, 280, 303
Kenny, A., 147
Kripke, S., 11, 109, 193, 281-282, 283, 285, 291, 301, 302, 304
Kuhn, T.S., 290, 304, 305
Locke, J., 11, 187
Lyotard, J-F., 290, 304
Massumi, B., 304
Mill, J.S., 169
Mitchell, B., xiii
Norris, C., xi, xiii, 292, 304
Owen, G.E.L., 85, 130, 146, 147
Plantinga, A., 147
Plato, 24, 116, 233, 244
Putnam, H., 293, 295, 297, 300, 305, 306
Quine, W.V.O., 3, 4, 6, 7, 9, 10, 11, 13, 14, 17, 20, 27, 37, 39, 84, 88, 101, 111-123, 123, 124, 125, 132, 141, 142, 145, 146, 147, 157, 169, 190, 193, 210, 211, 218, 249, 256-274, 275, 276, 277, 278, 290, 295, 295-299, 304, 305
Russell, B., 3, 4, 7, 9, 10, 11, 13, 14, 25, 27, 32, 36, 37, 39, 54, 56, 57, 60, 61, 70, 74, 75, 78, 81, 84, 86, 87, 100, 101, 102, 157, 160, 169, 181-193, 204, 207, 233, 253, 272, 278, 290, 291, 301, 303, 304
Ryle, G., 39, 266
Schiffer, S., 64, 65, 86, 272
Searle, J., 169, 174, 175, 179
Skilton, D., xiii

Sterelny, K., 283-284, 285, 290, 302, 304
Strawson, P.F., 3, 4, 7, 8, 9, 10, 11, 12, 18, 20, 21-24, 24, 37, 38, 47, 84, 85, 88-94, 94-102, 103, 104, 105, 106, 108, 111, 112, 113-123, 123, 130, 133, 136, 137, 145, 146, 156, 157, 186, 187, 189, 193, 209-219, 220-228, 229-255, 263-265, 266, 267, 272, 273, 275, 277, 278, 279, 284, 303
Trigg, R., xiii, xvi
Uehling, T. E., Jr., 306
Van Streaton, J., 12
Wettstein, H.K., 306
Wiggins, D., 81, 87
Winch, P., 39, 289, 303, 304
Wittgenstein, L., 28-30, 38, 54, 85, 86, 188, 262, 272, 273, 290, 293, 302, 303, 304, 305